# Universities, Knowledge Transfer and Regional Development

# NEW HORIZONS IN REGIONAL SCIENCE

**Series editor:** Philip McCann, *Professor of Economics, University of Waikato, New Zeeland and Professor of Urban and Regional Economics, University of Reading, UK*

Regional science analyses important issues surrounding the growth and development of urban and regional systems and is emerging as a major social science discipline. This series provides an invaluable forum for the publication of high quality scholarly work on urban and regional studies, industrial location economics, transport systems, economic geography and networks.

*New Horizons in Regional Science* aims to publish the best works by economists, geographers, urban and regional planners and other researchers from throughout the world. It is intended to serve a wide readership including academics, students and policymakers.

Titles in the series include:

Knowledge Externalities, Innovation Clusters and Regional Development
*Edited by Jordi Suriñach, Rosina Moreno and Esther Vayá*

Regional Knowledge Economies
Markets, Clusters and Innovation
*Philip Cooke, Carla De Laurentis, Franz Tödtling and Michaela Trippl*

Entrepreneurship, Industrial Location and Economic Growth
*Edited by Josep Maria Arauzo-Carod and Miguel Carlos Manjón-Antolín*

Creative Cities, Cultural Clusters and Local Economic Development
*Edited by Philip Cooke and Luciana Lazzeretti*

The Economics of Regional Clusters
Networks, Technology and Policy
*Edited by Uwe Blien and Gunther Maier*

Firm Mobility and International Networks
Innovation, Embeddedness and Economic Geography
*Joris Knoben*

Innovation, Agglomeration and Regional Competition
*Edited by Charlie Karlsson, Börje Johansson and Roger R. Stough*

Technological Change and Mature Industrial Regions
Firms, Knowledge and Policy
*Edited by Mahtab A. Farshchi, Odile E.M. Janne and Philip McCann*

Migration and Human Capital
*Edited by Jacques Poot, Brigitte Waldorf and Leo van Wissen*

Universities, Knowledge Transfer and Regional Development
Geography, Entrepreneurship and Policy
*Edited by Attila Varga*

# Universities, Knowledge Transfer and Regional Development

## Geography, Entrepreneurship and Policy

*Edited by*

Attila Varga

*Professor of Economics, Faculty of Business and Economics, University of Pécs, Hungary*

NEW HORIZONS IN REGIONAL SCIENCE

**Edward Elgar**
Cheltenham, UK • Northampton, MA, USA

Published by
Edward Elgar Publishing Limited
The Lypiatts
15 Lansdown Road
Cheltenham
Glos GL50 2JA
UK

Edward Elgar Publishing, Inc.
William Pratt House
9 Dewey Court
Northampton
Massachusetts 01060
USA

A catalogue record for this book
is available from the British Library

Library of Congress Control Number: 2009922752

ISBN 978 1 84542 931 7

Printed and bound by MPG Books Group, UK

# Contents

*v*

# Contributors

**Ács, Zoltán J.** School of Public Policy, George Mason University, Fairfax, Virginia, USA.

**Andersson, Martin** Jönköping International Business School, Sweden.

**Aspegren, Kevin** George Voinovich School for Leadership and Public Affairs, Ohio University, USA.

**Diez, Javier Revilla** Institute of Economic and Cultural Geography, University of Hannover, Germany.

**Faggian, Alessandra** School of Geography, University of Southampton, United Kingdom.

**Feser, Edward** Department of Urban and Regional Planning, University of Illinois at Urbana-Champaign, USA.

**Franzoni, Chiara** DISPEA-Politecnico di Torino, Italy.

**Frenken, Koen** Urban and Regional Research Centre Utrecht (URU), Faculty of Geosciences, Utrecht University, the Netherlands.

**Goldstein, Harvey A.** MODUL University Vienna, Austria.

**Gråsjö, Urban** Jönköping International Business School, Sweden.

**Hemer, Joachim** Fraunhofer Institute for Systems and Innovation Research, Germany.

**Karlsson, Charlie** Jönköping Business School, Sweden.

**Koschatzky, Knut** Fraunhofer Institute for Systems and Innovation Research, Karlsruhe, Germany.

**Lamb, William B.** Department of Management Systems, College of Business, Ohio University, USA.

**Lengyel, Imre** Faculty of Business and Economics, University of Szeged, Hungary.

**Lissoni, Francesco** University of Brescia and CESPRI, University of Bocconi, Italy.

**Maier, Gunther** Vienna University of Economics and Business Administration, Austria.

**McCann, Philip** Department of Economics, University of Waikato, New Zealand, and Department of Economics, University of Reading, UK.

**Mildahn, Björn** Institute of Geography, Kiel University, Germany.

**van Oort, Frank** Urban and Regional Research Centre Utrecht (URU), Faculty of Geosciences, Utrecht University and the Netherlands Environmental Assesment Agency, The Hague, the Netherlands.

**Parag, Andrea** Department of Economics and Regional Studies, Faculty of Business and Economics, University of Pécs, Hungary.

**Ponds, Roderik** the Netherlands Environmental Assesment Agency, The Hague, the Netherlands.

**Schiller, Daniel** Institute of Economic and Cultural Geography, University of Hannover, Germany.

**Sheppard, Stephen** Department of Economics, Williams College, USA.

**Sherman, Hugh D.** College of Business, University of Ohio, USA.

**Török, Ádám** University of Pannonia, Veszprém, Hungary and Hungarian Academy of Sciences, Hungary.

**Varga, Attila** Department of Economics and Regional Studies, Faculty of Business and Economics, University of Pécs, Hungary.

# 1. Introduction

## Attila Varga

Empirical research on academic knowledge transfers has been brought to the center of interest in economics and economic policy since the end of the 1980s for two main reasons. First, the emerging literatures of the new economic geography (Krugman 1991) and the endogenous growth theory (Romer 1986, 1990) pointed to the need of empirically testing the existence and significance of knowledge spillovers. The second reason is related to the growing interest in the mix of policies that are most suitable to generate 'university-based regional development' experienced first in Silicon Valley or in Route 128 (Isserman 1994; Reamer, Icerman and Youtie 2003). Recently emerging high technology centers in the US, Europe and Asia are strongly supported by regional economic development policies. Motivated by the above referred success stories, these programs put a significant weight on promoting knowledge transfers from universities to the local industry. Not only has the direct support for university research increased, but also a major portion of technology-related expenditures of regional governments is being spent on programs requiring different forms of university involvement. Moreover, about 70 percent of the total budget of US state technology programs is, in part, associated with some kind of university participation (Varga 2002).

Following Parker and Zilberman (1993) academic knowledge transfer is defined here as any process by which basic understanding, information and innovations move from a university to firms in the private sector. Academic knowledge transfer mechanisms can be classified into three broad categories: knowledge transmission via (formal or informal) networks of university and industry professionals (university-industry research collaborations, local labor market of graduates, faculty consulting, university seminars, conferences, student internships, local professional associations, continuing education of employees), diffusion of technology through formalized business relations (university spin-off companies, technology licensing), and knowledge transfers facilitated by the use of university physical facilities such as libraries, scientific laboratories, computer facilities and research parks located on university campuses (Varga 2000).

Within the academic knowledge transfers literature the geographical dimension has received particular attention. Studying localized knowledge spillovers as one type of agglomeration economies fits well into the research agendas of both theoretical and empirical economists while the potential of geographically constrained knowledge transfers to contribute to the development of regional economies makes the issue relevant for policy practitioners. To investigate the geography of university knowledge transfers two approaches have been developed: location studies and direct technology transfer studies (Varga 2002). Case studies, surveys, descriptive studies and econometric analyses show evidence that the effect of universities on the location of high technology activities is not constant over spatial entities and firms but varies according to industrial sectors, ownership status of firms, firm size, and city size (e.g., Florax 1992; Malecki and Bradbury 1992; Sivinatidou and Sivinatides 1995; Audretsch and Stephan 1996). Studies directly investigating the geography of knowledge transfers report that knowledge from universities tends to spill over locally with definite distance decay (Jaffe, Trajtenberg and Henderson 1993; Feldman 1994a; Audretsch and Feldman 1996; Varga 1998; Ács, Anselin and Varga 2002). This finding supports what is hypothesized about the localized nature of tacit knowledge transmission; however notable differences across industries are also reported.

Although a significant part of the literature on academic knowledge transfers focuses on the geographical aspects, several recently published papers raise the issue that besides pure spatial proximity to an academic institution some additional factors might also be instrumental for successful localized knowledge transmissions. Understanding the significance of those factors in regional economic development is at least as important for designing effective regional policies as improving our knowledge on the spatial proximity issue. Breschi and Lissoni (2007) argue that much of the localized transfers of knowledge are mediated by social networks of inventors. However as described in detail by Saxenian (1994) and supported among others also in Fischer et al. (2001) and Feldman and Desrochers (2004) cultural differences explain to a large extent the cross-regional variability in the intensity of localized innovation-related linkages. Somewhat related to culture, the level of entrepreneurship in the region is also a factor determining the extent to which the business opportunity in scientific knowledge is discovered and then the knowledge is transformed into innovation (Ács and Varga 2005;

Inzelt and Szerb 2006; Mueller 2006; Koo 2007). Agglomeration (the geographical concentration of innovative firms, research labs, business services and related industries) also influences the transfer of knowledge to regional industrial applications (Feldman 1994b; Koo 2005; Goldstein and Drucker 2006). Simulations in Varga (2000) show that the same levels of university research expenditures result in considerably more innovations in large concentrations of high technology activities than in smaller metropolitan areas in the US.

Three research questions have been investigated especially rigorously in the recent international literature on academic knowledge transfers. The first relates to the geography of knowledge flows from universities. What is the type of knowledge that tends to be transmitted more locally than globally? What is 'local'? In other words: what is the spatial extent of localized spillovers? Though these questions have already been investigated in earlier studies recent research reports new country- and region-specific evidences. An additional important advancement in the geography of academic knowledge transfers field is the increased application of the methodology of social network analysis (SNA) in empirical studies.

What are the channels of localized academic knowledge transfers? Are these knowledge flows mainly pure knowledge spillovers transmitted by mobility of professional labor or informal knowledge exchanges? What is the role of other mechanisms like student mobility or academic entrepreneurship? These questions, all related to the means of localized academic knowledge transfers, also receive particular attention in the most recent literature. The third field where similarly significant advancements can be observed is the account of the experience of regions where the 'recipe' of university supported regional development is taken seriously and implemented in the day-to-day practice of regional policy makers.

Chapters in this volume reflect the above areas of recent research. Individual contributions written by prominent scholars of the field are organized into four parts of this book. The first part 'sets the scene' in three reviews by outlining the most important advancements in theory, empirics and policy analysis. The second part gives a selection of research on the geography of academic knowledge transfers. Individual chapters in the third part have the commonality that all of them are related to two main channels of university knowledge flows: academic entrepreneurship and student mobility. The fourth part contains three case studies reporting the individual experiences in university-based development of three lagging regions of three continents.

Part One opens with the review by Harvey A. Goldstein on what we know and what we don't know about regional impacts of universities. In his survey he takes a practical view by pointing out that economic development policy officials and leaders in higher education often do not have solid empirical bases to decide what role(s) universities should play in economic development. Goldstein takes a stock of what has already been understood in the literature about the relationships between universities and regional economic development. In Chapter 3, Zoltán J. Ács provides a survey of the literature on knowledge spillovers from a perspective of how (in his words) the Jaffe-Feldman-Varga regional knowledge production function approach influences research in related fields such as the geography of innovation, entrepreneurship and development policy impact modeling. Turning to development policy analysis in Chapter 4, Edward Feser identifies some of the conceptual and practical issues that arise in the conduct of regional cluster studies focusing on joint university-industry strengths as a basis for policy action. He does it by grounding the discussion on a comparison of the methodologies of two recent applied cluster studies in the US.

Part Two presents three chapters on the geography of academic knowledge transfers. All the three contributions advance this field both by the methods applied and by their research results. In Chapter 5, Martin Andersson, Urban Gråsjö and Charlie Karlsson apply specific accessibility measures developed for Swedish spatial entities in order to empirically test the extent to which proximity to universities affects the location of business R&D. The underlying assumption of the study is that proximity to universities intensifies university-industry knowledge transfers and as such being closely situated to research done by universities is among the factors that drive the location of private research and development activities. In Chapter 6, Roderik Ponds, Frank van Oort and Koen Frenken take the geography of academic knowledge transfers field a step forward by comparatively studying the collaboration patterns of the three actors in scientific collaboration: universities, firms and governmental research institutions. Applying the methodology of social network analysis on Dutch data they test the hypothesis that research collaboration involving different kinds of organizations is more localized than collaboration in science between the same kinds of organizations. In Chapter 7, Attila Varga and Andrea Parag study the impact of international research collaboration on university patenting, applying recently collected data on selected departments of the University of Pécs. Instead of studying the impact of pure network size that has been common in earlier studies they focus on the research network structure effect. This is done by means of developing new partial and aggregate measures of network structure.

In Part Three two channels of academic knowledge transfers get closer investigation: academic spin-off firm formation and mobility of university graduates. In Chapter 8, Chiara Franzoni and Francesco Lissoni survey the notion of 'academic entrepreneur', as it emerges from a wide range of contributions to the economics and sociology of science. They suggest that the intensity and specific features of the entrepreneurial effort depend very much on the institutional characteristics of national academic systems. It is also emphasized that academic entrepreneurship results from the broad entrepreneurial agendas and not merely from the individual scientists' profit-seeking attitudes. Knut Koschatzky and Joachim Hemer in Chapter 9 report the results of 59 case studies among German academic spin-off firms. Their research questions relate first to the extent to which academic spin-offs are successful and second to the regional and non-regional factors that contribute to spin-off success. Koschatzky and Hemer also investigate the types of future promotional measures that could best contribute to the success of academic spin-offs and thus to economic and regional development.

Mobility of graduates is considered a main academic knowledge transmission channel in the literature. Effectiveness of this channel depends both on the quality of the inflows of students to universities and on the outflows of graduates. Accordingly two research questions are considered by the three chapters in the rest of Part Three. The first relates to competition among universities for students while the second to the size of the impact of graduate migration. In Chapter 10 Ádám Török approaches competition among universities from the perspective of competition in business. Török centers his contribution on two research questions. If competition is a fact in the university world, how can it be measured? Are often published university ranking lists a good measure of that competition? Taking an explicit spatial perspective, Günther Maier in Chapter 11 provides results of an empirical analysis of university competition in Austria. He frames the issue in terms of spatial competition between academic institutions and their corresponding market areas with the focus on the teaching functions of universities. The specific research question is whether universities in Austria act like spatial monopolists or like product differentiating suppliers in their competition for students. In Chapter 12, Alessandra Faggian, Philip McCann and Stephen Sheppard examine the relationship between tertiary educated human capital and regional performance. The way in which the study is being carried out is rather different to traditional analyses, which tend to focus on regional multiplier impacts. Instead, their chapter focuses on the extent to which the innovation dynamism of Britain's regions is interrelated with the interregional employment-migration behavior of university graduates.

In Part Four three studies report the experiences of university supported local economic development in three lagging regions of Asia, Europe and the United States of America. In Chapter 13 Daniel Schiller, Björn Mildahn and Javier Revilla Diez investigate the barriers that impede the transfer of knowledge between universities and enterprises in newly-industrialized countries, using the regional innovation system of Bangkok as an example. The chapter is structured along five hypotheses that are discussed theoretically and then analyzed empirically. Action recommendations for more efficient university-industry relationships in Thailand conclude the study. Imre Lengyel in Chapter 14 presents the bottom-up local development strategy of the Southern Great Plain region in Hungary. The strategy is centered around the University of Szeged. The peculiarity and the challenge of the case investigated by Lengyel lie in the fact that the university with outstanding scientific capacity is located in a lagging rural region of Hungary. The contribution is framed by the pyramid model of regional competitiveness. In the concluding study of Chapter 15 Hugh D. Sherman, William B. Lamb and Kevin Aspegren discuss the role universities can play in the care and feeding of high-growth businesses. The empirical sections of the chapter are based on information collected via interviews with development policy practitioners who actively participate in assisting high-growth businesses. Based on this information, Sherman, Lamb and Aspegren discuss the support needed by high-growth businesses, much of which can be facilitated by university programs. In particular, the chapter profiles the successful efforts of Ohio University's George Voinovich School for Leadership and Public Affairs in order to provide insight for future economic development initiatives.

I am most grateful to the colleagues who have agreed to share some of their results in this book. Earlier versions of most of the studies published in this volume were presented in a workshop organized at the Faculty of Business and Economics of the University of Pécs in October 2005. I would like to thank the Ohio-Pécs International Study Center, the University of Pécs and the Hungarian National Office for Research and Technology for supporting that event. Finally, I would like to thank Zita-Rózália Bedőházi for providing expert editorial assistance. Her endurance, care and attention in the production of the final copy have considerably enhanced the quality of this work.

# REFERENCES

Ács, Z.J. and A. Varga (2005), 'Entrepreneurship, agglomeration and technological change', *Small Business Economics*, 24, 323-334.

Ács, Z.J., L. Anselin and A. Varga (2002), 'Patents and innovation counts as measures of regional production of new knowledge', *Research Policy*, 31, 1069-1085.

Audretsch, D.B. and M. Feldman (1996), 'R&D spillovers and the geography of innovation and production', *The American Economic Review*, 86, 630-640.

Audretsch, D.B. and P. Stephan (1996), 'Company-scientist locational links: the case of biotechnology', *American Economic Review*, 86, 641-652.

Breschi, S. and F. Lissoni (2007), 'Mobility of inventors and the geography of knowledge spillovers. New evidence on US data', American Association of Geographers Annual Meeting, Special Session: *The Dynamic Geography of Innovation and Knowledge Creation*, San Francisco CA, April 17-21.

Feldman, M. (1994a), *The Geography of Innovation*, Kluwer Academic Publishers, Boston.

Feldman, M. (1994b), 'The university and economic development: the case of Johns Hopkins University and Baltimore', *Economic Development Quarterly*, 8, 67-77.

Feldman, M. and P. Desrochers (2004), 'Truth for its own sake: academic culture and technology transfer at Johns Hopkins University', *Minerva*, 42, 105-126.

Fischer, M., J. Diez and F. Snickars in association with A. Varga (2001), *Metropolitan Systems of Innovation. Theory and Evidence from Three Metropolitan Regions in Europe*, Springer, Berlin.

Florax, R. (1992), *The University: A Regional Booster? Economic Impacts of Academic Knowledge Infrastructure*, Avebury, Aldershot.

Goldstein, H. and J. Drucker (2006), 'The economic development impacts of universities on regions: do size and distance matter?', *Economic Development Quarterly*, 20 (1), 22-43.

Inzelt, A. and L. Szerb (2006), 'The innovation activity in a stagnating county of Hungary', *Acta Oeconomica*, 56, 279-299.

Isserman, A. (1994), 'State economic development policy and practice in the United States: a survey article', *International Regional Science Review*, 16, 49-100.

Jaffe, A., M. Trajtenberg and R. Henderson (1993), 'Geographic localization of knowledge spillovers as evidenced by patent citations', *Quarterly Journal of Economics*, 108, 577-598.

Koo, J. (2005), 'Agglomeration and spillovers in a simultaneous framework', *Annals of Regional Science*, 39, 35-47.

Koo, J. (2007), 'Determinants of localized technology spillovers: role of regional and industrial attributes', *Regional Studies*, 41, 1-17.

Krugman, P. (1991), 'Increasing returns and economic geography', *Journal of Political Economy*, 99, 483-499.

Malecki, E. and S. Bradbury (1992), 'R&D facilities and professional labour: labour force dynamics in high technology', *Regional Studies*, 26, 123-136.

Mueller, P. (2006), 'Exploring the knowledge filter: how entrepreneurship and university-industry relationships drive economic growth', *Research Policy*, 35, 1499-1508.

Parker, D. and D. Zilberman (1993), 'University technology transfers: impacts on local and US Economies', *Contemporary Policy Issues*, 11, 87-99.

Reamer, A., L. Icerman and J. Youtie (2003), 'Technology transfer and commercialization: their role in economic development', Economic Development Administration, US Department of Commerce.

Romer, P. (1986), 'Increasing returns and long-run growth', *Journal of Political Economy*, 94, 1002-1037.

Romer, P. (1990), 'Endogenous technological change', *Journal of Political Economy*, 98, S71-S102.

Saxenian, A. (1994), *Regional Advantage: Culture and Competition in Silicon Valley and Route 128*, Harvard University Press, Cambridge.

Sivitanidou, R. and P. Sivitanides (1995), 'The intrametropolitan distribution of R&D activities: theory and empirical evidence', *Journal of Regional Science*, 25, 391-415.

Varga, A. (1998), *University Research and Regional Innovation: A Spatial Econometric Analysis of Academic Technology Transfers*, Kluwer Academic Publishers, Boston.

Varga, A. (2000), 'Local academic knowledge spillovers and the concentration of economic activity', *Journal of Regional Science*, 40, 289-309.

Varga, A. (2002), 'Knowledge transfers from universities to the regional economy: A review of the literature', in A. Varga and L. Szerb (eds), *Innovation, Entrepreneurship and Regional Economic Development: International Experiences and Hungarian Challenges*, University of Pécs Press, Pécs, 147-171.

PART ONE

Setting the scene: analytical framework and knowledge inventory in theory, empirics and policy

# 2. What we know and what we don't know about the regional economic impacts of universities

## Harvey A. Goldstein

## 2.1 INTRODUCTION

Within the last 20 years there has been a burgeoning interest in the relationship between knowledge production and economic competitiveness at both the national and regional levels. Dramatic changes in regional economic conditions, globalization of both input and product markets, and the increase in the intensity of knowledge as an input to production across a wide range of products have contributed to institutions of higher education becoming key national and regional economic development actors.

In the context of the US, there have been additional factors driving university involvement in regional economic development, as an addition to the traditional, tripartite mission of teaching, research, and public service. These include: (1) changes in federal government research policy – most importantly the Bayh-Dole Act of 1980 – that give universities the intellectual property rights to patentable inventions arising from federal government sponsored research; (2) a long-term decline in federal government research funding in many disciplines; (3) growth in competing claims on state government spending and political limits on the ability of state governments to raise revenues through taxes; and (4) the competitive situation facing universities as they strive to increase their ranking – often measured by the total amount of research funding – and to attract 'star' faculty and the best graduate students. And while the role of universities as an 'engine' of regional economic development has been most vivid in the US, it has also become prominent in much of Europe and in parts of the Far East. Even in developing countries there is a great deal of interest by national governments, the World Bank, and other international donor agencies in investments in higher education as a tool for increasing national competitiveness.

There have been a large number of studies conducted and scholarly articles published in the broad area of how universities affect regional economic growth and development. They span a wide variety of specific research questions and methodological approaches, applied to many different study populations and temporal periods. Many studies have been motivated by policy interests, for example, assessing the efficiency or effectiveness of public investments in universities, *vis-à-vis* alternative instruments to stimulate economic development. Others have been motivated by more basic scientific interests such as how spillovers occur between universities and knowledge users among producers in the private sector, or by the effect of distance between where knowledge is produced and where it is adopted.

When reading this literature, however, there is a sense that we are often 'reinventing the wheel' on the one hand, and on the other hand, a low level of robustness of results. With results varying significantly depending upon data and methods used, economic development policy officials and leaders in higher education often do not have solid empirical bases to decide what role(s) universities should play in economic development, how universities can be effective in stimulating economic development, and how limited public resources for economic development might be most efficiently allocated between universities and other economic development institutions.

Given the current importance being placed upon universities in the US, Europe, and many parts of the developing world as engines of regional economic development, it may be propitious to take stock of what we know and what we do not know about the relationships between universities and regional economic development, including how these relationships can be estimated. After providing some definitions of concepts and categories to help fix ideas, the central part of the chapter identifies the key policy-relevant questions about the impacts of universities on regional economic development, and the extent to which we have reliable answers to these questions. We finish by addressing 'where should we go from here', by offering suggestions of the most fertile and valuable directions for further research.

## 2.2 SOME DEFINITIONS

Before beginning to address these questions, we first provide some clarification and definitions of some of the basic terms used here.

First, we restrict this discussion to those institutions of higher education that award bachelors, masters, and doctoral degrees. For the most part this means research universities. This definition eliminates, for example, community colleges in the context of the United States, though we acknowledge that community colleges can and do play an important role in economic development in human capital creation by providing specific training programs for skills in demand in the regional economy.

Second, we conceive of modern research universities as multi-product organizations (Goldstein et al. 1995). Their outputs include not only 'know-how' embodied in human capital, but also research, technology assistance and transfer, technology development, real estate investment and development, entertainment, and leadership. Universities allocate their resources among various activities to achieve some desired mix of outputs that meet their missions and institutional objectives.

Third, universities span both public and private in terms of ownership. The principal difference is that public universities are subject to many more state government regulations. It should be noted that the share of public university revenues from state governments has been steadily decreasing over time, though at the same time there has been an increase in autonomy. Thus a better term to use for public universities is publicly supported universities. Land-grant universities are a subset of publicly supported universities. Their traditional mission has been oriented to providing human capital, applied research, and technical assistance for the development of their state's agricultural and industrial sectors. Most states have one designated land grant university. Private universities are subject to all federal laws and regulations, but are not bound by most state regulations. However, they often still receive some revenues from state government through research funding or tuition grants for students.

Fourth, we use the term 'region' somewhat generically as a bounded geographic area that is smaller than a nation but is larger than a single city. In the context of the US, metropolitan statistical areas (MSAs) are the regions that come closest to an 'integrated' economic area. MSAs include the county in which the central city is located as well as suburban counties surrounding the central county. It should be noted, however, that in the policy discussion of how universities can be used to stimulate regional economic development, often states are the targeted geographical area since economic development policy is increasingly a state government function in the US.

## 2.3  THE KEY POLICY-RELEVANT QUESTIONS

Despite the eclectic nature of the literature concerning the relationship between higher education and regional economic development, we can identify six sets of questions that motivate researchers and focus the attention of policymakers.

### 2.3.1  What Kinds of Changes in Regional Economic Activity, or Outcomes, can we Reasonably Expect Universities to Stimulate?

The literature suggests that universities can stimulate the following kinds of economic impacts: increases in aggregate regional income and employment; productivity gains; increases in innovative activity; new business start-ups; and increased regional capacity for sustained economic growth.

There is no question that universities increase aggregate regional economic activity. Universities are typically large economic organizations that export educational services, employ a large number of workers, attract students (and their spending) to the region who would not be there otherwise, and purchase a variety of services and commodities from firms located in the region (Felsenstein 1994; Thanki 1999). Thus, increases in the size of the university will increase output, earnings, employment, and tax revenues within the region through a multiplier effect (Leslie and Brinkman 1988). The magnitude of the regional multiplier effect of a typical research university's spending will vary by the size and economic structure of the region, but generally will be in the range from 1.5 to 3.0 (Goldstein et al. 1995). While this may be a sufficient reason for policymakers to increase the budgets of publicly-supported universities, there are opportunity costs for using public funds for expansion of universities. The magnitude of the opportunity costs is difficult to assess using standard multiplier analysis such as regional input-output analysis.

There is less agreement about whether the other types of potential impacts at substantively important magnitudes can be attributable to universities. Empirical studies using regional production functions have shown that universities, on average, lead to increases in productivity growth and innovation in the region, controlling for other factors (e.g., Martin 1998). Using a knowledge production function approach, increases in regional productivity and innovative activity are most often interpreted as having stemmed from university research activity via knowledge spillovers to firms in the region (Varga 2000; Riddel and

Schwer 2003). Likewise, studies in the entrepreneurship literature have focused on the incidence of faculty and other university-based researchers creating new tech-based businesses as spin-offs from university-based research projects that in turn have led to commercial applications (Smilor et al. 1990; Steffensen et al. 2000; Feller et al. 2002). Yet other studies have found that aggregate regional productivity gains or the incidence of new firm formation stemming from university-based research may be substantively or statistically low because such impacts are contingent upon regional economic conditions (Varga 2000, 2001) and the internal organization and policies of the university itself (DiGregorio and Shane 2003). In the case of whether universities lead to increases in regional innovative activity, in addition to the issue of contingent conditions, there are considerable measurement issues. Innovative activity has been almost invariably measured by patents assigned to the university and to firms in the region. Certain types of innovations are not usually patentable, e.g. software, and the sources of patent data may not provide the actual location of the innovative activity because of multiple names and organizations to which the patent has been assigned.

Finally, increases in regional capacity for sustained economic development have been argued by some as perhaps the most important type of impact of universities. Yet empirically estimating its magnitude is fraught with conceptual and measurement problems. First, the concept of regional capacity itself has a number of variants that, while overlapping, are operationalized differently. Research on creative milieu and learning regions (e.g. Camagni 1991; Maillat and Lecoq 1992; Cooke 2001) have emphasized the role of universities in attracting bright, creative, and entrepreneurial individuals to a region because they like to live and work in areas where there is a concentration of similar people and because of the synergy that can result from the opportunities for serendipitous and informal interactions. Florida's (2002) work on creative regions argues that regions that will be economically competitive will be those can attract the 'creative class', and members of the creative class will be attracted to places that have institutions of higher education since such places tend also to be more tolerant, diverse, and 'cool'. Studies of successful regional innovative strategies (Landabaso et al. 2000; Cooke 2001) have keyed into the development of a region's knowledge infrastructure in which institutions of higher education are centerpieces. Still another concept of gain in regional capacity is the improvement in the structures of leadership and governance, and wiser public policy brought about by university leaders and prominent faculty engaged in regional problem-solving and helping to forge consensus among conflicting regional interests. Because of the inherent measurement difficulties in these concepts of regional capacity, almost all empirical studies to date have been case studies. While case studies as a method can provide rich contextual information about the role of universities, it is difficult to attribute impacts in the face of multiple factors and to generalize when conditions vary widely from case to case.

**2.3.2  How do Universities Stimulate Economic Development?**

A basic task in research about how universities can stimulate regional economic development is to distinguish different activities or functions within universities that have differential economic development impacts. We can identify at least seven different university activities that may potentially lead to economic development impacts (Goldstein et al. 1995):

1. Development of human capital (teaching)
2. Creation of knowledge (research)
3. Transfer of existing know-how (technical assistance)
4. Technological innovation
5. Capital investment
6. Regional leadership and governance
7. Co-production of knowledge infrastructure and creative milieu

While these activities are conceptually distinct, in practice they may overlap. Each type of activity, however, tends to have a distinct pattern of effects on the regional economy. The first two, teaching and research, constitute the traditional, or core, functions of the research university. Technical assistance through extension offices has been a third core function of land-grant universities in the US, but many public non-land-grant, and even some private universities have added it to their missions. Technological innovation is a relatively new institutionalized activity in many research universities, but perhaps has received the most attention in the popular media and clearly has been the most controversial in terms of its appropriateness as a university function.

Universities, whenever they have added to their physical facilities, invest capital in the built environment, which creates economic activity. What is new, though, is that a number of universities now invest in real estate in the region – office buildings, research parks, incubators, upscale housing – as a revenue producing activity (Wiewel 2004). As evidence, in 1983 there were 25 institutional members of the Association of University Real Estate Officials in the US and Canada. In 2007 there were over 190 institutions, representing most major universities. Likewise the number of university owned research parks has proliferated in the last 20 years (Luger and Goldstein 1991, 2006). The provision of regional leadership and governance has been discussed above as potential outcomes and is a form of university engagement in their respective communities. And the co-production of the region's knowledge infrastructure and creation of a particular kind of milieu is a passive activity that occurs by the very nature of a university existing in a region, but as discussed above, can lead to increased long-term capacity for sustainable economic development.

The mechanisms and magnitude of the regional economic development impacts of some of these activities are better understood than others. Given the extensive literature, we shall be selective in our discussion.

**Development of human capital (teaching)**
Teaching, or 'learning by learning', has been one of the two core missions of the modern universities, though within research universities it is often difficult at the graduate level to separate human capital creation from knowledge creation. While it has traditionally occurred in classrooms or labs on university campuses, more recently it occurs through distance learning, as part of industrial extension off-campus, or even in non-degree community education programs. This diversity in the way teaching now is conducted also complicates how the economic development impacts can be estimated.

Although much has been written about the economic impacts of higher education measured as social rates of return to investments, curiously, the regional economic development impacts of teaching have not been emphasized in the recent literature. We know there is no more important ingredient for sustained regional economic development than a supply of creative and highly skilled members of the labor force. Universities, of course, are the institutions that we rely upon most for providing advanced training in scientific, engineering, professional, and many technical areas. Yet this particular economic development role of universities has often been discounted, and sometimes perceived as a diversion of resources from direct economic development activities. One reason for this is that the region realizes an economic development impact from teaching only to the extent the students being trained take jobs within the regional labor market. Otherwise the human capital investment leaks out of the region.

The best measures for university activity in creating human capital are the number of degrees awarded annually, disaggregated by discipline and by type of degree, but only for those graduates who remain in the regional labor market. It is relatively easy to obtain data on the number of degrees awarded for each university, but the migration data are rather difficult to obtain. Universities do not track the location of their graduates in a systematic way, so either estimation based upon aggregate regional census data or special surveys would be needed. Which specific measures of university activity are best – for example, total degrees awarded, number of graduate degrees, or number of graduate degrees in science and technology fields – is not clear either. In econometric analyses we often are constrained by concerns about multi-collinearity and degrees of freedom so only one or two measures are ever included. Goldstein and Drucker (2006), after experimenting with a number of alternative measures, chose total number of science and engineering degrees, and the ratio of graduate science and engineering degrees to total degrees awarded. Their estimated model results indicated that the total number of science and engineering degrees awarded was negatively, though weakly, related to increases in average regional earnings, while the ratio of graduate science and engineering degrees was positively and strongly related to change in average regional earnings. The former result is interpreted to be because of a crowding effect in the local labor market.

**Creation of new knowledge (research)**
The second core mission of the modern university is the creation of new knowledge through basic scholarly or scientific research. The term basic refers here to research that focuses on answering general scientific questions or problems rather than on particular technological questions (Trajtenberg et al. 1992, p. 3). This definition does not imply that all research that leads to the uncovering of general theory necessarily starts with theoretical interests. Indeed in the history of science a number of important theoretical breakthroughs began with attempts to solve applied problems (Stokes 1997; Rosenberg 2002). Rather, this distinction between basic and applied research is made because it implies that basic research is difficult for individual firms or organizations to appropriate. For that reason the geographical impact of basic research is likely to be wide, with a flat spatial gradient. And because basic research is not directly appropriable, and has many of the qualities of a 'public good', it is most likely going to be funded by government agencies rather than by private industry. That the source of most basic research funding is government also has an important effect on the spatial distribution of research funds. To the extent that research funding now represents a major source of the overall budget and expenditures of research universities, government allocation of research funding has direct

implications for the magnitude of local and regional multiplier effects of universities.

Aside from the spending multiplier effect, university research activity induces economic development in the region to the extent that firms and other organizations within the region are able to use that knowledge to increase their productivity or innovativeness. This capacity clearly varies. We know that it depends upon the match between knowledge outputs and the technology area of the firm. A university, for e.g., can be world class in genetics and DNA research but if there are no biotechnology firms in the region, then the potential economic benefit will not be realized. But the literature also suggests that there are resource and cultural dimensions at work. Ács et al.'s (1994) empirical work suggests that larger firms may be more adept at exploiting knowledge produced within their own labs, but smaller firms, with fewer resources to invest in knowledge acquisition, focus on exploiting spillovers from university labs. But even here, many smaller firms may still be at a disadvantage because they may lack the resources and expertise to be able to assess which scientific outputs may lead to increases in productivity or product innovation. Similarly, the culture of some firms may inhibit communication between scientists in the university and personnel in the firm.

The most common measure for (basic) knowledge outputs of universities is the annual amount of research expenditures, all sources. Perhaps a better measure, at least in the context of the US, is the amount of federal government funded research, since it excludes private industry funding. Measuring the capacity of regional firms to utilize the research outputs of universities has been more difficult. The percentage of larger firms in particular industry categories (from size of firm distribution data) may serve as a proxy for this capacity. Goldstein and Drucker (2006) found that a region's aggregate university research expenditures was related to regional economic development outcomes, controlling for other factors including human capital creation and university involvement in technology development. The magnitude of the effect varied however, by size of the region: it was strongest in medium-sized regions employment of 75,000-200,000, weaker in the largest-sized regions where other non-university factors 'overwhelm' university research effects, and negative in the smallest regions, where there is a paucity of firms to absorb the spillovers from university research.

**Transfer of existing know-how**
This includes the outreach, extension, and public service functions originally associated with the mission of land-grant universities, but that

are now solidly established in most public and many private research universities in the US. It is an activity practiced through industrial and agricultural extension centers, small business assistance centers, economics and business research bureaus, and the clinical programs within the professional schools of universities including business, medicine, public health, social work, planning, public policy, and law. It is also carried out by individual faculty serving as consultants.

Among the best-known and most widely replicated programs for university-based technical assistance is the Manufacturing Extension Partnerships (MEPs). The MEPs are a loose federation of over 400 centers partially funded by the National Institute of Standards and Technology (NIST) of the US Department of Commerce and operated and staffed largely by universities and community colleges. MEPs focus on helping small and medium-sized manufacturing firms identify and adopt new and advanced technologies to become more competitive. Shapira (2001) found the net benefits of the MEP program to be positive, but also identified a number of organizational and institutional barriers to even greater effectiveness.

Another popular program in the US affiliated with a large number of universities is the small business development centers (SBDC), a national program partially funded by the US Small Business Administration. The SBDC program focuses on providing a wider variety of services and types of assistance compared to the MEPs, including helping to prepare business plans, assisting the development of marketing plans, financial analysis, accessing capital, and start-up assistance. A number of evaluations focused on the extent of the benefits of SBDCs have been conducted over 20 years (Chrisman and McMullan 2002).

Measures are often 'counts' of businesses and organizations receiving assistance, or person-days of assistance provided. Surveys of satisfaction of organizations receiving assistance are sometimes administered to indicate the value of services provided by universities. Public universities have incentives to generate these measures to show their engagement. Yet because the activities can be so disparate and spread among a number of units within universities, and there exist no standardized measures adopted among universities, it is difficult to measure their economic development effect statistically. As a result, the estimates we do have are based upon case studies and often employ perceptual measures of economic benefit.

**Technological innovation**
Technology development as a university activity refers to the production

of inventions that have commercial applications. While such outputs often stem from basic research conducted in university labs, inventions may also be generated as a direct and intentional outcome of industry funded research. This activity probably receives more popular and political attention than any other, as it has the potential to lead to huge publicity for the inventor/scientist, and windfall revenues for the university that owns the patent and sells licenses based on the invention. Yet the probability of hitting the 'jackpot' is rather small. Nevertheless, the increase in the number of research universities in the US that have now invested considerably in their infrastructure to support patenting and licensing has been dramatic. Prior to 1970, there were only five universities with technology transfer programs. Between 1970 and 1979, 15 additional universities initiated technology transfer programs. But in the 1980s the number of universities that started technology transfer programs was three times the number that had existed in total in 1979, and between 1990 and 2004 that number doubled again (AUTM 2004).

The most common measures of university technology development are annual number of patents issued and value of licenses sold. Other measures are number of spin-off firms formed based upon technology developed in the university and number of citations by private industry R&D labs of patents issued to a university. All of these measures have some limitations that have been amply cited in the literature. Among them, not all inventions are patentable, the incidence of spin-offs from university-based technology development is still rare, and the full extent of the economic value is difficult to trace since not all such 'knowledge' is appropriable. Thus utilizing the available measures leads to an underestimate of the impact of this activity. Agrawal and Henderson (2002), for example, found in their study of knowledge transfer from MIT's mechanical and electrical engineering departments that patenting represented only a small portion of modes of knowledge creation and knowledge transfer, and that patent counts are not representative of the extent and incidence of knowledge transfer from universities and to private firms. On the other hand, that technology development is a form of knowledge that generally is highly appropriable makes this activity the most controversial, as it runs counter to the tradition of a university in which knowledge creation is regarded as 'open science' (Merton 1973) and the products of research located within the public domain.

**Capital investment**
Universities spend considerable amounts of money each year building, renovating, and maintaining facilities to carry out their core academic

missions (Abramson 2006). But they also invest in research parks, incubators, and advanced technology labs, and off-campus housing for staff and students, as well as in infrastructure that also serves the larger region, including roads, power stations, water delivery systems, and athletic and recreational facilities. Reference was made earlier to the proliferation in the number of universities that have their own real estate offices for management of their property portfolios. Depending on regulations and restrictions, some universities invest in start-up business ventures, and some also invest generally in commercial real estate. Whatever the motivation to undertake capital projects, the economic impacts within the region just due to the multiplier effect can be considerable.

The economic development effect of capital investment is most commonly estimated by the use of regional input-output models. Many universities hire consultants (e.g. Urban Land Institute, 2006) to conduct such studies as they are often used to justify larger capital budgets in the case of public universities and more favorable tax treatment in the case of private universities. Calder and Greenstein (2001), Cortes (2004), Habily (2004), Sherry (2005), and Szatan (2004) provide descriptive accounts of universities as developers.

Because almost all studies have been in the form of individual case studies using disparate measures and methods, lack of comparable data has inhibited the development of rigorous cross-sectional econometric estimates of the effects of university investment impacts.

**Regional leadership and governance**
Top administrators and distinguished faculty of universities are increasingly asked to serve on local and regional government boards, committees, and commissions to provide leadership in addressing critical social, economic, and environmental problems or, as representatives of an important institutional stakeholder, to help plan for the region's future. The form of the leadership can come from providing specific expertise, say, in improving environmental quality or helping to frame an economic development strategy for the region, to providing a certain type of moral authority that comes from the tradition of universities having a fiduciary role in society that rises above narrow and self-serving interests. Of course, by helping to improve the region's quality of life, the university's self interests are indirectly being served.

Systematic measures of the incidence and amount of this type of activity do not exist, and estimates of the economic development impact are exceedingly difficult to quantify because of lack of measures and because often the effect is so indirect. Indeed, it is highly idiosyncratic, and may depend upon the culture of the university and the character of town-gown relations within the region. Again, we have to rely upon individual case studies in the form of narratives to document university activity and economic development outcomes. A study by the Initiative for a Competitive Inner City and CEOs for Cities (2002) provides a set of exemplary case studies of individual universities partnering with city and business leaders that led to successful urban economic revitalization programs in the US. While highly valuable in identifying 'best practice' and what is possible in often disadvantaged areas, these cases are not representative. Although the number of such narrative studies has been growing, it is risky to try to generalize their results.

## Co-production of knowledge infrastructure and milieu

Universities serve as repositories of stored knowledge in the form of faculty, non-faculty researchers, books and documents. They also serve as nodes along with other knowledge-generating organizations in the region, and provide linkages to these other nodes. Thus universities are critical actors in a particular type of infrastructure that helps increase the productivity of knowledge workers and firms through spillovers, and increases the attractiveness of the region to bright, creative, and entrepreneurial individuals, who may not have any direct economic relationship with the university or its faculty (Smith 1997; Goldstein 2005). The concentration of such individuals in the region often leads to a peculiar cultural and social milieu that, in turn, reinforces the attractiveness of the region for further rounds of in-migration of similar people.

This mostly passive activity of universities is felt to be an increasingly important way that universities stimulate regional economic development, but again, it is difficult to measure. A knowledge production framework can take into account these economic development effects, but can not separate them from direct knowledge spillovers stemming from human capital creation, knowledge creation, and technology development. Surveys of new technology-based firms have been used to estimate the effect of these activities by identifying those firms that have a direct economic relationship with the university from those that do not, and by asking how the presence of the university entered into their locational decision, and the specific benefits, if any, that the firm receives by being located within the region of a research university (Luger and Goldstein 1991).

### 2.3.3  What Regional Factors Condition the Magnitude of Universities' Impacts?

The literature suggests that certain regional conditions matter in the extent to which universities stimulate regional economic development. Which particular factors are most important?

Both the theoretical and empirical literatures posit that certain regional conditions matter in whether universities are significant drivers of economic development. The most consistently cited factors are size of region, presence of agglomeration economies, industrial structure, and the 'culture' of the region. There is some disagreement about the effect of size of region on magnitude of regional economic impact, however. There is also a lack of consensus about how these factors are best conceptualized and measured. In addition, some researchers have either hypothesized or found that size of firm distribution, firm ownership status, and entrepreneurial climate are important.

The literature considers regional size important for at least two reasons: (1) larger regions will likely have a larger number of firms, or 'critical mass' of high technology firms that can 'absorb' the outputs, or knowledge spillovers, from universities; and (2) larger regions will tend to be more diverse, offer a wider array of producer services, and have a more developed knowledge infrastructure. Yet these are both dimensions of agglomeration economies, where size of region serves as a proxy. In addition to the two mentioned above, there is also the Marshallian concept of agglomeration economies about density of firms. A concentration of firms producing the same or related products in close proximity to one another should have increased capability of absorbing knowledge outputs from universities and also sharing tacit knowledge among themselves. Once we separate the concepts of size and agglomeration, we can understand how some small and medium-sized regions with particular industry structures and firm characteristics in combination with the concentration of knowledge assets of a university, can provide agglomeration economies (Goldstein and Drucker 2006). The well-known case study of Johns Hopkins University and the city of Baltimore by Feldman (1994) and a follow-up by Feldman and Desrochers (2004) offer an excellent example of how the absence of producer services, and venture capital (plus an underdeveloped entrepreneurial culture) have been critical missing elements in the development of a concentration of technology-based firms in Baltimore, despite The Johns Hopkins University consistently having the largest research expenditures of all US universities, and Baltimore being a large, diverse metropolitan area.

Empirical studies (Jaffe 1989; Bania et al. 1993; Ács et al. 1994; Varga 2001) have each shown marked differences among industries, even among high technology industries, in the degree to which firms have both the capability and economic incentives to absorb the knowledge outputs of universities. Obviously, in cases of underdeveloped regions with little or no technology-based industries, the absorption of knowledge outcomes of universities will be low. There may also be a mismatch between the particular knowledge assets and strengths of the universities and the particular high technology industries in the region. Either way, the human capital created in the university in terms of its graduates will out-migrate in search of better career opportunities commensurate with skills and training, and knowledge outputs in the form of research products and technology will be absorbed, if at all, by firms located outside the region. Varga (1997) has argued that the data we use to represent regional industry structure also distinguish between routine and non-routine production activity. Finally, there is variation among industries in the degree to which high technology industries rely upon research outputs from universities *vis-á-vis* from their own research labs or from other industrial R&D labs (Lawton-Smith, 2006).

The 'culture' of the region as a mediating factor often refers to the region's entrepreneurial tradition and support for risk taking by firms. This quality might partially explain why some firms within particular sectors are more likely to seek out and absorb knowledge outputs from universities, as well as differences in the incidence of local spin-offs and start-ups by faculty and graduates from university-based research (Keane and Allison 1999). A related dimension of regional culture is the degree of collaboration, cooperation, and 'connectedness' among firms and between firms and researchers in universities and other knowledge-creating institutions. This factor has been emphasized more in the literature on 'learning' and 'creative' regions, and creating effective regional innovation systems. Yet it would seem to be pertinent to helping understand why in some cases there are close relationships and many opportunities for informal contacts between researchers in universities, and scientists, engineers, and entrepreneurs in private industry, and in other regions this does not exist (Dosi et al. 1988; Faulkner and Senker 1995; Storper, 1995; Cooke and Morgan 1998; Karlsson and Zhang 2001). Measurement issues represent a chronic challenge to incorporating concepts of regional culture into empirical studies of estimation of university impacts.

Other regional factors sometimes cited are size-of-firm distribution and firm ownership status. Size-of-firm seems to matter, but there is disagreement among studies about whether large or small firms are more likely to receive knowledge spillover benefits from universities. The results vary in part because we should distinguish more carefully between capacity to absorb – an advantage for larger firms – and greater dependency on knowledge spillovers as externalities – higher for smaller firms because of more limited R&D resources and smaller information span. There may also be a preference of university researchers to interact with researchers in larger firms (Howells 1986).

From studies of site-selection criteria (Galbraith and De Noble 1988), single-plant establishments are considered to be more likely to have interactions with both other firms and with university researchers in the region, because they are more dependent upon local sources of knowledge. Branch plants of multi-locational firms have access to the R&D facilities of the same firm, and are less constrained from entering into a cooperative relationship with only nearby universities.

Perhaps what is most noteworthy in reviewing the effect of regional conditions on university impacts is that the unexplained variation still tends to be relatively high, i.e., these may be the most important factors that influence absorption of knowledge spillovers from universities, but there are others that either we do not know about or are too difficult to measure, or are simply random.

### 2.3.4 What are the Influences of Universities' Internal Organization, Culture, and Regulations?

Universities vary in their missions, in the allocation of resources among different functions and outputs, and in the regulations they face that may constrain their activities in economic and business development. These variations are not only significant across national systems of higher education due to historical differences in the societal role of higher education, but also within nations, particularly those in which governance has been decentralized to the regional or state levels and where private universities are prominent, such as in the US. The question is: how much do these differences matter in the magnitude of university economic impacts?

The 'entrepreneurial turn' in higher education, mentioned in the introduction, describes research universities becoming more active in economic development functions rather than letting economic development outcomes stem mainly from their principal traditional functions of teaching and research. Economic and business development has been added to the mission statements of many public universities as an expansion of the more traditional 'public service' mission. Resources for the development or expansion of patenting and licensing offices, and university support for faculty spinning off new businesses has expanded rapidly on many campuses. At the same time in many publicly-supported universities, regulations prohibiting or restricting university employees from engaging in various forms of interaction with industry – from consulting to ownership – have been reduced. There is also widespread discussion and some movement to provide positive incentives to faculty to become engaged in activities that lead to economic development such as receiving credit for commercial applications in promotion and tenure decisions and in the award of merit salary increases. Yet despite the general entrepreneurial turn, there still exist significant differences among universities.

For the most part, studies that have attempted to assess the importance of these internal factors on university economic impact have been exploratory in purpose, and relied upon single or comparative case study designs (e.g. Bercovitz et al. 2001). Blumenthal et al. (1986) and more recently Renault (2003) have used cross-sectional designs to help control for other factors and to be able to generalize to larger populations. Renault chose to study the institutional policies and practices in 12 universities in the southeastern US and then drew a cross-sectional sample of faculty among the 12 universities in selected disciplines to identify relationships between organizational attributes and faculty attitudes towards engaging in entrepreneurial activities. She found that differences in university mission statements did not influence faculty attitudes toward patenting or spinning off companies. Faculty members at private universities were more likely to believe patenting was inappropriate for academia, or that their work was too basic or theoretical for commercialization, compared to public university faculty. A sizable percentage of faculty members felt that conflict of interest policies at their respective universities were inhibiting the spinning off of companies by faculty and other university researchers, and that the restrictions were becoming tighter compared to several years earlier. The greater the financial incentives provided by the university to faculty to patent (in terms of the revenue split), the greater the incidence of patenting activity. And finally, the reward structure at the departmental level in terms of tenure and promotion was closely associated with the decision of faculty of whether to engage in either patenting or spinning off companies.

Other studies have found that there is a relationship between university prestige and research and entrepreneurial capacity (Zucker et al. 1998; DiGregorio and Shane 2003). The latter found that the presence of university policies of making investments in start-ups and allocating a large share of royalties to the inventor are associated with high spin-off rates.

From the work that has been done there seem to be policies and practices at the university level that do affect faculty attitudes and behavior with regard to engaging in technology development, but that the norms of academia in general, differentiated somewhat among different disciplines, may be more important. Further research needs to be done.

### 2.3.5 How far do Knowledge Spillovers from Universities Extend?

The spatial dimension of knowledge spillovers from universities has become an important issue of the allocation of resources and funding for higher education by regional and state governments. When universities are seen as critical actors for regions to become or remain competitive in the knowledge economy, there is pressure to initiate or expand degree programs, research centers and institutes, and faculty expertise in a large number of locations and communities, often leading to inefficient and duplicative resources in some places, and insufficient resources to attain critical mass in others. Thus, having more precise knowledge of how far the positive externalities of universities extend could help improve the allocation of resources for higher education.

In terms of knowledge spillovers of university research, there is general agreement that (1) university research encourages industrial R&D within the same region (but not vice versa (Jaffe 1989)); and (2) firms close to major centers of academic research have a major advantage over those located at greater distances (Salter and Martin 2001). Yet there is still substantial variation in empirical results. First, spillovers from basic research within universities have been found to be much less localized than for applied research and technology development, because of differences in appropriability. Second, there are significant differences in the importance of private R&D labs' proximity to universities among industry sectors and technology areas. Third, the spatial gradient may be different among different types of universities.

Anselin et al. (1997) found that spillovers of university research leading to innovation in private industry in the US extended over a range of 75 miles from the university. Also in the US, Mansfield and Lee (1996) showed that firms invested their R&D in universities within 100 miles from the location of the firms' R&D laboratories. In Europe, however, a study by Beise and Stahl (1999) found that among 2,300 German firms, those that located near universities or Fachhochschule did not have a higher probability of using publicly-funded research, while Cooke et al. (2002) in their study of Tel Aviv, Belfast, and Cardiff found that universities have stronger linkages with national and international businesses than local and regional firms. Koo (2003) found that the degree to which knowledge spillovers are localized varies considerably by industry structure and the knowledge intensities of the industries in the region. Goldstein and Drucker (2006) experimented with different distances for measuring spatial spillovers of universities outwards beyond their MSA. They found spatial spillovers most significant at 60 miles from the centroid of the MSA with the research university. Finally, the spillovers from the most prestigious and highly ranked universities are likely to be much geographically wider than those from lesser ranked universities.

### 2.3.6 The Distribution of Economic Development Benefits from Universities

To the extent that universities stimulate regional economic development, how are the net benefits distributed? Do the positive externalities accrue to large corporations, or to small and medium-sized companies (SMEs)? With growth in employment and average earnings due to universities, how are these gains distributed between current residents and new in-migrants? Do the earnings and income gains accrue to only those with the highest skills and educational attainment, e.g. 'knowledge' workers, or do the benefits trickle down to production and lower-skilled service workers?

These are questions frequently asked by policymakers about a wide variety of economic development policies and strategies, and by and large they are rather difficult to answer. It is sufficiently hard to estimate the aggregate economic effects, let alone account for how they are distributed. The difficulties lie in the available secondary data and the general methodological problem of attribution when there are multiple factors leading to changes in income, jobs, and earnings. For these reasons, detailed case studies with the collection of primary data at the firm and employee level may offer the best research designs. Yet the sensitivity of the data will pose high risks of non-response.

Much of our extant knowledge about the distribution of economic benefits from universities comes from economic impact studies using regional input-output models (e.g. Luger et al. 2001; Walden 2001; ICF

Consulting, 2003). These provide estimates of how benefits are distributed among industry sectors, though changes in income distribution cannot be traced. And these results only take into account a limited set of activities by which universities affect regional economic development. In the course of conducting input-output studies of the spending impacts of universities, primary data about the location and type of firms from whom the university purchases equipment, supplies, and services can reveal how some benefits may be distributed geographically and by size of firm.

## 2.4  WHERE SHOULD WE GO FROM HERE?

Studies of the regional economic development impacts of universities have used a variety of methodological approaches, measures, data, time frames, and foci on particular functions of universities. In part because of this diversity, the body of results that we have to date lacks a high level of robustness. When policymakers ask about how universities might be most effectively used in achieving regional economic development goals, we often have to rely upon hunches rather than upon solid empirical evidence.

There seem to be a number of directions for future research that can prove to be fruitful. This author will briefly mention only a few that are biased in the direction of helping us understand how universities can be most effectively used in fashioning regional economic development policy.

The first is to expand the population of universities so as to include those institutions of higher education that are focused on teaching, training, and technology transfer, rather than focusing on just research. This would include community colleges and smaller universities in the US that have regional development missions, and Fachhochschule and polytechnics in Europe. The knowledge spillovers from university research may not be the most important way that institutions of higher education stimulate economic development in many regions, particularly those in declining industrial regions and perpherial areas and non-metropolitan areas.

Second, how universities can best contribute to their region's knowledge infrastructure, through interaction with other knowledge-producing organizations and intermediary institutions, is an underappreciated issue and underdeveloped part of the literature. A series of comparative case studies for exploring what factors produce the most conducive environments for collaboration and informal contacts between university and industry researchers (Meyer-Krahmer and Schmoch 1998) might hold the most methodological promise here.

Third, more work on helping us understand how far the regional economic development benefits of institutions of higher education extend, for different university functions, would be useful for more wisely deciding funding allocation when government budgets become increasingly tight and other government priorities crowd out expenditures for higher education.

Finally fourth, how important are internal university organization, reward structures, policies, regulations, and culture in affecting universities' regional economic development impacts? How important are they relative to regional economic conditions and strength of the university? This is mentioned in part because, in principle, policymakers have control over internal university policies, whereas regional economic conditions are much more difficult and take much longer to change. Studies here might be cross-national to take into account differences in national systems of higher education.

# REFERENCES

Abramson, P. (2006), '2006 college construction report', *College Planning and Management*, February.

Ács, Z.J., D.B. Audretsch and M. Feldman (1994), 'R&D spillovers and recipient firm size', *The Review of Economics and Statistics*, 76 (2), 336-340.

Agrawal, A. and R. Henderson (2002), 'Putting patents in context: exploring knowledge transfer from MIT', *Management Science*, 48 (1), 44-61.

Anselin, L., A. Varga and Z. Ács (1997), 'Local geographic spillovers between university research and high technology innovations', *Journal of Urban Economics*, 42 (3), 422-448.

Association of University Technology Managers (2004), AUTM US Licensing Survey, FY 2004.

Bania, N., R. Eberts and M. Fogarty (1993), 'Universities and the startup of new companies: can we generalize from Route 128 and Silicon Valley?', *The Review of Economics and Statistics*, 75, 761-766.

Beise, M. and H. Stahl (1999), 'Public research and industrial innovations in Germany', *Research Policy*, 28 (4), 397-422.

Bercovitz, J., M. Feldman, I. Feller and R. Burton (2001), 'Organizational structure as a determinant of academic patent and licensing behavior: an exploratory study of Duke, Johns Hopkins, and Pennsylvania State Universities', *Journal of Technology Transfer*, 26, 21-35.

Blumenthal, D., M. Gluck, K. Louis, M. Stoto and D. Wise (1986), 'University-industry research relationships in biotechnology: implications for the university', *Science*, 232 (4756), 1361-1366.

Calder, A. and R. Greenstein (2001), 'Universities as developers', *Land Lines*, July, 1-4.

Camagni, R. (1991), 'Local milieu, uncertainty and innovation networks: towards a dynamic theory of economic space', in R. Camagni (ed.), *Innovation Networks: Spatial Perspectives,* London, UK: Pinter.

Chrisman, J.J. and W.E. McMullan (2002), 'Some additional comments on the sources and measurement of the benefits of small business assistance programs', *Journal of Small Business Management*, 40 (1), 43-50.

Cooke, P. (2001), 'From technopoles to regional innovation systems: the evolution of localized technology development policy', *Canadian Journal of Regional Science*, 24 (1), 21-40.

Cooke, P. and K. Morgan (1998), *The Associational Economy: Firms, Regions, and Innovation*, Oxford, UK: Oxford University Press.

Cooke, P., C. Davies and R. Wilson (2002), 'Innovation advantages of cities: from knowledge to equity in five basic steps', *European Planning Studies,* 10 (2), 233-250.

Cortes, A. (2004), 'Estimating the impacts of urban universities on neighbourhood markets: an empirical analysis', *Urban Affairs Review*, January, 342-375.

DiGregorio, D. and S. Shane (2003), 'Why do some universities generate more start-ups than others?', *Research Policy*, 32, 209-227.

Dosi, G., C. Freeman, R. Nelson, G. Silverberg and L. Soete (eds) (1988), *Technical Change and Economic Theory*, Cambridge, UK: Cambridge University Press.

Faulkner, W. and J. Senker (1995), *Knowledge Frontiers: Public Sector Research and Industrial Innovation in Biotechnology, Engineering Ceramics, and Parallel Computing*, Oxford, UK: Clarendon Press.

Feldman, M. (1994), 'The university and economic development: the case of Johns Hopkins University and Baltimore', *Economic Development Quarterly*, 8, 67-77.

Feldman, M.P. and P. Desrochers (2004), 'Truth for its own sake: academic culture and technology transfer at Johns Hopkins University', *Minerva,* 42, 105-126.

Felsenstein, D. (1994), 'The university in local economic development: benefit or burden?' Unpublished paper, Department of Geography and Institute of Urban and Regional Studies, Hebrew University, Mt. Scopus, Jerusalem, Israel.

Feller, I., C.P. Ailes and J.D. Roessner (2002), 'Impacts of research universities on technological innovation in industry: evidence from engineering research centers', *Research Policy*, 31 (3), 457-474.

Florida, R. (2002), *The Rise of the Creative Class, and How It's Transforming Work, Leisure, Community, and Everyday Life*, New York: Basic Books.

Galbraith, C. and A. De Noble (1988), 'Location decisions of high technology firms: a comparison of firm size, industry type and institutional form', *Entrepreneurship: Theory and Practice*, 13, 31-47.

Goldstein, H. (2005), 'The role of knowledge infrastructure in regional economic development: the case of the Research Triangle', *Canadian Journal of Regional Science*, Summer, 199-220.

Goldstein, H. and J. Drucker (2006), 'The economic development impacts of universities on regions: do size and distance matter?', *Economic Development Quarterly*, 20 (1), 22-43.

Goldstein, H., G. Maier and M. Luger (1995), 'The university as an instrument for economic and business development', in D. Dill and B. Sporn (eds), *Emerging*

*Patterns of Social Demand and University Reform: Through a Glass Darkly*, Oxford, UK: Pergamon.

Habily, A.S. (2004), 'Revving up: universities and colleges as urban revitalization engines', *Economic Development America*, Winter, 6-7.

Howells, J. (1986), 'Industry-academic links in research and innovation: a national and regional perspective', *Regional Studies*, 20, 472-476.

ICF Consulting (2003), 'California's Future: It Starts Here. An Impact Study for the University of California', Report prepared for the UC Office of the President.

Initiative for a Competitive Inner City and CEOs for Cities (2002), 'Leveraging Colleges and Universities for Economic Revitalization: An Action Agenda'.

Jaffe, A. (1989), 'The real effects of academic research', *The American Economic Review*, 88, 957-970.

Karlsson, C. and W.-B. Zhang (2001), 'The role of universities in regional development: endogenous human capital and growth in a two region model', *The Annals of Regional Science*, 35, 179-197.

Keane, J. and J. Allison (1999), 'The intersection of the learning region and local and regional economic development', *Regional Studies*, 33, 896-902.

Koo, J. (2003), *When Technology Spillovers are Localized: Importance of Regional and Industrial Attributes*, Unpublished PhD dissertation, University of North Carolina at Chapel Hill, Chapel Hill, NC.

Landabaso, M., C. Oughton and K. Morgan (2000), 'Learning regions in Europe: theory, policy and practice through the RIS experience', in D. Gibson et al. (eds), *Systems and Policies for the Globalized Learning Economy*, Westport, CT: Quorum Books.

Lawton-Smith, H. (2006), 'Universities, innovation and territorial development: a review of the evidence', *Environment and Planning C: Government and Policy*, 25, 98-114.

Leslie, L. and P.T. Brinkman (1988), *The Economic Value of Higher Education*, New York: ACE and Macmillan.

Luger, M. and H. Goldstein (1991), *Technology in the Garden*, Chapel Hill, NC: The University of North Carolina Press.

Luger, M. and H. Goldstein (2006), 'Research parks redux: the changing landscape of the garden', Final report to the US Department of Commerce, Economic Development Administration, Award # 99-07-13827.

Luger, M. et al. (2001), 'The economic impact of the UNC system on the State of North Carolina', University of North Carolina, Office of Economic Development.

Maillat, D. and B. Lecoq (1992), 'New technologies and transformation of regional structures in Europe: the role of the milieu', *Entrepreneurship and Regional Development*, 4, 1-20.

Mansfield, E. and J.-Y. Lee (1996), 'The modern university: contributor to industrial innovation and recipient of industrial R&D support', *Research Policy*, 25, 1047-1058.

Martin, F. (1998), 'The economic impact of Canadian university R&D', *Research Policy*, 27, 677-687.

Merton, R.K. (1973), 'The normative structure of science', *The Sociology of Science*, Chicago: The University of Chicago Press.

Meyer-Krahmer, F. and U. Schmoch (1998), 'Science-based technologies: university-industry interactions in four fields', *Research Policy*, 27, 835-851.

Renault, C. (2003), *Increasing University Technology Transfer Productivity: Understanding Influences on Faculty Entrepreneurial Behavior*, Unpublished PhD dissertation, University of North Carolina at Chapel Hill, Chapel Hill, NC.

Riddel, M. and R.K. Schwer (2003), 'Regional innovative capacity with endogenous employment: empirical evidence from the US', *The Review of Regional Studies*, 33 (1), 73-84.

Rosenberg, N. (2002), 'Knowledge and innovation for economic development: should universities be economic institutions?', in P. Conceicao et al. (eds), *Knowledge for Inclusive Development*, Westport, CT: Greenwood Publishing Co.

Salter, A.J. and B.R. Martin (2001), 'The economic benefits of publicly funded basic research: a critical review', *Research Policy*, 30, 509-532.

Shapira, P. (2001), 'US manufacturing extension partnerships', *Research Policy*, 30, 977-992.

Sherry, B.A. (2005), 'Universities as developers; an international conversation', *Land Lines*, January, 11-20.

Smilor, R.W., D.V. Gibson and G.B. Dietrich (1990), 'University spin-out companies; technology start-ups from UT-Austin', *Journal of Business Venturing*, 5 (1), 63-76.

Smith, K. (1997), 'Economic infrastructures and innovation systems', in C. Edquist (ed.), *Systems of Innovation: Technologies, Institutions, and Organizations*, London, UK: Pinter, pp. 86-106.

Steffensen, M., E.M. Rogers and K. Speakman (2000), 'Spin-offs from research centers at a research university', *Journal of Business Venturing*, 15 (1), 93-111.

Stokes, D. (1997), *Pasteur's Quadrant: Basic Science and Technological Innovation*, Washington, DC: Brookings Institution Press.

Storper, M. (1995), 'The resurgence of regional economics, 10 years later: the region as a nexus of untraded interdependencies', *European Urban and Regional Studies*, 2 (3), 191-221.

Szatan, J. (2004), 'University-related development in Chicago invests billions', *Urban Land*, October, 79-80.

Thanki, R. (1999), 'How do we know the value of higher education to regional development?', *Regional Studies*, 33 (1), 84-89.

Trajtenberg, M., R. Henderson and A. Jaffe (1992), 'Ivory tower versus corporate lab: an empirical study of basic research and appropriability', *NBER Working Paper* No. 4146, National Bureau of Economic Research, Cambridge, MA.

Urban Land Institute (2006), *Universities and Real Estate/Campus Planning*, Washington, DC.

Varga, A. (1997), 'Regional effects of university research: a survey', Unpublished manuscript, West Virginia University, Regional Research Institute, Morgantown, W.V.

Varga, A. (2000), 'Local academic knowledge transfers and the concentration of economic activity', *Journal of Regional Science*, 40 (2), 289-309.

Varga, A (2001), 'Universities and regional economic development: does agglomeration matter?' in B. Johannsson, C. Karlsson and R. Stough (eds), *Theories of Endogenous Regional Growth: Lessons for Regional Policies*, Berlin: Springer-Verlag.

Walden, M. (2001), 'The net economic impact of North Carolina State University on the State of North Carolina, The Triangle Region, and Wake County', www.ncsu.edu/univ_relations/news_services/NCSUeconimpact.html

Wiewel, W. (2004), 'University real estate development; time for city planners to take notice', *Strategies*, Newsletter of the City Planning and Management Division of the American Planning Association, Winter 2004-05.

Zucker, L., M. Darby and M. Brewer (1998), 'Intellectual human capital and the birth of US biotechnology enterprises', *American Economic Review*, 88, 290-306.

# 3. Jaffe-Feldman-Varga: the search for knowledge spillovers

## Zoltán J. Ács

### 3.1 INTRODUCTION

In *The Structure of Scientific Revolutions*, Thomas Kuhn (1962) argued that 'normal science' shares two essential characteristics: (1) its achievement was sufficiently unprecedented to attract an enduring group of adherents away from competing modes of scientific activity; (2) it was sufficiently open-ended to leave all sorts of problems for the redefined group of practitioners to solve (Kuhn, 1962, 10). The shift from old growth theory to new growth theory represented one such transformation. In the words of Romer (1986, 204) this revolution 'removes the dead end in neoclassical theory and links microeconomic observations on routines, machine designs and the like with macroeconomic discussions of technology'. In the new growth theory a fundamental anomaly, which remains unresolved, is the identification and measurement of R&D spillovers, or the extent to which a start-up is able to exploit economically the investment in R&D made by another organization (Griliches, 1979).

David B. Audretsch and Zoltán J. Ács were attracted to the economics of technological change by the innovative prowls of new-technology-based firms in the 1980s. While the conventional wisdom was that large firms had an innovative advantage over small firms, in a 1988 article in the *American Economic Review*, they discovered an anomaly instead of solving a problem.

> A perhaps somewhat surprising result is that not only is the coefficient of the large-firm employment share positive and significant for small-firm innovations, but it is actually greater in magnitude than for large firms. This suggests that, ceteris paribus, the greater extent to which an industry is composed of large firms, the greater will be the innovative activity, but that increased innovative activity will tend to emanate more from the small firms than from the large firms. (Ács and Audretsch, 1988, 686)

The anomaly of where new-technology-based start-ups acquire knowledge was unresolved.

Building on the work of Griliches (1979), Adam Jaffe (1989) was the first to identify the extent to which university research spills over into the generation of commercial activity. Building on Jaffe's work, Maryann Feldman at Carnegie Mellon University expanded the knowledge production function to innovative activity and incorporated aspects of the regional knowledge infrastructure. Attila Varga at West Virginia University extends the Jaffe-Feldman approach by focusing on a more precise measure of local geographic spillovers. Varga approaches the issue of knowledge spillovers from an explicit spatial econometric perspective. The Jaffe-Feldman-Varga spillovers (from here on JFV) take us a long way toward understanding the role of knowledge spillovers in technological change. Building on this foundation the model has been recently extended to identify entrepreneurship as a conduit through which knowledge spillovers take place (Ács, Audretsch, Braunerhjelm and Carlsson 2009). Finally, the role of agglomerations in knowledge spillovers represents the final frontier in this scientific revolution (Clark, Feldman and Gertler 2000). The purpose of this chapter is to catalogue the contribution of JFV (two of them students of mine) that simultaneously and independently sparked a search for the mechanism of knowledge spillovers. Section 3.2 outlines the main contributions of Jaffe, Feldman and Varga. Section 3.3 examines extensions of the model by Jaffe, Trajtenberg and Henderson as well as recent criticisms of the model by Thomson and Fox-Kean. Section 3.4 examines spatialized explanations of economic growth by Ács and Varga, Fujita, Krugman and Venables, and Romer. Section 3.5 presents work on a knowledge spillover theory of entrepreneurship. Section 3.6 discusses agglomerations. Conclusions are in the final section.

## 3.2 JAFFE-FELDMAN-VARGA

In a 1989 paper in the *American Economic Review*, Adam Jaffe extended his pathbreaking 1986 study measuring the total R&D pool available for spillovers to identify the contribution of spillovers from university research to commercial innovation. Jaffe's findings were the first to identify the extent to which university research spills over into the generation of inventions and innovations by private firms. In order to relate the response of this measure to R&D spillovers from universities, Jaffe modifies the 'knowledge production function' introduced by Zvi

Griliches (1979) for two inputs: private corporate expenditures on R&D and research expenditures undertaken at universities.

In essence, this is a two-factor Cobb-Douglas production function that relates an output measure for 'knowledge' to two input measures: research and development performed by industry; and research performed by universities. Formally, this is expressed as:

$$\log(K) = \beta_{K1} \log(R) + \beta_{K2} \log(U) + \varepsilon_K \qquad (3.1)$$

where $K$ is a proxy for knowledge measured by patent counts, $R$ is industry R&D and $U$ is university research, with $\varepsilon_K$ as a stochastic error term. The analysis is carried out for US states for several points in time and disaggregated by sector. The potential interaction between university and industry research is captured by extending the model with two additional equations that allow for simultaneity between these two variables:

$$\log(R) = \beta_{R1} \log(U) + \beta_{R2} Z_2 + \varepsilon_R \qquad (3.2)$$

and

$$\log(U) = \beta_{U1} \log(R) + \beta_{U2} Z_1 + \varepsilon_U \qquad (3.3)$$

where $U$ and $R$ are as before, $Z_1$ and $Z_2$ are sets of exogenous local characteristics, and $\varepsilon_R$ and $\varepsilon_U$ are stochastic error terms.

Jaffe's statistical results provide evidence that corporate patent activity responds positively to commercial spillovers from university research. The results concerning the role of geographic proximity in spillovers from university research are clouded, however, by the lack of evidence that geographic proximity within the state matters as well. According to Jaffe (1989, 968), 'there is only weak evidence that spillovers are facilitated by geographic coincidence of universities and research labs within the state'. In other words, we know very little where knowledge spillovers go.

Maryann Feldman expands on the work of Jaffe in two ways (Ács, Audretsch and Feldman, 1992, 1994; Feldman, 1994; Feldman and Florida, 1994). First, she uses a new data source – a literature-based innovation output indicator developed by the US Small Business Administration that directly measures innovative activity (Ács and Audretsch, 1988) and she expands the knowledge production function

(Jaffe, 1989) to account for tacit knowledge and commercialization linkages.

Griliches (1979) introduced a model of technological innovation which views innovative output as the product of knowledge generating inputs. Jaffe (1989) modified this production function approach to consider spatial and technical area dimensions. However, Jaffe's model only considers what were previously defined as the elements of the formal knowledge base. Such a formulation does not consider other types of knowledge inputs, which contribute to the realization of innovative output. This is important since innovation requires both technical and business knowledge if profitability is to be the guide for making investments in research and development. Following the innovation knowledge base conceptual model, a more complete specification of innovative inputs would include

$$\log(K) = \beta_{K1} \log(R) + \beta_{K2} \log(U) + \beta_{K3} \log(BSERV) + \beta_{K4} \log(VA) + \varepsilon_K \quad (3.4)$$

where $K$ is measured by counts of innovations, and $R$ and $U$ are as before. $VA$ is the tacit knowledge embodied by the industry's presence in an area $BSERV$ stands for the presence of business services that represents a link to commercialization.

The last input in the knowledge base model is the most evasive. There are a variety of producer services that provide knowledge to the market and the commercialization process. For example, the services of patent attorneys are a critical input to the innovation process. Similarly, marketing information plays an important role in the commercialization process.

Substitution of the direct measure of innovative activity for the patent measure in the knowledge-production function generally strengthens Jaffe's (1989) arguments and reinforces his findings. Most importantly, use of the innovation data provides even greater support than was found by Jaffe: as he predicted, spillovers are facilitated by the geographic coincidence of universities and research labs within the state. In addition, there is at least some evidence that, because the patent and innovation measures capture different aspects of the process of technological change, results for specific sectors may be, at least to some extent, influenced by the technological regime. Thus, it is found that the importance of university spillovers relative to private-company R&D spending is considerably greater in the electronics sector when the direct measure of innovative activity is substituted for the patent measure.

However, the relative importance of industry R&D and university research as inputs in generating innovative output clearly varies between large and small firms (Ács, Audretsch and Feldman, 1994). That is, for large firms, not only is the elasticity of innovative activity with respect to industry R&D expenditures more than two times greater than the elasticity with respect to expenditures on research by universities, but it is nearly twice as large as the elasticity of small-firm innovative activity with respect to industry R&D. By contrast, for small firms the elasticity of innovative output with respect to expenditures on research by universities is about one-fifth greater than the elasticity with respect to industry R&D. Moreover, the elasticity of innovative activity with respect to university research is about 50 percent greater for small enterprises than for large corporations.

These results support the hypothesis that private corporation R&D plays a relatively more important role in generating innovative activity in large corporations than in small firms. By contrast, spillovers from the research activities of universities play a more decisive role in the innovative activity of small firms. Geographic proximity between university and corporate laboratories within a state clearly serves as a catalyst to innovative activity for firms of all sizes. However, the impact is apparently greater on small firms than on large firms.

There were two limitations in the Jaffe-Feldman research. First, the unit of analysis at the state level was too aggregate requiring a geographical coincidence index to control for co-location. Second, the research did not take into consideration the potential influence of spatial dependence that may invalidate the interpretation of econometric analyses based on contiguous cross-sectional data. Attila Varga extends this research by examining both the state and the metropolitan statistical area (MSA) levels and using spatial econometric techniques[1] (Anselin, Varga and Ács, 1997, 2000a, 2000b; Varga, 1998, 2000; Ács, Anselin and Varga, 2002).

These extensions yielded a more precise insight into the range of spatial externalities between innovation and R&D in the MSA and university research both within the MSA and in surrounding counties. He was able to shed some initial light on this issue for high technology innovations measured as an aggregate across five two-digit SIC industries and also at a more detailed industrial sector level. He found a positive and highly significant relationship between MSA innovations and university research indicating the presence of localized university research spillovers in innovation. In comparison to the effect of industrial knowledge spillovers (i.e., knowledge flows among industrial research laboratories) the size of the university effect is considerably smaller as it is one-third of the size of the industrial research coefficient. University knowledge spillovers follow a definite distance decay pattern as shown by the statistically significant albeit smaller size university research coefficient for adjoining counties within a 50-mile distance range from the MSA center.

There are notable differences among sectors with respect to the localized university effect as studied at the MSA level. Specifically for the four high technology sectors such as machinery, chemicals, electronics and instruments, significant localized university spillover impact was found only for electronics and instruments while for the other two industries the university research coefficient remains consistently insignificant.

Ács, Anselin and Varga (2002) test whether the patent data developed by the United States Patent and Trademark Office is in fact a reliable proxy measure of innovative activity at the regional level as compared to the literature-based innovation output indicator developed by the US Small Business Administration. This is important, since the patent data are readily available over time and can be used to study the dynamics of localized knowledge flows within regional innovation systems. Before this study, there was some evidence that patents provide a fairly reliable measure of innovative activity at the industry level (Ács and Audretsch, 1989) and some evidence that patents and innovations behave similarly at the state level (Ács, Audretsch and Feldman, 1992). However, this has not been tested at the sub-state level.

The correlation between the PTO patent and SBA innovation counts at the MSA level is reasonably high (0.79) and this could be taken as a first indication that patents might be a reliable measure of innovation at the regional level. However, this correlation coefficient value is not high enough to guarantee that the role of different regional actors in knowledge creation would turn out similar with both measures if applied in the same empirical model. Varga proceeds by replacing innovation counts with the patent measure in the same model as in Anselin, Varga and Ács (1997) to be able to directly compare the results of the two measures of new technological knowledge and assess the extent to which patents may be used as a reliable proxy.

Sizes of all the parameters in the estimated knowledge production function are smaller for innovation than for patents suggesting that firms in the product development stage rely on localized interactions (with universities as well as with other actors) less intensively than in earlier stages of the innovation process. The other important finding of this comparative study is that the importance of university knowledge spillovers (measured by the size of the university research parameter) compared to that of R&D spillovers among private firms is substantially less pronounced for patents than for innovations. Since patenting reflects more the earlier stages of innovation whereas the direct innovation measure accounts for the concluding stage of the innovation process, the relatively higher weight of local universities in innovation than in patenting appears to reflect the different spatial patterns of basic and applied research collaboration. To collaborate with universities in applied research, firms tend to choose local academic institutions whereas basic research collaboration can be carried out over larger distances.

## 3.3 EXTENSIONS OF THE JFV MODEL

Jaffe, Trajtenberg and Henderson (1993, 2005) expand on the above work to answer the question of whether knowledge externalities are localized. This is important since growth theory assumed that knowledge spills over to agents within the country, but not to other countries. This implicit assumption begs the question to what extent knowledge externalities are localized. Jaffe, Trajtenberg and Henderson extend the search for knowledge spillovers by using a matching method that found that knowledge spillovers are strongly localized. Their method matches each citing patent to a non-citing patent intended to control for the pre-existing geographic concentration of production. Using patent data, they came to two conclusions, one that spillovers are particularly significant at the local level, and that localization fades only slowly over time. These results and the large research issue are reproduced in Jaffe and Trajtenberg (2002).

Audretsch and Feldman (1996) explore the question of the geography of innovation and production. They provide evidence concerning the spatial dimension of knowledge spillovers. Their findings suggest that knowledge spillovers are geographically bounded and localized within spatial proximity to the knowledge source. Feldman and Audretsch (1999) further examine the question of knowledge spillovers by looking into the question of specialization versus diversity in cities. Their research supports the ideas that diversity leads to more innovation.

Recently, Thompson and Fox-Kean (2005a, 2005b) have challenged the findings of Jaffe, Trajtenberg and Henderson. They suggest that the Jaffe, Trajtenberg and Henderson method-matched case control methodology included a serious spurious component. Controlling for unobservables using matching methods is invariably a dangerous exercise because one can rarely be confident that the controls are doing their job. In some cases, imperfect matching may simply introduce noise and a corresponding loss of efficiency. They suggest at least two reasons why the matching method may not adequately control for existing patent activity. First, the level of aggregation might not be fine enough. Second, patents typically contain many distinct claims to each of which a technological classification is assigned. These two features of the control selection process mean that there is no guarantee that the control patent has any industrial similarity with the citing or to the originating patent. Of course, one of their conclusions that spillovers stop at the country level also needs explaining.

Empirical research done within the JFV framework and the extensions introduced so far were established and originally carried out in the United States with the use of state-, MSA- and county-level data sets. However, the issue of the geographic extent of knowledge spillovers has a definite international validity. Within the last decade the JFV model has been replicated and continuously refined to search for the geographical boundaries of knowledge flows in Europe, South America and Asia. Varga (2006) provides an assessment of the extent of this international literature.

## 3.4 'SPATIALIZED' EXPLANATION OF ECONOMIC GROWTH

Building on the JFV model of knowledge spillovers, Ács and Varga (2002) suggest a 'spatialized' theoretical framework of technology-led economic growth that needs to reflect three fundamental issues. First, it should provide an explanation of why knowledge-related economic activities concentrate in certain regions leaving others relatively underdeveloped. Second, it needs to answer the questions of how technological advances occur and what are the key processes and institutions involved with a particular focus on the geographic dimension. Third, it has to present an analytical framework where the role of technological change in regional and national economic growth is clearly explained. In order to answer these three questions Ács and Varga examine three separate and distinct literatures: the new economic geography, the new growth theory, and the new economics of innovation.

The three approaches focus on different aspects but at the same time are also complements of each other. The 'new' theories of growth endogenize technological change and as such interlink technological change with macroeconomic growth. However, the way technological change is described is strongly simplistic and the economy investigated is formulated in an aspatial model. On the other hand, systems of innovation frameworks are very detailed with respect to the innovation process but say nothing about macroeconomic growth. However, the spatial dimension has been introduced into the framework in the recently developed 'regional innovation systems' studies (e.g. Braczyk, Cooke and Hedenreich, 1998).

The idea behind the innovation systems approach is quite simple but as such extremely appealing. According to this in most cases, innovation is a result of a collective process and this process gets shaped in a systemic manner. The elements of the system are innovating firms and firms in related and connected industries (suppliers, buyers), private and public research laboratories, universities, supporting business services (like legal or technical services), financial institutions (especially venture capital), and the government. These elements are interconnected by innovation-related linkages where these linkages represent knowledge flows among them. Linkages can be informal in nature (occasional meetings in conferences, social events, etc.) or they can also be definitely formal (contracted research, collaborative product development, etc.). The effectiveness (i.e., productivity in terms of number of innovations) of the system is determined by both the knowledge already accumulated by the actors and the level of their interconnectedness (i.e., the intensity of knowledge flows). Ability and motivations for interactions are shaped largely by traditions, social norms, values and the countries' legal systems.

New economic geography models investigate general equilibrium in a spatial setting (Krugman, 1991). This means that they provide explanations not only for the determination of equilibrium prices, incomes and quantities in each market but also the development of the particular geographical structure of the economy. In other words, new economic geography derives economic and spatial equilibrium simultaneously (Fujita, Krugman and Venables, 1999; Fujita and Thisse, 2002). Spatial equilibrium arises as an outcome of the balance between centripetal forces working towards agglomeration (such as increasing returns to scale, industrial demand, localized knowledge spillovers) and centrifugal forces promoting dispersion (such as transportation costs). Until the latest developments in recent years, new economic geography models did not consider the spatial aspects of economic growth. However, even in the recent models explanation of technological change follows the same pattern as endogenous growth models and as such fail to reach the complexity inherent in innovation systems studies.

As emphasized by Ács and Varga (2002), each one of the above three approaches has its strengths and weaknesses but they could serve to create the building blocks of an explanatory framework of technology-led economic growth. They suggest that a specific combination of the Krugmanian theory of initial conditions for spatial concentration of economic activities with the Romerian theory of endogenous economic growth, complemented with a systematic representation of interactions among the actors of Nelson's innovation system, could be a way of developing an appropriate model of technology-led regional economic development.

Following Ács and Varga (2002), Varga (2006) develops an empirical modeling framework of geographical growth explanation. This framework is the spatial extension of the endogenous growth model in Romer (1990) and it integrates elements of the innovation systems and the new economic geography literatures. For a bit more formal treatment, Varga (2006) applies the generalized version of the Romer (1990) equation of macroeconomic level knowledge production developed in Jones (1995):[2]

$$dA = \delta \, H_A{}^{\lambda} A^{\varphi} \tag{3.5}$$

where $H_A$ stands for human capital in the research sector working on knowledge production (operationalized by the number of researchers), $A$ is the total stock of technological knowledge available at a certain point in time whereas $dA$ is the change in technological knowledge resulting from private efforts to invest in research and development. $\delta$, $\lambda$ and $\varphi$ are parameters.

Technological change is generated by research and its extent depends on the number of researchers involved in knowledge creation ($H_A$). However, their efficiency is directly related to the total stock of already available knowledge ($A$). Knowledge spillovers are central to the growth process: the higher $A$ is the larger the change in technology produced by the same number of researchers. Thus macroeconomic growth is strongly related to knowledge spillovers.

Parameters in the Romer knowledge production function play a decisive role in the effectiveness of macro level knowledge production. The same number of researchers with a similar value of $A$ can raise the level of already existing technological knowledge with significant differences depending on the size of the parameters. First, consider $\delta$ ($0 < \delta < 1$) which is the research productivity parameter. The larger is $\delta$

the more efficient is $H_A$ in producing economically useful new knowledge.

The size of $\varphi$ reflects the extent to which the total stock of already established knowledge impacts knowledge production. Given that $A$ stands for the level of codified knowledge (available in books, scientific papers or patent documentations), $\varphi$ is called the parameter of codified knowledge spillovers. The size of $\varphi$ reflects the portion of $A$ that spills over and, as such, its value largely influences the effectiveness of research in generating new technologies.

$\lambda$ is the research spillover parameter. The larger is $\lambda$ the stronger is the impact the same number of researchers plays in technological change. In contrast to $\varphi$ and $\delta$ that are determined primarily in the research sector and as such their values are exogenous to the economy, $\lambda$ is endogenous. Its value reflects the diffusion of (codified and tacit) knowledge accumulated by researchers. Technological diffusion depends on three types of interactions: first, on the intensity of interactions among researchers ($H_A$); second, on the quality of public research and the extent to which the private research sector is connected to it (especially to universities) by formal and informal linkages; and third, the development level of supporting/connected industries and business services and the integration of innovating firms into the system. The extensive innovation systems literature evidences that the same number of researchers contribute to different efficiencies depending on the development of the system. In the Romer equation, this is reflected in the size of $\lambda$.

Within the JFV framework, a series of papers demonstrate that a significant fraction of knowledge spillovers is bounded spatially. These findings imply that the geographic structure of R&D is a determinant of technological change and ultimately economic growth. Ceteris paribus, in an economy where the concentration of R&D institutions exceeds a critical level intensive knowledge spillovers result in a higher level of innovation than in a system where research is more evenly distributed over space. Thus $\lambda$ is also sensitive to the spatial structure of $H_A$. Even with the same number of researchers $\lambda$ can have different values depending on the extent to which research and development is concentrated in space.

## 3.5 A KNOWLEDGE SPILLOVER THEORY OF ENTREPRENEURSHIP

In this section, the JFV model of knowledge spillovers is extended by Ács and Audretsch who develop a Knowledge Spillover Theory of Entrepreneurship in order to answer the question: 'What is the conduit by which knowledge spillovers occur?' As a first step in this direction, the theory incorporates two of the above literatures, new growth theory (Romer) and the new economics of innovation (Nelson, 1991) to explain how entrepreneurship facilitates the spillover of knowledge.

A modern synthesis of the entrepreneur is someone who specializes in taking judgmental decisions about the coordination of scarce resources (Lazear, 2005). In this definition, the term 'someone' emphasizes that the entrepreneur is an individual. Judgmental decisions are decisions for which no obvious correct procedure exists – a judgmental decision cannot be made simply by plugging available numbers into a scientific formula and acting based on the number that comes out. In this framework, entrepreneurial activity depends upon the interaction between the characteristics of opportunity and the characteristics of the people who exploit them. Since discovery is a cognitive process, it can take place only at the individual level. Individuals, whether they are working in an existing organization or unemployed at the time of their discovery, are the entities that discover opportunities. The organizations that employ people are inanimate and cannot engage in discovery. Therefore, any explanation for the mode of opportunity discovery must be based on choices made by individuals about how they would like to exploit the opportunity that they have discovered (Hayek, 1937).

So where do opportunities come from? Today we know that the technology opportunity set is endogenously created by investments in new knowledge. The new growth theory, formalized by Romer (1986), assumes that firms exist exogenously and then engage in the pursuit of new economic knowledge as input into the process of generating endogenous growth. Technological change plays a central role in the explanation of economic growth, since on the steady state growth path the rate of per capita GDP growth equals the rate of technological change.

However, not only does new knowledge contribute to technological change, it also creates opportunities for use by third party firms, often entrepreneurial start-ups (Shane, 2001). The creation of new knowledge gives rise to new opportunities through knowledge spillovers, therefore entrepreneurial activity does not involve simply the arbitrage of opportunities (Kirzner, 1973) but also the exploitation of new opportunities created but not appropriated by incumbent organizations (Hellmann, 2007). Thus, while the entrepreneurship literature considers opportunity to exist exogenously, in the new economic growth literature opportunities are systematically and endogenously created through the purposeful investment in new knowledge. The Knowledge Spillover Theory of Entrepreneurship as suggested by Audretsch (1995, 48) 'proposes shifting the unit of observation away from exogenously assumed firms to individuals – agents confronted with new knowledge and the decision whether and how to act upon that new knowledge'.

The Theory relaxes two central (and unrealistic) assumptions of the endogenous growth model to develop a theory that improves the microeconomic foundations of endogenous growth theory (Ács, Audretsch, Braunerhjelm and Carlsson 2009). The first is that knowledge is automatically equated with economic knowledge. In fact, as Arrow (1962) emphasized, knowledge is inherently different from the traditional factors of production, resulting in a gap between knowledge ($K$) and what he called economic knowledge ($K^c$). The second involves the assumed spillover of knowledge. The existence of the factor of knowledge is equated with its automatic spillover, yielding endogenous growth. In the Knowledge Spillover Theory of Entrepreneurship, *institutions* impose a gap between new knowledge and economic knowledge ($0 < K^c / K < 1$) and results in a lower level of knowledge spillovers.

The model is one where new product innovations can come either from incumbent organizations or from entrepreneurial start-ups (Schumpeter, 1934). According to Baumol (2004, 9), 'the bulk of private R&D spending is shown to come from a tiny number of very large firms. Yet, the revolutionary breakthroughs continue to come predominantly from small entrepreneurial enterprises, with large industry providing streams of incremental improvements that also add up to major contributions.' We can think of incumbent firms that rely on the flow of knowledge to innovate focusing on incremental innovation, i.e. product improvements (Ács and Audretsch, 1988). Entrepreneurial start-ups that have access to knowledge spillovers from the stock of knowledge and entrepreneurial talent are more likely to be engaged in radical innovation that leads to new industries or completely replace existing products (Ács, Audretsch and Feldman, 1994). Start-ups played a major role in radical innovations such as software, semiconductors, biotechnology (Zucker, Darby and Brewer, 1998) and the information and communications technologies (Jorgenson, 2001). The presence of these activities is especially important at the early stages of the life cycle when technology is still fluid.

Equation (3.6) suggests that entrepreneurial startups ($E$) will be a function of the difference between expected profits ($\pi^*$) minus wages ($w$). Expected profits are conditioned by the knowledge stock ($K$) that positively affects start-ups and is negatively conditioned by knowledge commercialized by incumbent firms. Yet a rich literature suggests that there is a compelling array of financial, institutional and individual barriers to entrepreneurship, which results in a modification of the entrepreneurial choice equation:

$$E = \gamma \, [\pi^* \, (K^{\xi}) - w]/\beta \qquad\qquad (3.6)$$

where $\beta$ represents those institutional and individual barriers to entrepreneurship, spanning factors such as risk aversion, financial constraints, and legal and regulatory restrictions (Acemoglu, Johnson and Robinson, 2004). The existence of such barriers explains why economic agents might choose not to enter into entrepreneurship, even when confronted with knowledge that would otherwise generate a potentially profitable opportunity. Thus, this mode shows how local differences in knowledge stocks, the presence of large firms as deterrents to knowledge exploitation, and an entrepreneurial culture might explain regional variations in the rates of entrepreneurial activity. The primary theoretical predictions of the model are:

1. An increase in the stock of knowledge has a positive effect on the level of entrepreneurship.
2. The more efficient incumbents are at exploiting knowledge flows the smaller the effect of new knowledge on entrepreneurship.
3. Entrepreneurial activities are decreasing in the face of higher regulations, administrative barriers and governmental market intervention.

Thus, entrepreneurship becomes central to generating economic growth by serving as a conduit, albeit not the sole conduit, by which knowledge created by incumbent organizations spills over to agents who endogenously create new firms. The theory is actually a theory of endogenous entrepreneurship, where entrepreneurship is a response to opportunities created by investments in new knowledge that was not commercialized by incumbent firms. The theory suggests that, ceteris paribus, entrepreneurial activity will tend to be greater in contexts where investments in new knowledge are relatively high, since the start-ups will benefit from knowledge that has spilled over from the source actually producing that new knowledge. In a low knowledge context, the lack of new ideas will not generate entrepreneurial opportunities based on potential knowledge spillovers. In a recent series of studies Ács and Armington (2006), and Audretsch, Keilbach and Lehmann (2006) link entrepreneurship and economic growth at the regional level. Ács, Audretsch, Braunerhjelm and Carsson 2006 (2006) at the national level find that entrepreneurship does in fact offer an explanation for how knowledge spillovers occur.

Ács and Varga (2005) empirically test the theory within the JFV framework. They build their modeling approach on the interpretation of the Romerian equation (equation 3.5) provided in section 3.4. They start with the assumption that the value of l bears the influence of the level of entrepreneurship because the value of new economic knowledge is uncertain. While most R&D is carried out in knowledge-creating institutions (large firms and universities) it does not mean that the same individuals that discover the opportunity will carry out the exploitation. An implication of the theory of firm selection is that new firms may enter an industry in large numbers to exploit knowledge spillovers. The higher the rate of start-ups the greater should be the value of l because of knowledge spillovers.

The empirical model in which the parameter l in equation (3.5) is endogenized has the following form:

$$\log(NK) = \delta + \lambda\log(H) + \varphi\log(A) + \varepsilon \qquad (3.7)$$

$$\lambda = \beta_1 + \beta_2\log(ENTR) + \beta_3\log(AGGL) \qquad (3.8)$$

where *NK* stands for new knowledge (i.e., the change in *A*), *ENTR* is entrepreneurship, *AGGL* is agglomeration, *A* is the set of publicly available scientific-technological knowledge and $\varepsilon$ is stochastic error term. Implementation of (3.7) into (3.8) results in the following estimated equation:

$$\log(NK) = \delta + \beta_1\log(H) + \beta_2\log(ENTR)\log(H) + \\ \beta_3\log(AGGL)\log(H) + \varphi\log(A) + \varepsilon \quad (3.9)$$

In equation (3.9), the estimated value of the parameter $\beta_2$ measures the extent to which research interacted with entrepreneurship contributes to knowledge spillovers. Applying to European data, Ács and Varga (2005) found a statistically significant value of $\beta_2$ that is taken as a supporting evidence of the knowledge spillover theory of entrepreneurship.

## 3.6 AGGLOMERATION: THE FINAL FRONTIER

The JFV model can also be extended for empirically testing agglomeration effects in knowledge spillovers. Agglomeration forces are crucial in technological change and as such in economic growth explanation. Varga (2006) points out that in equation (3.5) the size of $\lambda$ is also influenced by agglomeration. Insights from the new economic geography can help understand the dynamic effects of the spatial structure of R&D on macroeconomic growth (Baldwin and Forslid, 2000; Fujita and Thisse, 2002; Baldwin, Forslid, Martin, Ottaviano and Robert-Nicoud 2003). If spatial proximity to other research labs, universities, firms and business services matters in innovation, firms are motivated to locate R&D laboratories in those regions where actors of the system of innovation are already agglomerated in order to decrease their costs to innovate.

Thus, spatial concentration of the system of innovation is a source of positive externalities and, as such, these externalities (as centrifugal forces in R&D location) determine the strength of the cumulative process that leads to a particular spatial economic structure. However, agglomeration effects could be negative as well. Increasing housing costs and travel time make innovation more expensive and might motivate labs to move out from the region. The actual balance between centrifugal and centripetal forces determines the geographical structure of the system of innovation. Through determining the size of $\lambda$ in equation (3.5), this also influences the rate of technological progress ($dA/A$) and, eventually, the macroeconomic growth rate ($dy/y$).

Within the JFV framework Varga (2000, 2001) estimates the magnitude of agglomeration effects. Based on a data set of 125 US metropolitan areas, he found that spatial concentration of high technology production and business services are in a definite positive relationship with the intensity of local academic knowledge transfers. Increasing returns resulting from the spatial concentration of economic activities were clearly demonstrated in the study. It was shown that the same amount of local expenditures on university research yields dramatically different levels of innovation output depending on the concentration of economic activities in the metropolitan area. It was found that a critical mass of agglomeration should be reached in order to experience substantial local economic effects of academic research spending. This critical mass was characterized with city population around three million, employment in high technology production facilities and business service firms about 160,000 and 4,000, respectively. In Varga (2001), agglomeration effects in university knowledge spillovers for two 'high technology' sectors (electronics and instruments) are also demonstrated.

How can the JFV framework contribute to study empirically the dynamism of agglomeration (i.e., the dynamism of $\lambda$) described in detail by the new economic geography? To model empirically the effects of centripetal and centrifugal forces on spatial structure, researchers develop spatial computable general equilibrium (SCGE) models (e.g. Thissen, 2003). These models are empirical counterparts of the new economic geography and are extremely powerful tools for explaining spatial distribution of economic activities under different starting assumptions.

Now it is technically possible to integrate the JFV approach into a combination of macroeconometric (Varga and Schalk 2004) and SCGE modeling in order to study the dynamic effects of knowledge spillovers on geography, technological change and macroeconomic growth. With this step the JFV model becomes a crucial bridge between academic research on the geography of innovation and policy analysis for studying different scenarios of economic development. Varga (2008) demonstrates that incorporating the lessons of the JFV framework into development policy analysis opens up the possibility of building 'new generation models' with such simulations where both regional, inter-regional and macro effects of different policy scenarios can be studied and compared to each other. The GMR-Hungary model (Varga, 2007) is the first one in this field.

## 3.7  CONCLUSIONS

In this review, we introduced the Jaffe-Feldman-Varga model and made an assessment as to its relevance for economics research. It is highlighted that this approach has become a widely applied tool for testing the spatial extent of knowledge spillovers in different countries, different sectors and at different spatial scales. In addition, this model became a workhorse of empirical studies of entrepreneurship, agglomeration and growth. The JFV approach has also proven to become a crucial element in 'new generation development policy modeling'. Thus the JFV model of knowledge spillovers and its extensions offer an avenue to re-explain the mechanism by which knowledge spillovers operate and open the door for a new understanding of regional and macroeconomic development. If this indeed happens, we would have experienced a paradigm shift in economic science.

## NOTES

1. When models are estimated for cross-sectional data on neighboring spatial units, the lack of independence across these units (or, the presence of spatial autocorrelation) can cause serious problems of model misspecification when ignored (Anselin, 1988). The methodology of spatial econometrics consists of testing for the potential presence of these misspecifications and of using the proper estimators for models that incorporate the spatial dependence explicitly (for a recent review, see Anselin, 2001).
2. The functional form corresponds to the Jones (1995) version, however, the interpretation of $\lambda$ and $\varphi$ is different in Varga (2006).

## REFERENCES

Acemoglu, D., S. Johnson and J. Robinson (2004), 'Institutions as the fundamental cause of long-run growth', in Philippe Aghion and Steven Durlauf (eds), *Handbook of Economic Growth*, New York, NY: Elsevier North Holland, Vol. 1, Chapter 6, pp. 405-472.

Ács, Z.J. and C. Armington (2006), *Entrepreneurship, Geography and American Economic Growth*, Cambridge, UK: Cambridge University Press.

Ács, Z.J. and D.B. Audretsch (1988), 'Innovation in large and small firms: an empirical analysis', *The American Economic Review*, 78 (4), 678-689.

Ács, Z.J. and D.B. Audretsch (1989), 'Patents as a measure of innovative activity', *Kyklos*, 42, 171-180.

Ács, Z.J. and A. Varga (2002), 'Geography, endogenous growth and innovation', *International Regional Science Review*, 25, 132-148.

Ács, Z.J. and A. Varga (2005), 'Entrepreneurship, agglomeration and technological change', *Small Business Economics*, 24 (3), 323-334.

Ács, Z.J., L. Anselin and A. Varga (2002), 'Patents and innovation counts as measures of regional production of new knowledge', *Research Policy*, 31, 1069-1085.

Ács, Z.J., D.B. Audretsch and M.P. Feldman (1992), 'Real effects of academic research: comment', *American Economic Review*, 82, 363-367.

Ács, Z.J., D.B. Audretsch and M.P. Feldman (1994), 'R&D spillovers and recipient firm size', *Review of Economic Statistics*, 99, 336-340.

Ács, Z.J., P. Braunerhjelm, D.B. Audretsch and B. Carlsson (2009), 'The knowledge spillover theory of entrepreneurship', *Small Business Economics*, 32 (1), 15-30.

Anselin, L. (1988), *Spatial Econometrics: Methods and Models*, Boston, MA: Kluwer Academic Publishers.

Anselin, L. (2001), 'Spatial Econometrics', in B. Baltagi (ed.), *A Companion to Theoretical Econometrics*, Oxford: Basil Blackwell, pp. 310-330.

Anselin, L., A. Varga and Z.J. Ács (1997), 'Local geographic spillovers between university research and high technology innovation', *Journal of Urban Economics*, 42, 422-448.

Anselin, L., A. Varga and Z.J. Ács (2000a), 'Geographic and sectoral characteristics of academic knowledge spillovers', *Papers in Regional Science*, 79, 435-445.

Anselin, L., A. Varga and Z.J. Ács (2000b), 'Geographic spillovers and university research: a spatial econometric approach', *Growth and Change*, 31, 501-515.

Arrow, K. (1962), 'The economic implications of learning by doing', *Review of Economic Studies*, 29, 155-173.

Audretsch, D.B (1995), *Innovation and Industry Evolution*, Cambridge, MA: MIT Press.

Audretsch, D.B. and M.P. Feldman (1996), 'R&D spillovers and the geography of innovation and production', *American Economic Review*, 86, 630-640.

Audretsch, D.B., M.C. Keilbach and E.E. Lehmann (2006), *Entrepreneurship and Economic Growth*, Oxford, UK: Oxford University Press.

Baldwin, R.E. and R. Forslid (2000), 'The core-periphery model and endogenous growth: stabilising and de-stabilizing integration', *Economica*, 67, 307-324.

Baldwin, R., R. Forslid, Ph. Martin, G. Ottaviano and F. Robert-Nicoud (2003), *Economic Geography and Public Policy*, Princeton, NJ: Princeton University Press.

Baumol, W. (2004), 'Entrepreneurial enterprises, large established firms and other components of the free-market growth machine', *Small Business Economics*, 23 (1), 9-21.

Braczyk, H., P. Cooke and M. Heidenreich (1998), *Regional Innovation Systems: The Role of Governances in a Globalized World*, London, UK: UCL Press.

Clark, G.L., M.P. Feldman and M.S. Gertler (2000), *The Oxford Handbook of Economic Geography*, Oxford, UK: Oxford University Press.

Feldman, M.P. (1994), *The Geography of Innovation*, New York, NY: Kluwer Academic Publishers.

Feldman, M.P. and D.B. Audretsch (1999), 'Innovation in cities: science-based diversity, specialization and localized competition', *European Economic Review*, 43, 409-429.

Feldman, M.P. and R. Florida (1994), 'The geographic sources of innovation: technological infrastructure and product innovation in the United States', *Annals of the Association of American Geographers*, 84, 210-229.

Fujita, M. and J. Thisse (2002), *Economics of Agglomeration. Cities, Industrial Location, and Regional Growth*, Cambridge, MA and London, UK: Cambridge University Press.

Fujita, M., P. Krugman and A. Venables (1999), *The Spatial Economy*, Cambridge, MA: MIT Press.

Griliches, Z. (1979), 'Issues in assessing the contributions of research and development to productivity growth', *The Bell Journal of Economics*, 10, 92-116.

Hayek, Frederick A. von (1937), 'Economics and knowledge', *Economica (New Series)*, 4, 33-54.

Hellmann, T. (2007), 'When do employees become entrepreneurs?', *Management Science*, 53 (6).

Jaffe, A. (1989), 'The real effects of academic research', *American Economic Review*, 79, 957-970.

Jaffe, A. and M. Trajtenberg (2002), *Patents, Citations and Innovations: A Window on the Knowledge Economy*, Cambridge, MA: MIT Press.

Jaffe, A., M. Trajtenberg and R. Henderson (1993), 'Geography, location of knowledge spillovers as evidence of patent citations', *Quarterly Journal of Economics*, 108, 483-499.

Jaffe, A., M. Trajtenberg and R. Henderson (2005), 'Patent citations and the geography of knowledge spillovers: a reassessment: Comment', *American Economic Review*, 95(1), 461-465.

Jones, C. (1995), 'R&D based models of economic growth', *Journal of Political Economy*, 103, 759-84.

Jorgenson, D.W. (2001), 'Information technology and the US economy', *American Economic Review*, 91, 1-32.

Kirzner, Isreal M. (1973), *Competition and Entrepreneurship*, Chicago, IL: University of Chicago Press.

Krugman, P. (1991), 'Increasing returns and economic geography', *Journal of Political Economy*, 99, 483-499.

Kuhn, T (1962), *The Structure of Scientific Revolutions*, Chicago, IL: The University of Chicago Press.

Lazear, E.P. (2005), 'Entrepreneurship', *Journal of Labor Economics*, 23 (4), 649-680.

Nelson, J.R. (1991), *National Innovation Systems*, Cambridge, MA: Harvard University Press.

Romer, P. (1986), 'Increasing returns and economic growth', *Journal of Political Economy*, 94, 1002-1037.

Romer, P. (1990), 'Endogenous technological change', *Journal of Political Economy*, 98, S71-S102.

Schumpeter, J.A. (1934), *The Theory of Economic Development*, Cambridge, MA: Harvard University Press.

Shane, S. (2001), 'Technological opportunity and new firm creation', *Management Science*, 47 (2), 205-220.

Thissen, M. (2003), 'RAEM 2.0 A Regional Applied General Equilibrium Model for the Netherlands', Manuscript, 19 pp.

Thompson, P. and M. Fox-Kean (2005a), 'Patent citations and the geography of knowledge spillovers: a reassessment', *American Economic Review*, 95 (1), 450-461.

Thompson, P. and M. Fox-Kean (2005b), 'Patent citations and the geography of knowledge spillovers: a reassessment: Reply', *American Economic Review*, 95 (1), 465-467.

Varga, A. (1998), *University Research and Regional Innovation: A Spatial Econometric Analysis of Academic Technology Transfers*, Boston, MA: Kluwer Academic Publishers.

Varga, A. (2000), 'Local academic knowledge spillovers and the concentration of economic activity', *Journal of Regional Science*, 40, 289-309.

Varga, A. (2001), 'Universities and regional economic development: does agglomeration matter?' in B. Johansson, C. Karlsson and R. Stough (eds), *Theories of Endogenous Regional Growth – Lessons for Regional Policies*, Berlin: Springer, pp. 345-367.

Varga, A. (2006), 'The spatial dimension of innovation and growth: empirical research methodology and policy analysis', *European Planning Studies*, 9, 1171-1186.

Varga, A. (2007), 'GMR-HUNGARY: A complex macro-regional model for the analysis of development policy impacts on the Hungarian economy', Final Report, Project No. NFH 370/2005.

Varga, A. (2008), 'From the geography of innovation to development policy analysis: the GMR-approach', *Annals of Economics and Statistics*, 87-88, 83-102.

Varga, A. and H.J. Schalk (2004), 'Knowledge spillovers, agglomeration and macroeconomic growth: an empirical approach', *Regional Studies*, 38, 977-989.

Zucker, L.G., M.R. Darby and M.B. Brewer (1998), 'Intellectual human capital and the birth of US biotechnology enterprises', *American Economic Review*, 88 (1), 290-306.

# 4. Detecting university-industry synergies: a comparison of two approaches in applied cluster analysis

## Edward Feser

## 4.1 INTRODUCTION

As the practice of formulating regional development strategies around various concepts of industry clusters has grown, so has the variety of methodologies for detecting appropriate areas of policy intervention.[1] A particular challenge for regions interested in facilitating the growth of technology-based industries and business start-ups is the conceptualization and measurement of the linkages between the regional science base, reflected most heavily in the region's universities but also in its non-academic research laboratories, and its industrial base (Paytas et al., 2004). According to most cluster theories, businesses are at the core of competitive clusters, with universities and other institutions forming a critical support infrastructure for continued industrial innovation and productivity growth. Following this view, many early applied regional cluster studies focused on detecting existing critical masses of businesses in related sectors, with the degree of 'relatedness' varying from membership in the same industry (a crude sector-based approach), to membership in buyer-supplier chains, utilization of similar technologies, or linkages via formal or informal innovation flows (Cortright, 2006). The implication in such studies is that strategies should aim to further strengthen large concentrations of related industries, effectively the region's industrial specializations, since by virtue of their large absolute and relative size they represent clear regional competitive advantages.

Some of the problems with this perspective are immediately apparent in smaller regions, regions with a preponderance of large declining industries, and regions particularly interested in fostering innovation-oriented economic development. Small regions may have few if any groups of related sectors that have achieved a meaningful degree of critical mass (a 'no clusters' problem). Some regions' existing clusters may be in decline or growing slowly if at all (a 'wrong clusters' problem). Well-known examples in the US are Detroit (in vehicle manufacturing) and the Carolinas (in textiles and apparel). Other regions may see the goal of economic development policy as not to further strengthen already large regional sectors, but to encourage the growth of new-technology-based industries that can represent the next economic base as larger industries inevitably decline (a 'new clusters' problem). Interestingly, few contributors to the literature conclude that cluster notions are uninformative for development planning purposes in such circumstances. Rather, the latest thrust in both the academic and applied literatures is that clusters follow life cycles, just as sectors do, and that studies should distinguish between existing (declining or regenerating) clusters and emerging and/or potential clusters (Chapman et al., 2004; Tödtling and Trippl, 2004; Dalum, et al., 2005; Wolter, 2005; Bergman, 2008).

In this chapter, I identify some of the conceptual and practical issues that arise in the conduct of regional cluster studies that focus on joint university-industry strengths as a basis for policy action. Rather than frame my arguments in the hypothetical, I ground the discussion in a comparison of the methodologies of two recent applied cluster studies in the US. The first is an analysis commissioned by the Appalachian Regional Commission (ARC) that sought to identify technology-based clusters across the ARC service area, an expansive territory that extends from northeastern Mississippi to central New York State, touching parts of 13 states in total. The second initiative, an effort pushed by university leaders but embraced and facilitated by a regional economic development marketing organization, focused on identifying current and potential clusters in a single region with a significant existing compliment of research universities and high technology businesses: the Research Triangle Park (RTP) region of North Carolina.

Both studies are useful cases to examine for two reasons. First, each is an example of 'best practice' in some sense. The ARC study is highly detailed in its geographic coverage and includes an unusually extensive and diverse array of quantitative indicators, some of which draw on unpublished, confidential federal data sources. The RTP project follows on the highly publicized five-region Clusters of Innovation initiative spearheaded by the US Council on Competitiveness.[2] The RTP study also sought to be a model for cluster-based economic development planning in other regions of North Carolina.

Second, the two studies are similar in that both sought to detect areas of existing or potential university-industry strength as sources of competitive advantage and future growth. Yet their differences in objectives, scope, design and methods help highlight some of the strengths and weaknesses of industry cluster analysis as a tool for economic development planning, particularly as it relates to understanding regional innovation. A comparison of the two methodologies exposes some of the 'fuzziness' inherent in the cluster concept as applied in practice and offers some guides about how cluster studies might best be designed and undertaken.

The chapter is organized as follows. Sections 4.2 and 4.3 briefly summarize the ARC and RTP projects in turn. I discuss each study's research design and approach to measurement, focusing specifically on how clusters are defined, how real (or hoped-for) linkages between local industries and research universities are conceptualized and measured, and how the studies are now being used to inform policy making. Section 4.4 then compares the two study approaches, focusing on two major issues: first, the degree to which each study approach can, in principle, inform typical regional economic development challenges in industrialized countries given their specific objectives and assumptions; and, second, whether each study approach is likely, in practice, to generate valid findings given definitional, data, and other empirical challenges. Section 4.5 concludes the chapter with a brief summary and discussion of an optimal approach to applied cluster analysis.

## 4.2 THE ARC STUDY

In late 2000 the Appalachian Regional Commission commissioned an inventory of high technology clusters in the Appalachian region. The analysis, which I was involved in designing and carrying out, was completed in 2002 (Feser et al., 2002). The study relied heavily on secondary data since the aim was to systematically inventory R&D, innovation, and technology specializations across all counties in the ARC service area – a total of 406 counties at that time – together with a buffer zone of border counties. The project examined technology-related assets within and near the region from two perspectives: the industrial base, defined as technology-based goods and services production and employment; and the knowledge base, defined as knowledge-creating institutions and programs. Technology clusters were defined as 'the overlap between the industry and the knowledge/innovation strengths' (Feser et al., 2002, p. 7). Multiple sources of evidence were assembled for each of the two dimensions 'given the myriad plausible ways that high-tech activity might be defined, measured (in terms of quantity), and assessed (in terms of quality)' (p. 7). Identified clusters were defined as places where a moderately to highly sophisticated knowledge infrastructure is joined with a substantial related industrial base.

### 4.2.1 Measurement of the High Technology Industrial Base

The study characterized Appalachia's industrial base by synthesizing findings generated with two kinds of related information: (1) the location of employment in eight high technology value chains; and (2) the location of science and engineering workers in thirteen occupational categories that mapped to the eight value chains. In the case of value chain employment, county-level data and a measure of spatial concentration were used to identify unique multi-county areas where technology-related activity was disproportionately clustered. The study used location quotients for the occupation analysis since data on science and engineering workers were available only for metropolitan areas. Because of the varying geographic units for which data could be obtained (e.g., zip codes, counties, metro areas), the study relied on graphic overlays in a GIS system to facilitate visual identification of those sub-regions within and along the border of Appalachia where technology-related activity is especially pronounced.

In the study, a given value chain is a group of technology-intensive Standard Industrial Classification (SIC) system sectors that are linked through buyer-supplier relationships (as revealed in a detailed analysis of national input-output patterns).[3] Each represents a technology-intensive buyer-supplier chain in the US economy as a whole, not Appalachia, and each is comprised of between 8 and 34-digit SIC codes.[4] The chains, which are not mutually exclusive, form the core classification around which the rest of the study is organized: they serve as a common reference category of high technology activity. Concordances between the chains and all other variable classifications (e.g., university disciplines, science and engineering occupations, industry utility patents, etc.) made possible a single consistent set of functional and geographic technology overlays from which spatial clusters could be identified.

The county employment data used in the study were from the confidential unsuppressed Unemployment Insurance Data Base (UDB) of the US Bureau of Labor Statistics, obtained with special permission. The UDB data, which contain employment and wage figures by establishment for all 50 states, permitted a fine-grained look at employment patterns even in very small counties. The metropolitan-level occupational employment data were from the US Bureau of Labor Statistics' Occupational Employment Statistics (OES) series. Fifty-six specific science, engineering, and engineering technician occupations were selected from the 709 occupations reported in the 1999 OES release and organized into 13 substantive groups that were then matched to the eight value chains.[5]

### 4.2.2 Measurement of the Knowledge Infrastructure Base

The study defined Appalachia's knowledge infrastructure as comprised of two major components: organizations conducting scientific research and applied innovation and the network of universities and colleges engaged in developing the region's human capital base. (In the case of major research universities, the two components came together.) The definition implied that Appalachia's science and innovation assets are principally based in its research universities and a limited number of non-academic research institutions (such as federal government laboratories), non-profit R&D organizations, state-sponsored technology agencies, and private sector businesses engaged in innovation. The region's higher education network consists of over 250 universities, colleges, and community colleges offering degree programs and specialized training in 15 science and engineering-related fields. There are 11 research-intensive universities located in ARC territory and an additional seven situated adjacent to or very nearby the ARC boundary. The seven adjacent schools were included on the assumption that their close spatial proximity yields a high potential spillover effect into the region.

The study used three measures to characterize university competitiveness or strength by discipline: (1) perceived faculty quality as judged by peers; (2) external research funding receipts; and (3) the number of enrolled full-time graduate students. The measures were converted into national rankings and a concordance between the set of value chain categories and the disciplines was created in order to establish a common scale for combining the disparate dimensions of research strength. The rankings were then averaged across the disciplines within each technology area. For example, the study defined Cornell University's rank for sponsored research relevant to the chemicals and plastics value chain as the arithmetic average of its national ranks for chemical engineering, materials engineering, and chemistry.

Given the rankings on the three indicators, Tier 1 universities were defined as those institutions with an average rank in the US top 20 for at least two out of the three measures. Tier 2 schools were defined as those with: (a) an average rank in the US top 20 for research funding or faculty quality; (b) an average rank in the US top 40 for all three measures; or (c) an average rank in the US top 20 for number of graduate students and a rank in the US top 40 for either (or both) faculty quality or research funding. The criteria effectively considered sponsored research and faculty quality as the leading barometer of a university's research capacity and output.

The study documented four additional indicators of R&D and/or educational capital in addition to university R&D. First, it identified all major non-academic research institutions located in or near the region (18 in total), documenting funding levels where data were available. Second, as a rough indicator of private sector innovation activity, the study measured patenting in industries in the eight value chain categories using data from the US Patent and Trademark Office (USPTO). Both location quotients and spatial autocorrelation-based concentration measures were used to identify geographic areas of patent clustering. Third, the project documented federal innovation funding program winners in all technology areas that matched to the value chains, mapping the winners by zip code.[6] Finally, the study used the US Department of Education's Integrated Postsecondary Education Data System (IPEDS) to track recent degrees granted in all postsecondary educational institutions in the region. The highly detailed degree data were aggregated into 15 disciplinary/program areas that paralleled the National Science Foundation discipline classification and then mapped to the eight industry value chains.

### 4.2.3 Joining the Information: Finding ARC's Technology Clusters

The final step in the methodology was to assemble the multiple variables specifying joint industrial and innovative strengths for each of the eight technology areas into a single GIS database. Mapping then revealed sub-regions where both high technology industry and related R&D and innovation activity were in evidence. The mapped data were also summarized using a simple index useful for analyzing broad patterns in technology clustering across the region.

Unlike the RTP study described in the next section, the ARC cluster analysis was not intended to inform a single specific policy initiative. Instead, its aim was to inventory technology clusters for subsequent use by ARC and local, regional and state economic development organizations in the region. The study showed that many of Appalachia's technology-related clusters are in traditional manufacturing (chemicals, motor vehicles, and industrial machinery), that the region's industrial base is oriented toward high-tech industries of moderate technology intensity, and that the joint spatial clustering of business and innovation/R&D in some very high-tech sectors (e.g., information technology, software, and aerospace) was quite limited as of the late 1990s. While some Appalachian universities boast significant existing or emerging R&D strengths in science and engineering disciplines, often those universities are not located nearby significant concentrations of industrial employment in related sectors. Likewise, while Appalachia has its share of federal laboratories and other non-university R&D institutions, the study found that they are often not spatially coincident with the technology-oriented industrial base. Indeed, the analysis found that a great many of the region's clusters are located on its periphery, making Appalachia's high-tech growth prospects heavily dependent on spillover (or 'spread') effects from neighboring cities and metropolitan areas.

## 4.3 THE RTP STUDY

The RTP region study originated with a project organized by the Council on Competitiveness, a non-profit organization formed in 1986 to advocate strategies to strengthen US competitiveness in the face of stiffening foreign competition. Historically the Council viewed US competitiveness issues through a national lens. However, in the late 1990s, influenced heavily by the ideas of Michael Porter, a member of the organization's Executive Board and an influential player in its genesis in the 1980s, the organization began to shift its focus to the regional scale. It came to see national innovation as contingent on regional innovation processes and the success of localized clusters:

> In healthy regions, competitiveness and innovation are concentrated in clusters, or interrelated industries, in which the region specializes. The nation's ability to produce high-value products and services that support high wage jobs depends on the creation and strengthening of these regional hubs of competitiveness and innovation. (Porter et al., 2001b, p. v)

In 2000 the Council began a multi-region cluster study that it called the 'Clusters of Innovation Initiative'. The project applied Porter's framework for analyzing clusters and regional competitiveness to five US regions: Atlanta, Pittsburgh, North Carolina's Research Triangle, San Diego and Wichita. The effort sought both to understand clusters in each individual region and to use the collective results to more rigorously test a variety of hypotheses about 'how innovative capacity is built' (Porter et al., 2001b, p. 1). Analysis for each case was based heavily on data from the Cluster Mapping Project at the Harvard Institute for Strategy and Competitiveness, supplemented with surveys and interviews in the host regions. The Cluster Mapping Project is a set of national industry cluster definitions – groups of SIC codes defined as clusters – matched to publicly available secondary data sources (County Business Patterns). The Clusters of Innovation Initiative documented trends in clusters in the study regions using the secondary employment data and then used interviews and survey-based information to help interpret those trends. Each of the five case reports follows a similar boilerplate. The executive summaries offer two major types of findings: first, general recommendations regarding the role of clusters in regional economies and implications for policy and the role of government; and, second, empirical findings regarding basic economic indicators, industrial composition, innovation assets, and identified clusters.

The Council's RTP study (Porter et al., 2001a) identified 14 existing and emerging clusters in the Triangle, some traditional (e.g., textiles, tobacco, heavy machinery) and some technology-oriented (e.g., communications equipment). The study called attention to weaknesses in the region's infrastructure, primary and secondary education systems, technology commercialization capacity, lack of industrial diversity, branch plant dominance, lack of collaboration among companies in different clusters, and high degree of interjurisdictional competition (i.e., absence of regional cooperation). One of its major recommendations was that regional officials seek to build a consensus around a shared, innovation-oriented strategy for regional economic development.

Regional government, academic, and business leaders were highly receptive to implementing the suggestions of the Council's report, in large part due to the leadership of the University of North Carolina. The UNC System president, Molly Corbett Broad, was a member of the Council on Competitiveness Executive Committee at the time, and she actively raised the financial and political support for the Triangle's inclusion as one of the Clusters of Innovation case regions. In 2002, regional leaders formed a 37-member Future Cluster Competitiveness Task Force, chaired by former North Carolina governor Jim Hunt and staffed by the Research Triangle Regional Partnership (RTRP), the region's state-designated economic development marketing organization. The Task Force set to work developing an economic development plan, relying heavily on planning and facilitation support from the Small Busi-

*Table 4.1  Summary of RTP study: analytical elements and plan*

| 1/2002 | Clusters of Innovation: Research Triangle | Identified 14 industry clusters in the region, from agricultural products to heavy machinery to pharmaceuticals/biotechnology. Argued for support for existing clusters, upgrading traditional clusters and promoting growth opportunities in areas that cross-cut several clusters (e.g., in environmental sciences, biotechnology and information technology, telecommunications and medicine, and biotechnology and agribusiness). | Council on Competitive-ness, Monitor Group and onthe-Frontier |
|--------|-------------------------------------------|-----------------------------------------------------------------|----------------|
| 2/2003 | A Blueprint for Life Sciences Industry Growth in the Research Triangle Region | Documents the size and potential growth of the region's life sciences industry. Developed primarily to influence debate over a biotechnology training initiative eventually funded by the NC legislature in mid-2003. Later listed as part of the RTP cluster project. | PMP Public Affairs Consulting |
| 9/2003 | R&D Inventory and Analysis of Growth Opportunities in the Research Triangle Region | Documents the size and potential growth of the region's life sciences industry. Developed primarily to influence debate over a biotechnology training initiative eventually funded by the NC legislature in mid-2003. Later listed as part of the RTP cluster project. | RTI International |
| 10/2003 | Recommended Actions from Focus Groups | Report of findings of focus groups around each of the eight application areas. A total of 68 people participated in the focus groups (not including project consultants or staff), some 33 from businesses in the region. | RTI International |
| 11/2003 | Identifying 'Targets of Opportunity': Competitive Clusters for RTP and its Sub-regions | Using secondary data and benchmark cluster definitions identified major clusters in the 19 counties including and around the Triangle, in the three core metro counties of the region, in six non-core metro counties, and in ten non-metro counties. Recommended 13 possible targets for the region, including two that were eventually added to the eight proposed by RTI: motor vehicle components and transportation/shipping/logistics. | Office of Economic Development University of North Carolina at Chapel Hill |
| 3/2004 | Staying on Top: Winning the Job Wars of the Future | Lays out a five-year, $5 million action research agenda 'to generate 100,000 new jobs and increase employment in all 13 counties' of the region. Calls for five major strategies: 'promote the growth of industry clusters where the region has a competitive advantage; use a balanced approach of targeted recruitment, global branding, business creation and existing business retention; integrate higher education into economic development efforts; develop creative, inclusive approaches to rural prosperity; create agile leadership networks to respond to market challenges, changes and opportunities' (p. 4). | Research Triangle Regional Partnership |

ness Technology Development Council (SBTDC), a unit of the University of North Carolina; consulting support from RTI International, a non-profit research organization formed in 1959 with the establishment of the Research Triangle Park; research assistance from the Office of Economic Development at the University of North Carolina at Chapel Hill; and staff support from RTRP. Upon completion of the plan, RTRP took on direction of its implementation, with continued support from SBTDC. Table 4.1 summarizes the major elements of the RTP cluster study.

Associated with each element is a document, culminating with the five year strategic plan: *Staying on Top: Winning the Job Wars of the Future* (RTRP, 2004). At the heart of the cluster analysis was an R&D inventory and technology growth opportunity study conducted by RTI International (RTI, 2003a). Eight technology areas identified by RTI, along with two rural-based industrial clusters highlighted in a study conducted by UNC's Office of Economic Development (OED) (Luger et al., 2003), became the clusters prioritized in the strategic plan. Table 4.2 lists the ten clusters. The next two sections provide a brief summary of the RTI and OED methodologies.

*Table 4.2  Clusters prioritized in RTP strategic plan*

| Cluster | Study |
|---|---|
| Pharmaceuticals | RTI |
| Biological agents and infectious diseases | RTI |
| Agricultural biotechnology | RTI |
| Pervasive computing | RTI |
| Advanced medical care | RTI |
| Analytical instrumentation | RTI |
| Nanoscale technologies | RTI |
| Informatics | RTI |
| Vehicle component parts | OED |
| Logistics and distribution | OED |

*Source*:  RTRP (2004).

### 4.3.1  Inventorying the Region's R&D Assets and Growth Opportunities

The RTI inventory and growth study sought to 'provide information on the research and development (R&D) strengths of the Research Triangle region and to identify opportunities for capitalizing on those strengths in "hot technology" areas over the next 10-20 years' (RTI, 2003a, p. 1).

Figure 4.1 documents the major steps in the analysis along with key information sources and variables.

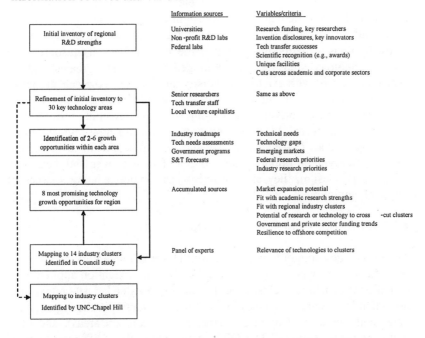

*Figure 4.1 Overview of RTI growth opportunity study methodology*

The RTI team began by using mostly secondary data on R&D spending, facilities, and technology transfer from universities, non-profit labs, and federal labs to create an initial inventory of research strengths in the region. Actual industrial R&D spending and activity in the Triangle received little attention at this stage, in large measure because of the absence of appropriate secondary data. RTI used interviews of key informants – senior university researchers, technology transfer staff, and local venture capitalists – to refine the initial list to 30 key technology areas. Two to six specific growth opportunities within each of the 30 areas were then identified based on information from existing industry roadmaps, technology needs assessment studies, government program documents, and science and technology forecasts. RTI's assessment of the research and market opportunities across the 30 areas is extraordinarily detailed, constituting 150 pages of the 179-page final report. With the help of a panel of experts, RTI also developed a rough concordance of the 30 areas to the 14 clusters identified in the Council's initial 2002 Research Triangle study.[7]

The RTI team used the two major pieces of accumulated evidence – the multiple identified growth opportunities within the 30 technology areas and the mapping of those technology areas to the 2002 Council study findings – to recommend that the region focus its resources on eight technology areas that were judged to have the greatest market potential, best fit with existing university and industry research strengths, highest relevance for multiple clusters, good funding support from government and/or private sector sources, and maximum resilience to offshore competition. In short, the eight areas are new and emerging technologies that RTI believes have the most potential to drive private sector employment opportunities in the Triangle over the next two decades. By design, they are not necessarily major existing sources of regional employment.

### 4.3.2  Identifying Cluster Targets of Opportunity

The OED study sought to complement RTI's R&D inventory by focusing on industrial activity (Luger et al., 2003). The investigation applied a set of national benchmark cluster definitions to an extended 19-county area and included separate analyses for core metro counties in the region, non-core metro counties (roughly interpreted as suburban areas), and non-metro or rural counties.[8] The OED team argued that 'different types of clusters make sense in different types of locations, and different parts of clusters are appropriate in different places' (Luger et al., 2003, p. 2). Using employment data for two time periods, measures of absolute and relative size (the latter a location quotient), and separate benchmarks for clusters as a whole and their technology-based components, the study identified 11 'appropriate target clusters' for the region, distinguishing the relevance of each for different sub-areas. Two major criteria filtered the targets from a total of 14 benchmarks: (1) 'indications of competitive local advantage, either by a sizable presence already in the region, or rapid changes in location quotient'; and (2) evidence that 'businesses that are adding jobs ... or at least not losing employment if the businesses are becoming more productive' (p. 19).

The OED analysis highlighted several traditional industry clusters with growth potential in rural areas of the region, but in the end, only two target clusters – logistics and distribution and vehicle component parts – were eventually included in the strategic plan. Even then, the plan makes clear that the two OED-identified clusters are of secondary importance and were included primarily to address concerns about the future of the region's rural areas. In *Staying on Top* it is noted:

> RTI identified eight areas that hold the highest potential for boosting economic growth in the near term ... These eight draw on the region's most competitive and innovative R&D assets. They represent the region's best opportunities for strong and sustainable job creation and business investment and growth ... Although universities and businesses will continue to pursue growth opportunities in all of the areas identified by RTI, the institutional partners working on this initiative will collaborate to focus new resources on these eight growth priorities ... In addition, the region will focus on two additional industry sectors that are important for job creation in more rural parts of the region: vehicle component parts and logistics and distribution. (p. 9)

Unlike the case for the eight technology clusters, there is no discussion in the plan of specific growth opportunities and strengths within the vehicle component parts and logistics and distribution clusters; the two clusters receive only brief mention.

### 4.3.3 The Regional Competitiveness Plan

Three other projects in addition to the RTI and OED analyses informed the development of the strategic plan: a series of focus groups organized for each of the eight technology areas (RTI, 2003b); a brief assessment of the region's life sciences industry (Pellerito, 2003); and the organization of six working groups made up of a diverse selection of business, academic and government officials and leaders. *Staying on Top* outlines five strategies and 30 action steps and calls for raising $5 million over five years to support an effort to 'generate 100,000 new jobs and increase employment in all 13 counties of the Research Triangle Region' (RTRP, 2004, p. 4).

## 4.4  LESSONS FOR TECHNOLOGY-FOCUSED CLUSTER STUDIES

What can be learned from the ARC and RTP projects that might inform the conduct of similar studies in other regions that seek to formulate development strategies around university-industry clusters? At the heart of this question are two major issues. First, how appropriate are the objectives of the studies, given what we know about the relationship between university research activities, private sector R&D, business competitiveness, and industrial growth? Underlying the objectives are implicit or explicit definitions of what clusters are and how they drive regional economic outcomes. Second, if the objectives of each project are taken at face value, how valid is the actual empirical work? Are there internal validity threats – factors that drive a wedge between study objectives and empirical results – that may ultimately lead to the formation and adoption of misleading policy actions or the neglect of issues essential to the strengthening of regional innovation systems? The validity threats may derive from data constraints, poor measures, resource constraints, or data collection procedures. The two issues might be reduced to the following: do we have the right objectives and can we legitimately meet those objectives with the tools at our disposal?

### 4.4.1 Appropriate Objectives?

The ARC study defines technology clusters as locations – specified as counties or multi-county areas – where there is current relative concentration of industrial activity in an identified high technology value chain as well as demonstrated university research and training strength in disciplines related to that value chain. From a policy standpoint, the study assumes that areas of joint university-industry specialization represent the appropriate building blocks of technology-based economic development. Moreover, specializations are detected relative to national norms, not local trends. A university specialization in a given discipline is based on its ranking against universities nationwide, not institutions within the given location or broader ARC region. Likewise, industrial specializations based on employment numbers and patent counts are identified from national share comparisons. The best academic program in the region is not a specialization in the ARC study if it is not highly ranked nationwide. Similarly, a large volume of local activity in a given value chain is an industrial concentration not because it is large locally, but because it is an outlier in a national spatial distribution. In those respects, the ARC study is consistent with Porter's diamond model and its emphasis on corporate activity and demonstrated existing competitive advantage.

In the RTP project, clusters are hybrids between current and prospective economic strengths, and the basis of the current strength for most of the identified clusters is academic rather than corporate, found mostly in strong university disciplines, productive scholars, projected national research priorities on which area universities believe they can capitalize, locally-based federal laboratories, and perhaps some evidence of private sector entrepreneurial activity. Implicit in the model is that the academic research strengths and cutting-edge science will translate to new economic engines in the future (the potential strengths), either via connections to existing technology industries or through university spin-offs or independent new business creation. That is clear in the strategic plan, which lists the key regional resources for each of the eight core clusters. Aside from the pharmaceuticals cluster, which represents an existing and widely-recognized industrial strength in the Triangle, most of the resource compendia highlight academic or non-profit research capabilities before mentioning any corporate presence or even private sector R&D. In essence, the RTP project identified areas where the Triangle is most likely to be a science and innovation leader and is gambling that, with the right strategies, innovation can be turned into job growth in offices and factories, not just federal laboratories, universities, and colleges.

The validity of each study varies depending on the policy concerns and context at hand. The ARC study suffers from what is, in one respect, a major weakness: identifying current clusters means focusing on areas of industrial and academic strength that may very well be in decline, or at least have already reached their peak. Indeed, the elements of the RTP project that are most similar to the ARC effort, namely the original Triangle region cluster analysis by the Council on Competitiveness and the benchmark industry cluster study by OED, identified some Triangle industries that are not likely to be the source of significant job and income growth over the next decades (e.g., textiles, heavy machinery, tobacco). The ARC study found that most of Appalachia's technology clusters are in chemicals and plastics, industrial machinery, motor vehicles, and household appliances, not exactly industries that are projected to expand rapidly in the US or the region.

On the other hand, the ARC study approach produces results that can effectively inform two key areas of economic development practice: (1) business retention activities focused on industrial modernization and (2) workforce development concerned with transitioning workers from declining sectors. Both are areas that are especially important for regions with a history in heavy industry and a modest complement of nationally competitive research institutions and major federal laboratories.

The RTP study's key strength is its emphasis on future growth and the fact that it builds on what is arguably the Triangle region's true locally-based competitive advantage: its three world class universities. The RTP approach addresses a major limitation of much applied industry cluster analysis head-on, namely, its tendency to document past industrial successes rather than future sources of growth. Among the RTP project's weaknesses are that it offers little understanding of the needs of existing industry, aside from those of a few selected sectors, and that it adopts a narrow view of human capital formation as a foundation of the region's competitiveness. One of the more consistent findings of the literature on the impact of universities on regional economies is the role of the university as educator of the next-generation workforce (Goldstein and Renault, 2004; Goldstein and Drucker, 2006). In focusing its economic development efforts around eight cutting edge areas of science, the RTP project implicitly downplays the importance of developing more basic training and curricula that can serve the current mix of businesses and jobs in the region more effectively. Admittedly, there are other initiatives underway in North Carolina to address university and college curricula issues in the context of workforce preparation, but they receive scant mention in what is labeled as the official 'Competitiveness Plan for the Research Triangle Region'.

There are two other issues related to each study's objectives that deserve consideration. The first is the nature of the connection between universities and innovation-led industry growth. It is hard to imagine that the design of the RTP project and resulting strategic plan were not influenced by local observation of the tight connections between the region's universities and its pharmaceuticals and biotechnology industries. The Triangle is one of a few legitimate biotechnology hubs in the US (Cortright and Mayer, 2002), and the close linkages between the academy and the private sector in biotechnology are well documented. The vision of university-industry collaborative science, small business entrepreneurship, university spin-offs, and technology transfer that characterizes the biotechnology industry permeates much of the RTP's plan. How likely is that vision – a kind of nirvana of university-industry synergies – to extend to other areas of technology, such as pervasive computing, analytical instrumentation, informatics, and nanotechnology? The weight of the evidence to date is mixed. While authors like Mowery and Sampat (2005) believe such synergies are increasing, they are hard pressed to find clear empirical evidence of it in the US. Betts and Lee (2004) argue that the role universities play in supporting the growth of technology clusters is overstated, and a careful historical analysis of efforts to translate the Silicon Valley model to other regions cautions against uni-dimensional explanations of the link between university research prowess and industrial growth (Leslie and Kargon, 1996). The conceptualized nexus between universities and industries in the ARC study is problematic as well, if current areas of joint university-industry specialization reflect a past economic history where universities acted as last-resort substitutes for low levels of technical capability in local firms

(Rosenberg and Nelson, 1994). Clearly, whether and how university-business connections lead to the formation of new regional industries is not obvious.

The second issue relates to cluster policy implementation and its likely success. One might make a case that the RTP study, despite some of the drawbacks noted above, conceptualizes clusters in a manner best suited to what might be realistically achieved from a policy standpoint. There is an interesting paradox in what might be described as 'industry cluster practice'. On the one hand, many development officials are most interested in how cluster ideas might be used to promote the growth of next-generation industries, that is, clusters that do not currently exist in any sense of critical mass. On the other hand, most academics and cluster consultants argue clusters cannot be built from scratch because of the overwhelming dominance of market imperatives over which policy makers have little control (Porter, 2003; Ketels, 2004). The RTP model sees the region's future primarily in its science base. That base is strengthened with continued investments in university programs and laboratories, along with other relatively modest entrepreneurship support initiatives (e.g., incubators, technology transfer, R&D tax incentives, and marketing). Building a science base, particularly in higher education, is something that the public sector has proven it can do reasonably well. The strategy has the additional and not insignificant advantage of a ready external funding source: federal research dollars. A cluster strategy that is tilted toward the university's role may or may not generate thousands of new private sector jobs, but it is certainly capable of supporting the development of quality educational institutions and programs that have their own public interest benefits. The woeful record of strategies that seek to influence the behavior of businesses directly may indeed favor an academic science-led cluster approach, other things equal.

### 4.4.2 Industry Cluster Empirics

The discussion above presumes the two studies have the right analytical methods and data to produce findings that satisfy their stated objectives. In actuality, there are a number of reasons to interpret and apply the results of applied industry cluster analysis cautiously (Feser and Luger, 2003). Selected similarities and differences in the approaches of the ARC and RTP studies help to highlight some of the key empirical limitations of analysis aimed at identifying industry clusters in general, and university-industry clusters in particular.

Perhaps the most significant challenge facing any applied industry cluster study is data availability. That is especially true of studies focused on innovation activity or emerging markets. There are a wide variety of sources of data on industry trends, but they are often limited to employment series at the regional level. Employment is a reasonable measure of overall industry size from a labor market standpoint, but it is prone to giving a misleading view of the true presence, impact or growth of an industry in a region. Some industries are highly capital-intensive, generating significant sales, and often regional interindustry linkages, with few employees. Other industries are growing in sales even as they cut employees. Substitutes for employment data are few. Readily available secondary data on industry R&D activities do not exist in the US at the sub-state level, and even at the state level the data are not reliable since they are based on very small samples in surveys administered by the National Science Foundation. That is one reason why the RTP study did not attempt to systematically assess industrial innovation trends. The ARC study used patent data as a proxy measure of industrial innovation by sector in Appalachia. While that avoided ignoring private sector innovation entirely, patent data are not always a good barometer of the inputs in the research process, and industries also vary substantially in the degree to which they use patents to protect intellectual property (Griliches, 1990).

The data challenge is not simply availability, but also whether or not available data series have meaningful levels of detail. Those levels may pertain to industry, geography, university and college disciplines, occupation, and areas of innovation. In the case of industry data, even the most disaggregated sectoral categories may be insufficient to distinguish important firm and market specializations. The problem is even more significant for trend data on university research. In the US, the National Science Foundation compiles university research expenditures by discipline but the categories utilized are often too broad to document activity in the narrow scientific specialties that would provide the best hint of entrepreneurship opportunities. A given institution may have a scientific strength that is completely obscured if researchers from different disciplines are contributors. The artificial nature of conventional disciplinary classification schemes is only going to increase as research becomes more multi-disciplinary.

The ARC and RTP studies address data limitations in different ways. Since the objective of the ARC analysis – i.e., to document clustering across hundreds of counties – necessitated reliance on secondary data sources, the project utilized maximum sectoral and disciplinary detail and employed a variety of different measures of industrial and innovation activity. On the industry side, that led to the request for access to confidential federal employment data and the special run of patent data. The RTP study relied much more heavily on primary data collection, principally via review of the academic and trade literatures and interviews with key informants. Studying only a single region, the RTI technology assessment was able to utilize an iterative strategy whereby it produced an initial snapshot of regional academic, federal laboratory, and private non-profit research strengths; presented the assessment to local experts; revisited contested aspects of the assessment in a subsequent review; presented a revised snapshot for review; and so forth. The RTP study also substantially reduced the scale of its data needs by conceptualizing clusters as research strengths based mainly in academic science. That implied that local private sector R&D trends, which are much harder to objectively assess given the paucity of secondary data and the reluctance of firms to divulge their R&D expenditures, could be safely ignored.

Another major challenge for any assessment of real or potential academic-business linkages is the need to somehow map the scientific activity in the university to relevant business sectors, and vice versa. Both the ARC and RTP study utilize academic-industry concordances that substantially influence their findings. In the ARC study, eight national high technology value chains form the basis of the results; disciplines, patents by sectors, occupational employment, innovation awardees, and other variables are all assigned to one or more value chains. There is much that is a 'leap of faith' in the mapping. First, the value chains themselves may or may not accurately distinguish technology specializations in the US or Appalachia. Certainly eight chains appears to be a very small number for an economy as diverse as that of the US; they may very well be too aggregated. The origin of the value chains in an objective analysis of input-output trends lends them credibility, but it is also possible that the input-output data are in themselves insufficient for specifying related technology industries in proper detail. Second, the mapping of some of the indicators to the value chains involves an element of guesswork, as good data on the connections are not available. The RTI technology assessment used information from 'experts' to map 30 initial technology areas to the Triangle industry clusters documented by the Council on Competitiveness. Unfortunately, it is not evident that such an approach involved any less informed guesswork than in the case of the ARC study.

The sheer scope of the ARC study probably makes it more prone to either overlooking legitimate specializations that offer the best opportunities for technology-based economic development or failing to provide the level of detail necessary to formulate real development plans. For example, the study identified a 'pharmaceuticals and medical technologies' cluster in Pittsburgh based on a revealed concentration of industry employment in the pharmaceuticals value chain, second tier rankings of Carnegie Mellon University (CMU) and the University of Pittsburgh (Pitt) in related health sciences disciplines, and a concentration of federal innovation research award winners in related health fields. Although this is useful, more detailed information on the source of the overall pharmaceuticals and medical technologies specialization is required to formulate actionable plans.

For example, a close look at the Pittsburgh region indicates that one of its key health sciences strengths is tissue engineering, a discipline that crosses several medical and technical specialties, including cell biology, molecular biology, biomaterial engineering, computer-assisted design, microscopic imaging, robotics engineering, and scientific equipment manufacturing. The McGowan Institute for Regenerative Medicine (MIRM), established by the Pitt School of Medicine and the Pitt Medical Center health system, is considered one of the nation's most ambitious tissue engineering programs. In 2002 Congress established the National Tissue Engineering Center (NTEC) in Pittsburgh to facilitate related Department of Defense research among the top military and civilian scientists in the field. NTEC unites the clinical and research facilities of MIRM, Pitt, the Pitt Medical Center, research at CMU, and the Windber Research Institute with the US Army's Institute for Surgical Research and the Walter Reed Army Medical Center. The region recognizes the economic development potential of tissue engineering and the biosciences in general. Current development initiatives include the Pittsburgh Tissue Engineering Initiative (PTEI) and Pittsburgh Life Sciences Greenhouse (PLSG), the latter a public-private collaborative effort aimed at supporting the growth of regional life sciences companies in the areas of bioinformatics, bio-nanotechnology, diagnostics, medical devices, medical robotics, therapeutics, and tools and services. There are currently five biotech companies in a PLSG-run incubator space.

The Pittsburgh example illustrates a key difference between the ARC and RTP studies. Because of its heavy reliance on secondary data, the ARC cluster analysis searched for broadly defined and already well-developed clusters. Its overall findings are useful for gaining a sense of the overall pattern of technology-based economic activity in Appalachia, but more detail on the strength and vintage of specializations within identified clusters is needed in any particular region to formulate useful development strategies. Because a region has a lot of pharmaceuticals activity does not mean that more pharmaceuticals and biomedical employment or drugs research is the appropriate policy target. It is possible there are newer emerging areas of science and private sector growth that should be fostered within a broader health sciences cluster. By focusing on emerging science, the RTP study emphasizes innovation and new markets. Overall, if informing specific policy options in a single location is the aim, the internal validity of the RTP study is stronger. Scientific specialties in the university are easier to assess than innovation strengths in industry and the small scope of the RTP study meant that it could use primary data collection to rectify the limitations of secondary data. That meant more detail and greater narrowing of possible areas of development intervention.

## 4.5  SUMMARY AND IMPLICATIONS

In this chapter I have compared two major recent industry cluster studies that focus on the intersection of industrial development and the regional science base. The different objectives and methodologies of the two projects illustrate some of the major challenges in applied industry cluster analysis as a basis for economic development policy and planning in general, and technology-based economic development in particular.

Both the ARC and RTP studies demonstrate how important it is to utilize primary and secondary data on industry and innovation strengths cautiously. Excessively broad industry categories, aggregated disciplinary classification schemes, the absence of private sector R&D expenditure data, and the limited geography of many secondary data series make it hard to pinpoint narrower industrial and scientific specializations. Primary data collection, particularly via key informant interviews, offers a solution, but not one without its own risks. The strengths and weaknesses of qualitative methods in applied cluster analysis deserve much greater attention in the research literature. Too many applied analyses appeal to 'experts' as the primary data source without explaining how such experts were identified, their advice solicited, and their opinions documented and analyzed.

Both the ARC and RTP studies make assumptions about the importance of the regional science base for regional industry growth. In the case of the RTP study, in particular, there is much implicit faith that university-based science lays the groundwork for regional industries of the future. In an increasingly knowledge-based economy, that is not an assumption that policy officials find especially hard to swallow. However, while studies have found a connection between universities and regional economic growth, the preponderance of evidence suggests the link is modest and, most likely, based mainly in the traditional human capital function of higher education (Goldstein and Renault, 2004). It probably makes sense to foster the growth of academic science-based clusters as an element of a broader regional development plan in some instances, just not the sole element.

Clearly the value of different approaches varies depending on the region concerned. It is hard to imagine how the RTP study approach could be applied to a lagging region with a very modest science base, no top-notch universities, and an industrial structure dominated by branch plants conducting no localized R&D. In such cases, analysis of the fit between existing industrial concentrations, possible new growth industries (either via local growth or inward investment), and university and community college curricula is probably more useful since workforce development concerns are likely to well outweigh the potential for technology-based development. Likewise, the ARC study approach would be quite limited in its scope in an innovation-rich region like North Carolina's Research Triangle, where the prospects for growth emerging out of local R&D are good.

In the end, is there an optimal approach to industry cluster analysis aimed at detecting regional innovation strengths? The answer is yes, if we conceive of cluster analysis as a flexible methodology and not a set of fixed definitions and measures (Feser and Luger, 2003). The ideal industry cluster study begins with a careful articulation of project goals and objectives, which themselves are based on the types of policies the study is intended to inform. The assumed conceptual relationship between industry clusters and regional development outcomes should be explicitly outlined in the goals articulation phase of any study. In the RTP analysis, clusters are more or less scientific fields presenting economic growth opportunities given the existing critical mass of R&D. That makes sense if the policy aims are centered on technology-based economic development and entrepreneurship and the region is seeking to leverage academic innovation into business formation. The RTP study falls short as a cluster analysis if the intention is to inform workforce development and traditional industry modernization programs. Likewise, the ARC study offers insights into the existing technology mix in the Appalachian region, but it is far less useful as a source of evidence of emerging technology fields. Unfortunately, neither the RTP nor the ARC study are as explicit about goals as they might be, and therefore both are subject to the criticism that they leave significant dimensions of the technology economy in their respective regions unexplored. A pitfall common to industry cluster analysis is the tendency to try to be too comprehensive, to produce a study that can become the basis for all regional economic development planning efforts. The optimal cluster study – if optimality is defined as the capacity to reliably inform policy making – is designed to address information needs required for specific policy objectives. The need for the analysis should drive the design of the project, not the other way around.

## NOTES

1. Special thanks to Shannon Landwehr for research assistance.
2. See http://www.compete.org/nri/clusters_innovation.asp.
3. The eight chains are chemicals and plastics; information technology and instruments; industrial machinery; motor vehicles; aerospace; household appliances; communications services & software; and pharmaceuticals and medical technologies.
4. The chains were derived from a factor analysis of US input-output data following the approach in Feser and Bergman (2000). A detailed description of the methodology is available in Feser and Koo (2000).
5. For example, chemists/chemical engineers and materials engineers/scientists occupations were matched to the chemicals and plastics value chain.
6. The study documented grantees of the Small Business Innovation Research (SBIR) program, Small Business Technology Transfer (STTR) program, and Advanced Technologies Program (ATP).
7. The concordance includes a four-level qualitative scoring of the 'relevance' of the technology areas to each cluster.
8. The benchmark cluster definitions are from Feser and Koo (2000) and are derived using the same basic methodology as the high technology value chain clusters used in the ARC study (see also Feser and Bergman, 2000).

## REFERENCES

Bergman, E.M. (2008), 'Cluster life-cycles: an emerging synthesis', in Charlie Karlsson (ed.), *Handbook of Research on Cluster Theory*, Cheltenham, UK and Northampton, MA, USA, Edward Elgar, pp. 114-132.

Betts, J.R. and C.W.B. Lee (2004), 'Universities as drivers of regional and national innovation: an assessment of the linkages from universities to innovation and economic growth', San Diego, CA, University of California at San Diego.

Chapman, K., D. MacKinnon and A. Cumbers (2004), 'Adjustment or renewal in regional clusters? A study of diversification amongst SMEs in the Aberdeen oil complex', *Transactions of the Institute of British Geographers*, 29, 382-396.

Cortright, J. (2006), *Making Sense of Clusters: Regional Competitiveness and Economic Development*, Metropolitan Policy Project, Brookings Institution, Washington, DC.

Cortright, J. and H. Mayer (2002), 'Signs of Life: The Growth of Biotechnology Centers in the US', Brookings Institution, Washington, DC.

Dalum, B., C.Ø.R. Pedersen and G. Villumsen (2005), 'Technological life-cycles: lessons from a cluster facing disruption', *European Urban and Regional Studies*, 12, 229-246.

Feser, E.J. and E.M. Bergman (2000), 'National industry cluster templates: a framework for applied regional cluster analysis', *Regional Studies*, 34, 1-19.

Feser, E.J. and J. Koo (2000), *High Tech Clusters in North Carolina*, North Carolina Board of Science and Technology, Raleigh, NC.

Feser, E.J. and M.I. Luger (2003), 'Cluster analysis as a mode of inquiry: its use in science and technology policymaking in North Carolina', *European Planning Studies* 11 (1), 11-24.

Feser, E., H. Goldstein, H. Renski and C. Renault (2002), 'Regional Technology Assets and Opportunities: The Geographic Clustering of High-Tech Industry, Science and Innovation in Appalachia, Appalachian Regional Commission', *Report Paper*, Center for Urban and Regional Studies, University of North Carolina at Chapel Hill, Chapel Hill.

Goldstein, H. and J. Drucker (2006), 'The economic development impacts of universities on regions: do size and distance matter?' *Economic Development Quarterly*, 20 (1), 22-43.

Goldstein, H. and C. Renault (2004), 'Contributions of universities to regional economic development: a quasi-experimental approach', *Regional Studies*, 38, 733-746.

Griliches, Z. (1990), 'Patent statistics as economic indicators: a survey', *Journal of Economic Literature*, 28, 1661-1707.

Ketels, C. (2004), 'European clusters', *Structural Change in Europe*, 3, 1-5, Habarth Publications, Bollschweil, Germany.

Leslie, S.W. and R.H. Kargon (1996), 'Selling Silicon Valley: Frederick Terman's Model for regional advantage', *Business History Review*, 70, 435-472.

Luger, M.I., L.S. Stewart and G. Androney (2003), *Identifying 'Targets of Opportunity': Competitive Clusters for RTRP and its Sub-Regions*, Office of Economic Development, University of North Carolina, Chapel Hill.

Mowery, D.C. and B.N. Sampat (2005), 'Universities in national innovation systems', in J. Fagerberg, D.C. Mowery and R.R. Nelson (eds), *Oxford Handbook of Innovation*, Oxford University Press, pp. 209-239.

Paytas, J., R. Gradeck and L. Andrews (2004), *Universities and the Development of Industry Clusters*, Economic Development Administration, US Department of Commerce, Washington, DC.

Pellerito, P. (2003), *A Blueprint for Life Sciences Industry Growth in the Research Triangle Region*, PMP Public Affairs Consulting, Chapel Hill, NC.

Porter, M.E. (2003), 'The economic performance of regions', *Regional Studies*, 37, 549-578.

Porter, M.E., Monitor Group, ontheFrontier and Council on Competitiveness (2001a), *Clusters of Innovation Initiative: Research Triangle*, Washington, DC.

Porter, Michael E., Monitor Group, ontheFrontier and Council on Competitiveness (2001b), *Clusters of Innovation: Regional Foundations of US Competitiveness*, Washington, DC.

Rosenberg, N. and R.R. Nelson (1994), 'American universities and technical advance in industry', *Research Policy*, 23, 323-348.

RTI (2003a), *Future Cluster Competitiveness: R&D Inventory and Analysis of Growth Opportunities in the Research Triangle Region*, Research Triangle Park, NC, RTI International.

RTI (2003b), *Future Competitiveness Effort: Recommended Actions from Focus Groups*, RTI International, Research Triangle Park, NC.

RTRP (2004), *Staying on Top: Winning the Job Wars of the Future*, Research Triangle Regional Partnership, RDU Airport, NC.

Tödtling, F. and M. Trippl (2004), 'Like Phoenix from the ashes? The renewal of clusters in old industrial areas', *Urban Studies*, 41, 1175-1195.

Wolter, K. (2005), 'A life cycle for clusters? Agglomerations, bifurcations and adaptation', Working paper, Institute for Industrial Economics and Strategy, Frederiksberg, Denmark.

PART TWO


The geography of academic knowledge
transfers: recent developments

# 5. The role of higher education and university R&D for industrial R&D location

## Martin Andersson, Urban Gråsjö and Charlie Karlsson

### 5.1 INTRODUCTION

The rapid globalization in recent years has created a radically new competitive situation for the rich industrialized countries. They have experienced increased penetration of their domestic markets not only by 'simple' products such as textiles, clothes and shoes, but also to an increasing degree of more advanced industry products such as computers, cars and mobile phones.

The increased import penetration has generated different reactions in different countries. In the US and the European Union (EU), voices have been raised for introducing various protectionist measures. Another reaction heard, for example in Sweden, is that the rich industrialized countries must increase the knowledge content and the degree of sophistication of their products to retain their comparative advantages in their export markets and to find new export markets.

How this shall be achieved in practical terms is much more of an open question. Of course, there has been a strong argument for increasing the share of young people going to university as well as for increasing the public R&D investments. However, higher education by itself does not make old products more knowledge intensive or generate new products or processes, nor do public R&D investments to any major extent. Public R&D investments are mainly directed towards basic research.

The upgrading of old products, the generation of new products and the development and upgrading of processes is the object of industry research mainly controlled by private industry. The possibilities for governments to directly control private R&D is limited in market economies and the options to subsidize private R&D are also limited within for example the EU, since such subsidies distort competition. There are, however, indirect measures, such as investments in higher education and public R&D, to influence industry R&D. The question is: how responsive is private industry to these kinds of indirect measures?

The answer to that question is by no means simple. To take Sweden as an example, we can observe that about 90 percent of all private R&D is controlled by multinational enterprises (MNEs) and about 40 percent of that is performed by foreign-owned MNEs. Swedish MNEs on the other hand perform almost half of their R&D abroad (Lööf, 2005). Given that a very limited number of MNEs, of which a high share is foreign owned, control so much of the industry R&D it is an open question to what extent national and regional policies can influence industry R&D in a country or region. MNEs, for instance, optimize their operations on a global scale.

Nevertheless, a possible way in which national and regional policies can affect industry R&D is to make regions attractive as locations for R&D by private firms. In an earlier paper, two of the authors (Andersson and Karlsson, 2004) of this chapter studied among other things how the location of industry R&D is influenced by the location of public, i.e. university, R&D. The empirical results pointed in the direction of a positive influence. However, that paper did not consider the possible influence of higher education in terms of students and graduates. A recent study by Faggian and McCann (2006), for instance, shows that migration effects of embodied human capital in students and graduates is of greater importance than (informal) university-industry spillovers. This suggests that the effect of students and graduates at institutions of higher education on the location of industry R&D needs to be considered. In view of this, the purpose of this chapter is to analyse to what extent the location and extent of public R&D and higher education, in terms of potential recruitment of graduates, influence the location and the extent of industry R&D in Sweden.

The chapter is organized as follows. In Section 5.2 we discuss some general characteristics of innovation processes and their dependence upon knowledge flows, and give an overview of the general spatial conditions for innovation processes, which is followed by an elaboration over the factors determining the location of company R&D in general and its dependence on closeness to university R&D and higher education, in particular. The data, the models and the econometric techniques used in the empirical analysis are presented in Section 5.3. Our empirical analysis can be found in Section 5.4 and our conclusions in Section 5.5.

# 5.2 INNOVATION PROCESSES, AGGLOMERATION AND THE LOCATION OF INDUSTRY R&D

### 5.2.1 Innovation Processes and Knowledge Flows

Knowledge and information are critical inputs in innovation processes. However, there is an important distinction. Knowledge is a critical input in innovation processes, which is used to generate new or improved products and/or new production techniques. This knowledge input may come from different sources: (1) scientific knowledge in the form of basic scientific principles, (2) technological knowledge in the form of technical solution and blueprints, and/or (3) entrepreneurial knowledge, that is knowledge about customer preferences and willingness to pay, market conditions, business concepts and business methods, etc. (Karlsson and Johansson, 2006). These types of knowledge can be developed in R&D processes over time and can be accessed through various types of knowledge sources. This explains why firms are active in many different knowledge networks with many economic actors possessing different types of knowledge (Batten, Kobayashi and Andersson, 1989).

Except for knowledge sources internal to firms, they derive information and knowledge from external knowledge sources. The literature on innovation networks and innovation systems frequently refers to the fact that firms gain knowledge from customers, suppliers, other firms in the industry, universities, R&D institutes, etc.

For innovative firms, interaction with the scientific community is generally considered as crucial. Innovative firms are often assumed to be highly dependent upon knowledge generated in nearby research universities. In this chapter we are particularly interested in studying the role of physical accessibility to research universities and institutions of higher education for innovative firms. We do so by analysing the role played by physical accessibility to research universities and institutions of higher education for the location of industry R&D.

### 5.2.2 The Spatial Conditions for Innovation Processes

It is a common observation that innovations tend to be agglomerated, i.e. to be clustered in particular in large urban regions. Large urban regions offer proximity advantages, which facilitate information and knowledge flows (Artle, 1959; Vernon, 1962; Feldman and Audretsch, 1999;

Glaeser, 1999) and create a proximity-based communication externality (Fujita and Thisse, 2002).

Why do innovations tend to be clustered in large urban regions? What are the proximity and accessibility benefits these regions offer? Already Marshall (1920) identified the exchange of information as an externality leading to the agglomeration of economic actors. In more recent years, Henderson (1974) has used information exchange in the form of technology spillovers as an explanation for the agglomeration of economic agents across space. Proximity is seen as critical for all the different types and forms of knowledge flows essential in innovation processes (Karlsson and Johansson, 2006).

Within large functional urban regions,[1] firms have normally high accessibility to pertinent actors, such as customers, suppliers, research universities and other institutions of higher education, R&D institutes, etc. These actors are all important sources for intellectual capital and innovation-relevant information and knowledge. Large functional urban regions also offer greater opportunities for labor mobility, which is an important channel for knowledge transfers and for the maintenance and growth of knowledge capital.[2]

There has been a steady progression of theoretical frameworks and methods to investigate the influence of spatial factors on innovation processes. These range from general knowledge production functions that embody broad forms of distance-sensitive knowledge flows, including tacit as well as formal knowledge inputs, to more precisely specified models of knowledge flows and spillovers including so-called localized knowledge spillovers through patents, patent citations and product innovations (Varga, 2002). Studies of localized knowledge spillovers occasionally, although not always, distinguish clearly between pecuniary and technological spillovers, their public and club good features, and various forms of private intellectual property.

The size of a functional region is critical also to its ability to attract, support and maintain innovative firms. Large functional urban regions offer special advantages to firms by offering many clusters of specific industries as well as a broad range of industries. They combine the specialization advantages of the Marshall-Arrow-Romer type with the diversity advantages advocated by Jacobs (1969). Hence, large functional urban regions offer innovative firms advantages in the form of both network and agglomeration externalities, which stimulate knowledge transfers.

Large functional urban regions are also important since they represent an arena at which different geographical scales meet. They represent a nexus between local, regional, national and international networks of different kinds including knowledge networks (Lagendijk, 2001).

### 5.2.3 The Location of Industry R&D

There are plenty of evidences in the literature that industry R&D is substantially more concentrated spatially than industry production.[3] For example, Kelly and Hageman (1999) show that innovation exhibits strong geographical clustering, independently of the distribution of employment. Sectors locate their R&D not where they are producing but near to where other sectors do their R&D. However, Feldman and Audretsch (1999) found that there are substantial differences across sectors in spatial clustering with some industries, like computers and pharmaceuticals, displaying a higher degree of concentration compared to all manufacturing. Similar conclusions were drawn by Breschi (1999) after an examination of patent data for the period 1978-1991 from the European Patent Office (EPO).

Theoretical arguments concerning localized knowledge flows suggests that knowledge production and innovative activities within a company will tend to be more efficient in agglomerations containing research universities and other R&D performing companies, since the access to knowledge flows and potential knowledge externalities is greater. The knowledge production and the innovative activities will be more productive in such agglomerations because there is a high probability that companies can access potentially useful external knowledge at a cost that is lower than producing this knowledge internally or trying to acquire it externally from a geographic distance (Harhoff, 2000). The cost of transferring such knowledge is a function of geographic time distance. This is one of the arguments why R&D agglomerations are likely to give rise to localized knowledge externalities (Siegel, Westhead and Wright, 2003). Thus, given the character of knowledge flows, it seems natural to assume that the spatial dimension is a key factor explaining the location of R&D activities of companies. The location of R&D activities of companies is influenced by the potential knowledge externalities from knowledge flows from university R&D and R&D in other companies.

There is a rich literature regarding various aspects of the relationship between university R&D and industry R&D and innovation, and to a certain extent also about the relationship between higher education and industry R&D and innovation. Some studies focus on the ability of companies to utilize knowledge flows from universities (Cohen and Levinthal, 1989, 1990; Cockburn and Henderson, 1998; Ziedonis, 1999;

Lim, 2000). Another strand of literature studies the characteristics of universities that generate knowledge flows of interest for industry R&D and innovation (Henderson, Jaffe and Trajtenberg, 1998; Jensen and Thursby, 1998; Feldman et al., 2002; Thursby and Thursby, 2002; Di Gregorio and Shane, 2000). A third set of studies analyze the channels through which knowledge flows from universities to industry (Agrawal and Henderson, 2002; Cohen, Nelson and Walsh, 2002; Colyvas et al., 2002; Shane, 2002). These channels include:

- Personal networks of academic and industry researchers (Liebeskind et al., 1996; MacPherson, 1998)
- University researchers consulting to industry or serving on company boards
- University researchers leaving university to work for industry
- Technological spillovers of newly created knowledge from universities to industry
- Purchases by industry of newly created university knowledge or intellectual property
- Spin-offs of new firms from universities, i.e. academic entrepreneurship (Slaughter and Lesley, 1997; Stuart and Shane, 2002)
- Universities creating incubators, enterprise centers and science parks to improve interaction with industry, and to facilitate university knowledge transfers
- Participation in conferences, seminars and presentations
- Flows of fresh graduates at the master and PhD level to industry (Varga, 2000)

However, there seem to be fewer studies that explicitly study the influences of university R&D on companies in general and on company R&D in particular. The specific and relative role played by each of these different links for the development of industry is not well understood (Karlsson and Manduchi, 2001). Moreover, the actual links between universities and industry in many cases have proved difficult to detect. However, recent studies have shown strong evidence of knowledge transfers and spillover flows, as demonstrated by the joint distributions of university capacity and high technology sectors (Varga, 1997, 2002).

At a general level research universities have been identified as a location factor of growing importance (Henderson, Jaffe and Trajtenberg, 1995; Zucker and Darby, 2001, 2005; Adams, 2002; Hall, Link and Scott, 2003). It has been suggested that regions with strong research universities have better opportunities to attract and support innovative firms than regions without such universities. Regionally-based science parks can be seen as an institutional set-up to integrate the resources of research universities and innovative firms (Luger and Goldstein, 1991). Also, network type interactions among innovative firms and private and public research institutions seem to be of growing importance (Lundvall, 1992; Nelson, 1993; Etzkowitz and Leyersdorff, 2000; Charles, 2003).

Zucker, Darby and Brewer (1998) examine the location decisions of companies relative to the location of star university scientists. Mariani (2002), in a study of Japanese investments in Europe, showed that geographical proximity to the local science base is an important factor for locating only R&D laboratories compared to R&D and production and production only. Agrawal and Cockburn (2002) use data on scientific publications and patents as indicators of university R&D and industry R&D and find strong evidences of geographic concentration in both activities at the level of metropolitan statistical areas (MSAs) in the US. They also find strong evidences of co-location of upstream and downstream R&D activities. Agrawal and Cockburn (2003) report that high levels of university publishing in metropolitan areas in the United States and Canada tend to be matched by high levels of company patenting in the same technology field and metropolitan area, suggesting co-location of research activities. Other empirical studies suggest a strong correlation between the specialization of the regional R&D infrastructure and the innovative activities conducted by industry (Feldman, 1994a; Felder, Fier and Nerlingar 1997a, b; Harhoff, 1997a, b; Nerlinger, 1998). These results can be interpreted as indicating that knowledge externalities from the R&D infrastructure can be best used in innovation activities in companies in the same or closely related scientific and technological field(s). The correlation tends to increase with the complexity of the R&D and innovation activities and the more specific the demand for technological know-how (Feldman, 1994a; Feldman and Florida, 1994). Results presented by Bade and Nerlinger (2000) indicate strong correlations between the occurrence of new technology-based firms and the proximity to R&D facilities comprising universities, technical colleges and non-university R&D institutes as well as private R&D.

Griliches' (1979) 'knowledge production function approach' did not acknowledge that knowledgeable persons and knowledge production activities are spread out in geography and at the same time to a high degree concentrated to agglomerations. However, the original 'knowledge production function approach' has later been modified to also accommodate the spatial dimension (Jaffe, 1989; Audretsch and Feldman, 1994, 1996; Feldman, 1994a, b). The inputs and outputs considered in these studies vary from study to study and so does the geographic unit of analysis. With a few exceptions (Henderson, Jaffe and Trajtenberg, 1994; Beise and Stahl, 1999), empirical research suggests that knowledge flows from public science to companies decline with geographical distance.

### 5.2.4 Accessibility and Industry R&D Location across Regions

The preceding sections suggest that investments in industry R&D are attracted to locations with high accessibility to university R&D and students and graduates at institutions of higher education. This section illustrates how this kind of accessibility potentially affects firms' evaluation of alternative locations for their R&D investments. Given the discussion in the introduction about the dominating role of multinational national enterprises (MNEs) in industry R&D, it seems natural to take an MNE as an example.

Consider an MNE that is about to choose a location for its R&D investments. Since such a firm operates at a global level, we may conjecture that the MNE's first decision is which part of the world it should choose. This can be choice among, e.g., Western Europe, Asia or North America, and is likely to depend on which part of the world constitutes the main (or home) market for the MNE. Given the choice of continent, the question is then which country to choose.[4] Conditional on the choice of Europe, for instance, this involves a choice between Germany, France, Sweden or the UK, etc. Once a country has been chosen, the question is which region to locate in. An MNE certainly evaluates the alternative regions in an economic territory (such as a country) based on a set of pertinent attributes of the regions. A main conjecture here is that accessibility to university R&D and institutions of higher education are two such pertinent attributes of regions.

Against the background above we now consider an MNE that has decided to locate its R&D investments in a country which hosts $N$ $=\{1,\ldots,n\}$ regions. How is a region's accessibility to university R&D and institutions of higher education related to the attractiveness of the region as regards location of an R&D unit? Here, we employ the basic set-up in Johansson, Klaesson and Olsson (2002) and relate accessibility to the preference structure in random choice theory, which starts from a stochastic specification of the preference value associated with different alternatives. We start with institutions of higher education in terms of students and graduates.

An R&D unit $r$ located in region $i \in N$ by an MNE faces the set of $N$

alternatives as regards recruitment of graduates from institutions of higher education. The set $N$ denotes the set of regions in the national economy considered. Thus, each alternative pertains to students and graduates at institutions of higher education in a specific region. We might now ask: what determines the preference value of R&D unit $r$ as regards potential recruitment of (and collaboration with) graduates in region $j \in N$? This preference value, denoted by $U_{ij,r}$ is specified as consisting of two parts:

$$U_{ij,r} = u_{ij,r} + \varepsilon_{ij} \tag{5.1}$$

where $u_{ij,r}$ denotes the deterministically known part and $\varepsilon_{ij}$ denotes random influence from non-observed factors. $u_{ij,r}$ is in turn assumed to be a function of: (i) the size of the institution of higher education (in terms of graduates), $H_j$; (ii) the quality differential of the graduates between region $i$ and $j$, $q_j$; (iii) the wage differential of graduates (cost differential) in region $j$ and $i$, $(p_j - p_i)$; (iv) interaction costs between $i$ and $j$, $c_{ij}$; and (v) travel time between region $i$ and $j$, $t_{ij}$. $u_{ij,r}$ is specified in Equation (5.2):

$$u_{ij,r} = \ln H_j + \delta(q_j - q_i) - \alpha(p_j - p_i) - \beta c_{ij} - \gamma t_{ij} \tag{5.2}$$

where $\delta$, $\alpha$, $\beta$ and $\gamma$ are parameters. In Equation (5.2) the size of the institution of higher education in region $j$ can be interpreted as the attraction factor in region $j$.

Assuming that the $\varepsilon_{ij}$ are i.i.d. and extreme value distributed (the standard assumption in discrete choice theory), the probability that an R&D unit $r$ will hire (or collaborate with) graduates in region $j$, given that the R&D unit is located in region $i$, $P_{ij,r}$, is given by:[5]

$$P_{ij,r} = \frac{e^{\{u_{ij,r}\}}}{\sum_{j \in N} e^{\{u_{ij,r}\}}} \tag{5.3}$$

Equation (5.3) is the general expression for the choice probabilities in the multinomial logit (MNL) model (see e.g. Anderson, de Palma and Thisse, 1992). In Equation (5.3), the numerator is the preference value for graduates from institutions of higher education in municipality $j$ whereas the denominator is the sum of such preference values.

Now consider the denominator in (5.3) and assume that (i) the quality of the graduates and the wages are equal in all regions[6] and (ii) the monetary travel costs are proportional to the time distance such that $c_{ij} = \eta_{ij}$. Using these assumptions, the denominator in (5.3) can be expressed as:

$$T_i^H = \sum_{j \in N} H_j e^{\{-\lambda \eta_{ij}\}} \tag{5.4}$$

where $\lambda = (\beta\gamma + \gamma)$. The expression in (5.4) gives the sum of the preference values, conditional on a location in region $i$. Observe that the set of $N$ alternatives is the same in any region in the set $N = \{1,\dots,n\}$. Thus, an R&D unit located in region $s \in N$ faces the same set of alternatives, i.e. alternative regions, as an R&D unit located in $i \in N$, but the sum of the preference values from that location is different. Hence, locations where the sum of such preference values is high can thus be interpreted as location with high attractiveness.[7] $T_i^H$ in Equation (5.3) is also a standard measure of accessibility with exponential distance decay. This type of accessibility measure belongs to the family of accessibility measures that satisfies criteria of consistency and meaningfulness, as has been shown in Weibull (1976). Both the size of the attractor and time distances in (5.4) are arguments in the preference function in (5.2). An analogous reasoning can be applied to university R&D. Region $i$'s accessibility to university R&D can thus be expressed as:

$$T_i^U = \sum_{j \in M} U_j^{R\&D} e^{\{-\lambda \eta_{ij}\}} \tag{5.5}$$

where $U_j^{R\&D}$ denotes the size of the investments in university R&D in region $j$.

A national economy can be divided into functional regions that consist of one or several localities. Such localities are here labeled municipalities. Functional regions are connected to other functional regions by means of economic, knowledge and infrastructure networks. The same prevails for the different municipalities within a functional region. Moreover, each municipality can also be looked upon as a number of nodes connected by the same type of networks. With reference to such a structure, it is possible to separate between (i) intra-municipal accessibility, (ii) intra-regional accessibility.[8] Letting $R$ denote the set of municipalities belonging to functional region $R$, such that $R \subset N$, the different accessibilities can be expressed as (accessibility to $H$ is here taken as an example):

$$\text{Intra-municipal} \Rightarrow T_{iM}^{H} = H_{i}\,e^{\{-\lambda_{ii}\}}$$

$$\text{Intra-regional} \quad \Rightarrow T_{iR}^{H} = \sum_{j \in R, i \neq j} H_{j}\,e^{\{-\lambda_{ij}\}}$$

The advantage of this separation is that the effect of each type of accessibility can be estimated directly. The subsequent sections of the chapter analyse the relationship between the location of industry R&D and accessibility to university R&D and accessibility to students and graduates at institutions of higher education using data at the municipality level in Sweden.

## 5.3 EMPIRICAL MODEL, VARIABLES AND DATA

The empirical model used to assess the relationship between the location of industry R&D and accessibility to university R&D and student enrollment at institutions of higher education, respectively, is presented in Equation (5.6) below, where $t$ refers to 2001 and $t - \tau$ to 1995.

$$I_{i,t}^{R\&D} = \alpha + \beta_1 I_{i,t-\tau}^{R\&D} + \beta_2 MP_{i,t} + \beta_3 T_{iM,t}^{H} + \beta_4 T_{iR,t}^{H} + \dots$$

$$\dots + \beta_5 T_{iM,t}^{U} + \beta_6 T_{iR,t}^{U} + \varepsilon_{i,t} \quad (5.6)$$

The variables in the model are explained in Table 5.1. Industry R&D in municipality $i$ in 2001 is thus explained by: (i) industry R&D in 1995, (ii) market potential of municipality $i$ in 2001, (iii) intra-municipal and intra-regional accessibility to students enrolled at institutions of higher education in 2001, (iv) intra-municipal and intra-regional accessibility to university R&D in 2001. The industry R&D in previous periods is a natural control variable to include. R&D activities are often path dependent in the sense that the R&D intensity across space changes in slow processes. The existence of R&D units in a region can also be an attractor for new R&D units as it indicates availability of industry R&D experience. The available information about R&D projects and activities can also be expected to be larger in regions where R&D units are located. The market potential of a municipality – proxied by total accessibility to population – can be seen as a 'catch-all' variable and is related to the theoretical discussion in previous sections on the advantages of large regions as locations for R&D. It pertains to the potential for agglomeration economies, encompassing knowledge and information flows, transportation facilities, legal and commercial services. The accessibility to university R&D and student enrollment at universities and institutions for higher education pertain to the potential for knowledge flows from university R&D activities to R&D units and access to qualified personnel.

*Table 5.1   Description of variables in Equation (5.6), excluding regional dummies*

| Variable | Description | Motivation | Source |
|---|---|---|---|
| $I_{i,t}^{R\&D}$ | Industry R&D in terms of man-years in municipality $i$ 2001. | Dependent variable | Statistics Sweden (SCB) |
| $I_{i,t-\tau}^{R\&D}$ | Industry R&D in terms of man-years in municipality $i$ 1995. | Path dependence phenomena. Availability of industry R&D experience and information of R&D activities in municipality. | Statistics Sweden (SCB) |
| $MP_{i,t}$ | Market potential of municipality $i$ in 2001. Measured as the municipality's accessibility to population. | Agglomeration phenomena. | Statistics Sweden (SCB) |
| $T^H$ | Physical accessibility to student enrollment at institutions of higher education in 2001. | Access to qualified labor and large potential for knowledge flows to R&D units. | Statistics Sweden (SCB) and Swedish Road Administration (SRA) |
| $T^U$ | Physical accessibility to university R&D in terms of man-years in 2001. | Access to university R&D. Large potential for knowledge flows to R&D units | Statistics Sweden (SCB) and Swedish Road Administration (SRA) |

The data used in this chapter are constructed based on secondary material from Statistics Sweden (SCB). The R&D data originates from SCB and are collected by SCB via questionnaires that are sent out to firms and universities. The R&D data is measured in man-years. One man-year is the amount of work a full-time employee performs during a year. This means that a full-time employee who only spends 50 percent of her work on R&D counts as 0.5 man-years. The data on students and graduates at institutions of higher education also comes from SCB. These data report the number of students enrolled at institutions of higher education (i.e. university students and graduates) in different regions in different municipalities in Sweden.

Accessibility is calculated for university R&D and student enrollment at institutions of higher education. These calculations are based on a Swedish travel time-distance matrix, which reports the minimum travel time by car: (i) between zones within municipalities and (ii) between municipalities. This matrix is provided by the Swedish Road Administration (SRA). The time distance sensitivity parameter was set to 0.017 in these calculations. This is the sensitivity parameter found by Hugosson and Johansson (2001) in a study of the spatial extent of business trips across Swedish regions.

In order to provide the reader with a feel for the data of main interest in the chapter, descriptive statistics for industry and university R&D and

*Table 5.2  Descriptive statistics for industry and university R&D and student enrollment at institutions of higher education across Swedish municipalities in 2001*

|                | Industry R&D | University R&D | Student enrollment* |
|----------------|--------------|----------------|---------------------|
| Min            | 0            | 0              | 0                   |
| Max            | 11 912.35    | 3 452.03       | 44 578              |
| Mean           | 171.87       | 68.94          | 1 006.89            |
| Median         | 7.92         | 0              | 6                   |
| Std. deviation | 894.58       | 380.57         | 4 135.90            |
| Skewness**     | 10.45        | 7.08           | 7.04                |
|                | (0.14)       | (0.14)         | (0.14)              |
| Kurtosis**     | 123.27       | 52.31          | 59.56               |
|                | (0.29)       | (0.29)         | (0.28)              |
| No. obs.       | 286          | 286            | 286                 |

*Notes*:  * Refers to student enrollment at institutions of higher education.
         ** Standard errors presented within brackets.

*Source*:  Statistics Sweden (SCB).

student enrollment at institutions of higher education are presented in Table 5.2. As can be seen from the difference between the mean and median, the data are highly skewed to the right.

Figure 5.1 compares the spatial concentration of industry and university R&D and student enrollment at institutions of higher education with population in 2001. Municipalities were ranked in ascending order according to their share of the total population. Then, the cumulative percentage of population, industry R&D and university R&D were calculated.

Cumulative share (%)

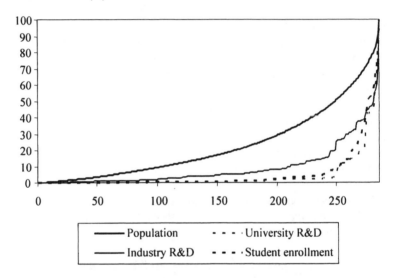

*Note:* Municipalities ranked in ascending order according to population share.

*Figure 5.1  The spatial concentration of industry and university R&D and
            student enrollment at institutions of higher education relative to
            population*

As is evident from the figure, industry and university R&D as well as student enrollment at institutions of higher education are more concentrated than population. Moreover, university R&D and student enrollment show a similar concentration pattern whereas industry R&D is relatively less concentrated in space.

## 5.4 THE LOCATION OF INDUSTRY R&D: AN EMPIRICAL ANALYSIS

Parameter estimates of the variables in Equation (5.6) are presented in Table 5.3. Three models are tested; one including accessibility to both university R&D and students and graduates, and two with each accessibility variable separately. This is so because $T_{iM,t}^{H}$ and $T_{iM,t}^{U}$ as well as $T_{iR,t}^{H}$ and $T_{iR,t}^{U}$ are correlated. Such a correlation might distort the parameter estimates (a correlation matrix is presented in the appendix). The econometric analysis will indicate the relative importance of the two types of accessibility (intra-municipal and intra-regional). Also, differences between accessibility to university R&D and student enrollment at institutions of higher education can be assessed. The standard errors reported in Table 5.3 are bootstrapped using 1000 replications, which reduces biases from potential heteroscedasticity which is common in analyses on regional data. The bootstrap often provides a tractable way to reduce or eliminate finite-sample distortions of the levels of statistical tests (Horowitz, 1995).

Since the dependent variable is bottom-censored, i.e. cannot be below zero, and several municipalities have zero R&D man-years in the industry, the parameters are estimated using a Tobit model after Tobin (1958). The Tobit model assumes a latent dependent variable, $Y_i^*$, that is not directly observed. $Y_i$ is however observed and is defined as:

$$Y_i = Y_i^* \text{ iff } Y_i^* > c$$
$$Y_i = c \text{ iff } Y_i^* \le c \tag{5.7}$$

where $c$ denotes the lower bound below which $Y_i^*$ is censored. $Y_i^*$ is assumed to be generated by the canonical linear regression model such that:

$$Y_i^* = \beta' X_i + \varepsilon_i \tag{5.8}$$

where $X_i$ denotes a vector of regressors and $\beta$ a corresponding vector of parameters. The error terms are assumed to be i.i.d. $\varepsilon_i \sim N(0,\sigma^2)$. In our case, the lower bound is zero, $c = 0$. Neglecting such a lower bound can lead to biased and inconsistent parameter estimates.

The estimates reported in the table shows that industry R&D in

*Table 5.3  Tobit estimates of the parameters in Equation (5.7),*
       *(bootstrapped standard errors, 1000 replications)*

| Variable | Estimates (Tobit model) 1 | Estimates (Tobit model) 2 | Estimates (Tobit model) 3 |
|---|---|---|---|
| $I_{i,t-\tau}^{R\&D}$ | 1.05 (0.05) | 1.06 (0.05) | 1.06 (0.05) |
| $MP_{i,t}$ | 0.0001 (0.00) | 0.0001 (0.00) | 0.0001 (0.00) |
| $T_{iM,t}^{H}$ | 0.02 (0.01) | 0.03 (0.01) | – |
| $T_{iR,t}^{H}$ | n.s | −0.002 (0.00) | – |
| $T_{iM,t}^{U}$ | n.s | – | n.s |
| $T_{iR,t}^{U}$ | n.s. | – | −0.03 (0.01) |
| Pseudo $R^2$ | 0.28 | 0.27 | 0.26 |
| # censored obs. | 106 | 106 | 106 |
| # obs. | 286 | 286 | 286 |

*Notes*:
\* only parameter estimates significant at the 0.05 level are shown in the table (n.s denotes insignificance at the 0.05 level).
\*\* bootstrapped standard errors presented within brackets.
\*\*\* region dummies not shown.

previous time periods is a good predictor of the R&D in present time periods. The parameter estimate is significant and positive in all specifications. Likewise, the parameter estimate of the market potential variable comes out as significant and positive. Thus, consistent with previous literature and theory, industry R&D seems to be attracted to larger regions with a larger potential for agglomeration economies. The intra-municipal accessibility to students and graduates at institutes of higher education comes out as positive and significant in the specifications where it is included, whereas intra-municipal accessibility to university R&D comes out as insignificant. Interestingly, the estimated parameter for the intra-regional accessibility variable in both model 2 (student enrollment) and model 3 (university R&D) is negative and significant. This can be interpreted as a competition effect. If the intra-regional accessibility is high it means that there are plenty of resources in other surrounding localities in the region which thereby are attractive locations for industry R&D units.

Overall, the results point toward that the main attractor for industry R&D is not the university R&D as such, but rather the human capital embodied in students and graduates at institutions of higher education. This cannot, however, be interpreted as that university R&D is unimportant. The supply of students and graduates are certainly large in regions where university R&D is large. Rather, as aforementioned, the results suggest that the main attractor is the supply of students and graduates. Thus, consistent with other recent studies (c.f. Faggian and McCann, 2006), the results indicate that university R&D is mainly accessed indirectly via labor-market transactions.

## 5.5 CONCLUSIONS

We have endeavored in this chapter to analyse how the location of industry R&D is related to accessibility to students and graduates at institutions of higher education and university R&D. Two types of accessibilities were considered in the chapter: (i) intra-municipal accessibility and (ii) intra-regional accessibility.

By relating the location of industry R&D across Swedish municipalities to a set of municipal characteristics, including intra-municipal and intra-regional accessibility to university R&D and students and graduates at institutions of higher education, it is shown that the location of industry R&D can be related to accessibility to students and graduates at institutions of higher education. However, we find no significance as regards accessibility to university R&D. This indicates that the main attractor for industry R&D on average is not the university R&D as such, but rather the human capital embodied in students and graduates at institutions of higher education. However, this is not interpreted as that university R&D is unimportant. The supply of students and graduates are certainly large in regions where university R&D is large. Rather, as aforementioned, the results suggest that the main attractor is the supply of students and graduates. Thus, consistent with other recent studies (c.f. Faggian and McCann, 2006), the results indicate that university R&D is mainly accessed indirectly via labor-market transactions.

The methodology applied in the chapter, however, does admittedly not allow for causative interpretations. An avenue for further research is therefore to use methodologies that allow for causative interpretations.

Moreover, the chapter has used aggregate R&D data across industries. Controlling for heterogeneity across industries requires disaggregate R&D data and is a further topic for additional research.

## NOTES

1. Functional urban regions are delimited by labor and housing market perimeters.
2. One hour's travel time by car from a given location seems to set the limit for the positive proximity externalities associated with agglomerations (Johansson, Klaesson and Olsson, 2002).
3. However, there are authors who claim that R&D-intensive and high-tech industries do not necessarily agglomerate (Devereux, Griffith and Simpson, 1999; Shaver and Flyer, 2000; Kalnins and Chung, 2004; Barrios, Bertinelli, Strobl and Teixeira, 2003; Alecke, Alsleben, Untiedt and Scharr, 2006). In her study of Japanese investments in Europe Mariani (2002) found that R&D tends to locate close to production activities.
4. It should here be recognized that ample evidence suggests that MNEs are particularly likely to locate their R&D activities to their home country compared to other activities (Patel and Pavitt, 1995).
5. This condition is derived in several texts, see *inter alia* Train (1993), Anderson, de Palma and Thisse (1992).
6. This is a gross simplification. However, data on quality and wages for different graduates are not available. This assumption can be motivated by that quality and wages outweigh each other.
7. Compare Ben-Akiva and Lerman (1979).
8. It is possible to also define a third type of accessibility – extra-regional accessibility (see e.g. Johansson, Klaesson and Olsson, 2002). However, this accessibility is not considered in this chapter. There is now a set of papers using municipalities as the observational unit, which have used all three types of accessibility. These papers have shown that extra-regional accessibility is insignificant in a large number of alternative settings (see e.g. Gråsjö, 2005a, b; Andersson and Karlsson, 2007; Andersson and Gråsjö, 2009).

## REFERENCES

Adams, J.D. (2002), 'Comparative localization of academic and industrial spillovers', *Journal of Economic Geography*, 2, 253-278.

Agrawal, A. and I.M. Cockburn (2002), 'University research, industrial R&D and the Anchor Tenant Hypothesis', *NBER Working Paper*, 9212, Cambridge, MA.

Agrawal, A. and I.M. Cockburn (2003), 'The Anchor Tenant Hypothesis: exploring the role of large, local, R&D intensive firms in regional innovation systems', *International Journal of Industrial Organisation*, 21, 1227-1253.

Agrawal, A. and R. Henderson (2002), 'Putting patents in context: exploring knowledge transfer from MIT', *Management Science*, 48, 44-60.

Alecke, B., C. Alsleben, G. Untiedt and F. Scharr (2003), 'New empirical evidence on the geographic concentration of German industries: do high-tech clusters really matter?', *The Annals of Regional Science*, 40, 19-42.

Anderson, S.P., A. de Palma and J.-F. Thisse (1992), *Discrete Choice Theory of Product Differentiation*, Cambridge, MA: The MIT Press.

Andersson, M. and U. Gråsjö (2009), 'Spatial dependence and the representation of space in empircal models', *Annals of Regional Science*, 43 (1), 159-180.

Andersson, M. and C. Karlsson (2007), 'Knowledge in regional economic growth – the role of knowledge accessibility', *Industry and Innovation*, 14, 129-149.

Artle, R. (1959), *The Structure of the Stockholm Economy – Toward a Framework for Projecting Metropolitan Community Development*, Stockholm: School of Economics.

Audretsch, D.B. and M. Feldman (1994), 'Knowledge spillovers and the geography of innovation and production', *Discussion Paper 953*, Centre for Economic Policy Research, London.

Audretsch, D.B. and M. Feldman (1996), 'R&D spillovers and the geography of innovation and production', *The American Economic Review*, 86, 630-640.

Bade, F.-J. and E.A. Nerlinger (2000), 'The spatial distribution of new technology-based firms: empirical results for West-Germany', *Papers of Regional Science*, 79, 155-176.

Barrios, S., L. Bertinelli, E. Strobl and A.C. Teixeira (2003), 'Agglomeration economies and the location of industries: a comparison of three small European countries', *CORE Discussion Paper*, 2003-2067.

Batten, D.F., K. Kobayashi and Å.E. Andersson (1989), 'Knowledge, nodes and networks: an analytical perspective', in Å.E Andersson, D.F Batten and C. Karlsson (eds), *Knowledge and Industrial Organization*, Berlin: Springer Verlag, pp. 31-46.

Beise, M. and H. Stahl (1999), 'Public research and industrial innovation in Germany', *Research Policy*, 28, 397-422.

Ben-Akiva, M. and S. Lerman (1979), 'Disaggregate travel and mobility choice models and measures of accessibility', in D.A. Henscher and P.R. Stopher (eds), *Behavioral Travel Modelling*, London: Croom Helm, pp. 654-679.

Breschi, S. (1999), 'Spatial patterns of innovation: evidence from patent data', in Gambardella and Malerba (eds), *The Organization of Economic Innovation in Europe*, Cambridge: Cambridge University Press, pp. 71-102.

Charles, D. (2003), 'Universities and territorial development: reshaping the regional role of UK universities', *Local Economy*, 18, 7-20.

Cockburn, I. and R. Henderson (1998), 'Absorptive capacity, co-authoring behaviour and the organisation of research in drug discovery', *Journal of Industrial Economics*, XLVI, 157-182.

Cohen, D.W. and D.A. Levinthal (1989), 'Innovation and learning: the two faces of R&D', *The Economic Journal*, 99, 569-596.

Cohen, D.W. and D.A. Levinthal (1990), 'Absorptive capacity: a new perspective on learning and innovation', *Administrative Science Quarterly*, 35, 128-152.

Cohen, D.W., R.R. Nelson and J. Walsh (2002), 'Links and impacts: the influence of public research on industrial R&D', *Management Science*, 48, 1-23.

Colyvas, J., M. Crow, A. Gelijns, R. Mazzoleni, R.R. Nelson, N. Rosenberg and B.N. Sampat (2002), 'How do university inventions go into practice?', *Management Science*, 48, 61-72.

Devereux, M.P., R. Griffith and H. Simpson (1999), 'The geographic distribution of production activity in the UK', *IFS Working Paper*, W99/26, Institute for Fiscal Studies.

Di Gregorio and S. Shane (2000), 'Why do some universities generate more spillovers than others?', University of Maryland (mimeo).

Etzkowitz, H. and L. Leyersdorff (2000), 'The dynamics of innovation: from national systems and "Mode 2" to a triple helix of university-industry-government relations', *Research Policy*, 29, 109-123.

Faggian, A. and P. McCann (2006), 'Human capital and regional knowledge assets: a simultaneous equation model', *Oxford Economic Papers*, 58(3), 475-500.

Felder, J., A. Fier and E. Nerlinger (1997a), 'High Tech in den neuen Ländern: Unternehmensgründungen und Standorte', *Zeitschrift für Wirtschaftsgeographie*, 41, 1-17.

Felder, J., A. Fier and E. Nerlinger (1997b), 'Im Osten nichts Neues? Unternehmensgründungen in High-Tech Industrien', in D. Harhoff (ed.), *Unternehmensgründungen – Empirische Analysen für die alten and neuen Bundesländer*, 7, Baden-Baden, pp. 73-100.

Feldman, M. (1994a), *The Geography of Innovation*, Dordrecht: Kluwer.

Feldman, M. (1994b), 'Knowledge complementarity and innovation', *Small Business Economics*, 6, 363-372.

Feldman, M. and D.B. Audretsch (1999), 'Innovation in cities: science-based diversity, specialisation and localised competition', *European Economic Review*, 43, 409-29.

Feldman, M.P. and R. Florida (1994), 'The geographic sources of innovation: technological infrastructure and product innovation in the United States', *Annals of the Association of American Geographers*, 84, 210-229.

Feldman, M., I. Feller, J. Bercovitz and R. Burton (2002), 'Equity and the technology transfer strategies of American research universities', *Management Science*, 48, 105-121.

Fujita, M. and J.-F. Thisse (2002), *Eonomics of Agglomeration – Cities, Industrial Location and Regional Growth*, Cambridge: Cambridge University Press.

Glaeser, E. (1999), 'Learning in cities', *Journal of Urban Economics*, 46, 254-277.

Gråsjö, U. (2005a), 'Accessibility to R&D and patent production', *Working Paper Series in Economics and Institutions of Innovation*, 37, Royal Institute of Technology, CESIS.

Gråsjö, U. (2005b), 'Human capital, R&D and regional exports', *Working Paper Series in Economics and Institutions of Innovation*, 50, Royal Institute of Technology, CESIS.

Griliches, Z. (1979), 'Issues in assessing the contribution of research and development to productivity growth', *Bell Journal of Economics*, 10, 92-116.

Hall, B.H., A.N. Link and J.T. Scott (2003), 'Universities as research partners', *Review of Economics and Statistics*, 85, 485-491.

Harhoff, D. (1997a), 'Innovation in German Manufacturing Enterprises: Empirical Studies on Productivity, Externalities, Corporate Finance, and Tax Policy', Habilitationsschrift, Universität Mannheim.

Harhoff, D. (ed.) (1997b), *Unternehmensgründungen – Empirische Analysen für die alten and neuen Bundesländer*, 7, Baden-Baden.

Harhoff, D. (2000), 'R&D spillovers, technological proximity, and productivity growth – evidence from German panel data', *Schmalenbach Business Review*, 52, 238-260.

Henderson, J.V. (1974), 'The size and type of cities', *American Economic Review*, 89, 650-656.

Henderson, R.M., A. Jaffe and M. Trajtenberg (1994), 'Numbers up, quality down? Trends in university patenting 1965-1992', Presentation to CEPR/AAAS Conference University Goals, Institutional Mechanisms and the Industrial Transferability of Research, Stanford University, March 18-20.

Henderson, R.M., A. Jaffe and M. Trajtenberg (1995), 'Universities as a source of commercial technology: a detailed analysis of university patenting 1965-1988', *NBER WP* 5068.

Henderson, R.M., A. Jaffe and M. Trajtenberg (1998), 'Universities as a source of commercial technology: a detailed analysis of university patenting, 1965-1988', *Review of Economics and Statistics*, 80, 119-127.

Horowitz, J.L. (1995), 'Boostrap methods in econometrics – theory and numerical performance', Mimeograph, University of Iowa.

Hugosson, P. and B. Johansson (2001), 'Business trips between functional regions', Mimeograph, Jönköping University

Jacobs, J. (1969), *The Economy of Cities*, New York: Vintage.

Jaffe, A.B. (1989), 'Real effects of academic research', *American Economic Review*, 79, 957-970.

Jensen, R. and M. Thursby (1998), 'Proofs and prototypes for sale: the tale of university licensing', *NBER Working Paper*, 6698, Cambridge, MA.

Johansson, B., J. Klaesson and M. Olsson (2002), 'Time distances and labor market integration', *Papers in Regional Science*, 81, 305-327.

Kalnins, A. and W. Chung (2004), 'Resource-seeking agglomeration: a study of market entry in the lodging industry', *Strategic Management Journal*, 25, 689-699.

Karlsson, C. and B. Johansson (2006), 'Towards a dynamic theory for the spatial knowledge economy', in B. Johansson, C. Karlsson and R.R. Stough (eds), *Entrepreneurship and Dynamics in the Knowledge Economy*, London and New York: Routledge, pp. 12-46.

Karlsson, C. and A. Manduchi (2001), 'Knowledge spillovers in a spatial context – a critical review and assessment', in M.M. Fischer and J. Fröhlich (eds), *Knowledge, Complexity and Innovation Systems*, Berlin: Springer-Verlag, pp. 101-123.

Kelly, M. and A. Hageman (1999), 'Marshallian externalities in innovation', *Journal of Economic Growth*, 4, 39-54.

Lagendijk, A. (2001), 'Scaling knowledge production: how significant is the region?', in M.M. Fischer and J. Fröhlich (eds), *Knowledge, Complexity and Innovation Systems*, Berlin: Springer-Verlag, pp. 79-100.

Liebeskind, J.P., A.L. Oliver, L. Zucker and M. Brewer (1996), 'Social networks, learning and flexibility: sourcing scientific knowledge in new biotechnology firms', *Organizational Science*, 7, 428-443.

Lim, K. (2000), 'The many faces of absorptive capacity: spillovers of copper interconnect technology for semiconductor chips', MIT (mimeo).

Lööf, H. (2005), 'Den växande utlandskontrollen av ekonomierna i Norden. Effekter på FoU, innovation och produktivitet', *ITPS A* 005.

Luger, M. and H. Goldstein (1991), *Technology in the Garden: Research Parks and Regional Development*, Chapel Hill: University of North Carolina Press.

Lundvall, B.-Å. (ed.) (1992), *National Systems of Innovation*, London: Pinter.

MacPherson, A.D. (1998), 'Academic-industry linkages and small firm innovation: evidence from the scientific instruments sector', *Entrepreneurship and Regional Development*, 10, 261-276.

Mariani, M. (2002), 'Next to production or to technological clusters? The economics and management of R&D location', *Journal of Management and Governance*, 6, 131-152.

Marshall, A. (1920), *Principles of Economics*, 8th edition, London: Macmillan.

Nelson, R.R. (ed.) (1993), *National Innovation Systems: A Comparative Study*, New York: Oxford University Press.

Nerlinger, E. (1998), 'Standorte und Entwicklung junger innovativer Unternehmen: Empirische Ergebnisse für West-Deutschland', *ZEW-Schriftenreihe*, 27, Baden-Baden.

Patel, P. and Pavitt, K., 1995, 'Patterns of technological activity: their measurement and interpretation', in Paul Stoneman (ed.), *Handbook of the Economics of Innovation and Technological Change*, Oxford: Blackwell.

Shane, S. (2002), 'Selling university technology: patterns from MIT', *Management Science*, 48, 122-137.

Shaver, J.M. and F. Flyer (2000), 'Agglomeration economies, firm heterogeneity, and foreign direct investments in the United States', *Strategic Management Journal*, 21, 1175-1193.

Siegel, D.S., P. Westhead and M. Wright (2003), 'Assessing the impact of university science parks on research productivity: exploratory firm-level evidence from the United Kingdom', *International Journal of Industrial Organisation*, 21, 1357-1369.

Slaughter, S. and L.L. Leslie (1997), *Academic Capitalism: Politics, Policies, and the Entrepreneurial University*, Baltimore and London: The Johns Hopkins University Press.

Stuart, T.E. and S. Shane (2002), 'Organisational endowments and the performance of university start-ups', *Management Science*, 48, 151-170.

Thursby, J. and M. Thursby (2002), 'Who is selling the ivory tower? Sources of growth in university licensing', *Management Science*, 48, 90-104.

Tobin, J. (1958), 'Estimations of relationships for limited dependent variables', *Econometrica*, 26, 24-36.

Train, K. (1993), *Qualitative Choice Analysis – Theory, Econometrics and an Application to Automobile Demand*, Cambridge, MA: MIT Press.

Varga, A. (1997), *University Research and Regional Innovation: A Spatial Econometric Analysis of Technology Transfers*, Boston: Kluwer Academic Publishers.

Varga, A. (2000), 'Local academic knowledge transfers and the concentration of economic activity', *Journal of Regional Science*, 40, 289-309.

Varga, A. (2002), 'Knowledge transfers from universities and the regional economy: a review of the literature', in A. Varga and L. Szerb (eds), *Innovation, Entrepreneurship and Regional Economic Development: International Experiences and Hungarian Challenges*, Pécs: University of Pécs Press, pp. 141-171.

Vernon, R. (1962), *Metropolis 1985*, Cambridge, MA, Harvard University Press.
Weibull, J. (1976), 'An axiomatic approach to the measurement of accessibility', *Regional Science and Urban Economics*, 6, 357-379.
Ziedonis, A. (1999), 'Inward technology transfer by firms: the case of university technology licences', University of California, Berkeley (mimeo).
Zucker, L.G. and M.R. Darby (2001), 'Capturing technological opportunity via Japan's star scientists: evidence from Japanese firm's biotech patents and products', *The Journal of Technology Transfer Springer*, 26(1-2), 37-58.
Zucker, L.G. and M.R. Darby (2005), 'Socio-economic impact of nanoscale science: initial results and NanoBank', *NBER Working Paper*, 11181.
Zucker, L., M. Darby and M. Brewer (1998), 'Intellectual human capital and the birth of US biotechnology enterprises', *American Economic Review*, 88, 290-306.

## APPENDIX

*Table 5A.1  Correlations between independent variables*

| | $I_{i,t-\tau}^{R\&D}$ | $MP_{i,t}$ | $T_{iM,t}^{H}$ | $T_{iR,t}^{H}$ | $T_{iM,t}^{U}$ | $T_{iR,t}^{U}$ |
|---|---|---|---|---|---|---|
| $I_{i,t-\tau}^{R\&D}$ | 1 | – | – | – | – | – |
| $MP_{i,t}$ | 0.236* | 1 | – | – | – | – |
| $T_{iM,t}^{H}$ | 0.835* | 0.181* | 1 | – | – | – |
| $T_{iR,t}^{H}$ | 0.040 | 0.689* | −0.056 | 1 | – | – |
| $T_{iM,t}^{U}$ | 0.743* | 0.203* | 0.919* | 0.011 | 1 | – |
| $T_{iR,t}^{U}$ | 0.078 | 0.685* | −0.150 | 0.988* | 0.030 | 1 |

*Note*:  * Indicates significance at the 0.05 level.

# 6. Internationalization and regional embedding of scientific research in the Netherlands[1]

## Roderik Ponds,[2] Frank van Oort and Koen Frenken

## 6.1 INTRODUCTION

The relationship between scientific research, technological innovation and regional economic development has been an important theme in innovation studies and economic geography for many years now. The literature indicates that, in particular in the so-called science-based industries, the interaction between research institutes and firms is a crucial factor in innovation processes. A number of scholars have focused on the role of geography in these interaction processes and have found evidence for localized knowledge spillovers from universities and other academic organizations (see amongst others; Jaffe et al. 1993; Varga 1998; Anselin et al. 2000). An important source of knowledge spillovers is research collaboration, which can be indicated by co-authorships in patent or publication data (Schmoch 1997). However, the precise role of geography with regard to research collaboration is still unclear. Little is known about the role of spatial proximity in scientific collaboration and about how this affects the nature and probability of networking among research institutes and firms. Since collaboration in scientific knowledge production has become a central policy issue, it is surprising that only a few researchers have tried to understand the geography of these research collaborations.

The majority of research collaborations occur between universities. From previous research, we know that research collaboration in academia has internationalized rapidly (Katz and Martin 1997; Wagner 2005). An important part of research collaboration, especially within applied sciences, takes place in heterogeneous networks including universities, firms and governmental institutes (Etzkowitz and Leydesdorff 2000). In this context, it has been argued that the regional scale is more relevant as heterogeneous actors require geographical and institutional proximity to reduce uncertainties inherent in risky collaborative endeavours (Cooke et al. 1997). Here, we test the hypothesis that research collaboration involving different kinds of organizations (firms, universities, governmental research institutes) is more localized than collaboration in science between the same kind of organizations. Collaboration patterns in the field of biotechnology and semiconductors will be analysed for the period 1988-2004 at different spatial levels in the Netherlands in relation to international collaborations.

## 6.2 GEOGRAPHY OF SCIENCE VERSUS GEOGRAPHY OF INNOVATION

Consensus has grown among economists and economic geographers that knowledge production and knowledge spillovers are to an important extent geographically localized (Jaffe 1989; Audretsch and Feldman 1996; Feldman 1999). Economists point to agglomeration economies as a source of concentration of industrial activity and innovation. Agglomeration economies of various sorts arise when people and firms engaged in production are geographically concentrated. These 'Marshallian' externalities concern labour market pooling, access to specialized suppliers of intermediate goods and business services, and knowledge spillovers (Feser 2002). To test for knowledge spillovers, most scholars apply a knowledge production function approach to explain the regional production of patents or innovations as a result of public and private R&D inputs. In more than one case, and for different spatial levels, scholars have been able to indicate that such spillovers turn out to be statistically significant, that is, exert a significant and positive effect on knowledge output as measured by patents or innovations. In particular, the money spent on university research in a region is said to be very beneficial for innovation in that region (Jaffe 1989). Knowledge spillovers from universities and other academic research institutions seem to be spatially bounded, as shown by Jaffe et al. (1993), who found that the large majority of citations to US patents stem from the same state as the one from which the cited patent originated, even when corrected for differences in regional sector distributions. Importantly, there exist significant differences between industries and between technologies (Ács 2002; Anselin et al. 2000). The extent and importance of localized knowledge spillovers is dependent on the structure of the industry, the stage of the lifecycle of firms and sectors, and the underlying nature of the knowledge base. This underlines the importance of analysing innovation at the level of a sector or a technology, rather than at the level of the economy as a whole (below, we will also adopt a sector-based approach).

Theoretically, the insights drawn from the geography of innovation are to a limited extent relevant for understanding the geography of research collaboration. Scientific research is fundamentally different from industrial innovation (Dasgupta and David 1994). Gittelman and Kogut (2003, p. 367) state it like this: 'the logic of scientific discovery does not adhere to the same logic that governs the development of new technologies'. Scientific research and (research for) industrial innovation take place in different socio-economic structures (Dasgupta and David 1994). Because of these differences, the world of science and the world of technology can be seen as two different communities with their own set of rules and behaviour.

The socio-economic structures of these communities differ in at least two structural ways. The first main difference is the goal of the research and as a consequence the underlying incentive structure. The main goal in science, and of scientific publishing, is to add new knowledge to the existing 'stock of knowledge' and to diffuse this new knowledge as widely as possible, whereas industrial research and innovation is concerned with 'adding to the streams of rents that may be derived from possession of (rights to use) private knowledge' (Dasgupta and David 1994, p. 498). As a result the incentive structure regarding knowledge production in academia and industry is conflicting: in academia actors want to maximize diffusion of their knowledge, while actors in industry want to minimize diffusion of their knowledge. When universities and industries collaborate in R&D, the differences in incentive structure give rise to complex institutional arrangements. The complexity of these collaborations render it generally impossible to encode all contingencies in a contract, and, as a consequence, these networks have to rely at least partially on less formal institutions that reduce the risk of opportunism. One may therefore argue that in the case of collaboration between academic and non-academic organizations (like university-industry relations), as stressed by the regional innovation system literature, geographical proximity may be supportive to establish successful partnerships between organizations with structural different backgrounds. Collaboration between academic organizations and non-academic organizations like firms is especially important and frequently occurring in the so-called science-based industries and technologies (Pavitt 1984; DeSolla Price 1984).

The question remains why firms do scientific research and publish (some of) the results in scientific journals. The answer is that production and publication of scientific knowledge can be part of a firm's strategy to realize profits. Benefits of basis research can be first-mover advantages; advantages for a firm being the first to have new knowledge thereby creating an unique position to competitors (Pavitt 1984; Rosenberg 1990). Collaboration with academia can play an important role in this context, because it allows firms to access critical human resources and physical infrastructures. A second reason to invest in scientific research is absorptive capacity: by doing research, a firm is better able to reap the benefits from research done outside the firm (Cohen and Levinthal 1990). These arguments explain why firms do scientific research but not why they publish their results of this research in scientific journals. Rosenberg (1990) sees the publication of the results of a firm's scientific research as 'a ticket of admission to an information network' (p. 170). Cohen and Levinthal (1989) state that internal capability to generate knowledge, and external collaboration to acquire knowledge or to learn from external knowledge, are not substitutes but complementary to each other. Internal scientific knowledge production brings new knowledge and creates an ability to learn form external sources. External collaboration provides access to new knowledge that cannot be generated inside the firm (Lundvall 1992). Especially in industries like science-based industries with a complex knowledge base, consisting of a combination of knowledge from different fields, it is impossible for an individual firm to generate this knowledge by itself and to keep up with the development in all fields. To learn from external sources one has to collaborate with external actors and to be active in a network of research institutes, universities and other companies. To become a member of these networks, a non-academic organization has to be part of the scientific community and by publication of the outcomes of scientific research in this community the firm becomes 'a member' (Cockburn and Henderson 1998). In particular, when firms collaborate with universities or governmental research institutes, publication is almost inevitable.

The second important difference between academia and industry is the nature of knowledge produced, although this difference should not be understood as a clear-cut dichotomy. Scientific knowledge production is organized around a global discourse and new knowledge is produced and diffused by publications in journals within an international epistemic community. An epistemic community can be defined as a group of agents sharing a common goal of knowledge creation and a common framework in order to communicate and to create mutual understanding (Cowan et al. 2000; Cohendet and Meyer-Krahmer 2001). Codification of new knowledge by a common codebook is the way new knowledge becomes accessible for all members. Another feature of an epistemic community is the fact that the knowledge created is also accessible for the outside world, to the extent that outsiders are able to understand and use the codified knowledge. Outsiders without knowledge of the codebook have difficulties understanding and interpreting this knowledge, although it is codified. The understanding of the codebook discriminates between those who can understand and learn from the knowledge and those who cannot. In this way science can be seen as an international community bounded by a common codebook and driven by the goal of creating and adding knowledge around a global discourse. Collaboration within an epistemic community is not bounded in space. The major determinant for collaboration is the understanding of the codebook and the membership of the community, which are not so much influenced by geographical proximity.

Industrial knowledge production is, in general, rooted in a more 'local' problem solving context and the knowledge it produces is often more specific to the context of one or a few individual actors. The need for codification is less and therefore the tacit component large. Collaboration and the exchange of tacit knowledge is therefore often more bounded in space because of the importance of mutual trust and face-to-face contacts which are both supposed to be positively influenced by geographical proximity (Gertler 2003). Scientific knowledge production is therefore expected to be less localized in space than innovation.

The main thesis underlying our study holds that geographical proximity can facilitate collaboration between organizations with different socio-economic structures. In such heterogeneous collaboration networks, problems typically arise from conflicts of interest or from cognitive differences. Geographical proximity may help to overcome these problems since it eases the necessary trust by enabling frequent face-to-face contacts and the presence of a common institutional environment (e.g. language, law). However geographical proximity is not a prerequisite nor a sufficient condition for collaboration (Boschma 2005) and other forms of proximity may be at least as important. Technological proximity for example is generally being seen as an important other form of proximity for knowledge spillovers to occur (see for example Greunz 2003). Two establishments of the same multinational firm can be seen as organizationally close and can therefore easily collaborate over longer distances. Organizations with the same institutional background are less dependent on geographical proximity for successful collaboration in research due to a higher similarity in the goals and incentives. Following Boschma (2005), geographical proximity can compensate for the lack of institutional and cognitive proximity in interaction. And, inversely, institutional and cognitive proximity facilitates interaction over long geographical distances.

In the following sections, the spatial characteristics of collaboration in scientific knowledge production between various organizations will be analysed. The main goal is to find out what the spatial patterns of different forms of collaboration in scientific knowledge production are. In particular we are interested in possible differences in spatial scales in collaboration between organizations with the same institutional background and the so-called heterogeneous collaborations. The main hypothesis is that heterogeneous collaborations are more regionalized than homogeneous collaborations.

## 6.3 DATA

Because of the importance of scientific knowledge for technological innovation in science-based industries, two of the most high-tech, highly dynamic science-based technologies were selected; biotechnology and semiconductors. Within these technologies and their underlying scientific fields collaboration between academic and non-academic organizations is frequently occurring, which make them perfect cases to test our hypothesis. Biotechnology is one of the fastest growing industries and developments within these industries are considered to have major impacts on industries like pharmaceuticals, chemistry and agricultural industry. The semiconductor industry is one of the major drivers of technological change in industries like ICT and electronics. A comparison between these two is interesting because the 'science base' of both industries consists of different scientific fields. Biotechnology is considered as the prime example of a life-science-based industry whereas the semiconductor industry is a typical physical-sciences-based industry (Marsili 2001).

Within the Netherlands, biotechnology is regarded as a crucial technology by policymakers (as in many countries) for future economic development and therefore stimulated in various ways. An important part of these policies consists of the stimulation of collaboration in scientific knowledge production. This is also the case at the EU level, as exemplified in the EU framework programmes. The semiconductor technology is a crucial technology within the electronics industry, which is an important industry in the Netherlands, especially in the south where for example the research labs of Philips are located.

The main data source in scientometrics in general (and used in this study) is the Web of Science, a product offered by the Institute of Scientific Information (ISI, http://www.isinet.com/). Web of Science contains information on publications in all major journals in the world for 1988 onwards. We analysed publications for those disciplines that contributed the most to technological innovation in biotechnology and semiconductors. Based on the EU report 'linking science to technology' (Verbeek et al. 2002), that analyses the importance of different science disciplines for different technologies, the relevant scientific sub-disciplines for respectively biotechnology and semiconductors were selected and grouped together. In Table 6.1 the selected science fields are shown.

*Table 6.1 The relevant science fields\* for technological innovation in biotechnology and semiconductors*

| Biotechnology | Semiconductors |
| --- | --- |
| Biochemistry and molecular biology | Applied physics |
| Microbiology | Electrical and electronical engineering |
| Genetics and heredity | Physics-condensed matters |
| Immunology | Crystallography |
| Virology | Material science |
| Biophysics | Nuclear science and technology |
| Biotechnology and applied microbiology | |

*Note*: \*as defined by the Institute for Scientific Information (ISI).

Collaboration among research institutions is defined as the co-occurrence of two addresses in the same publication record. This means that a single-author paper with two or more affiliations is also counted as collaboration whereas a multi-authored paper with one address (i.e. an intra-organization collaboration) is not regarded as collaboration (Katz and Martin 1997). The use of institutions rather than persons to indicate collaborations is necessary if one is interested in the geography of collaboration. Addresses refer to institutional affiliation and not to single persons per se.

All semiconductors publications with at least one address in the Netherlands in the period 1988-2004 are downloaded for biotechnology and semiconductors. From this data, all publications with at least two addresses were selected. In Figures 6.1 and 6.2 the number of co-publications (at the organizational level), single publications and total publications are shown. From these figures it can be concluded that

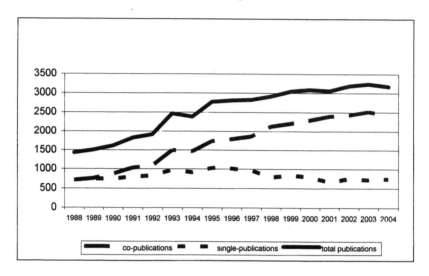

*Figure 6.1 Biotechnology: number of publications per year*

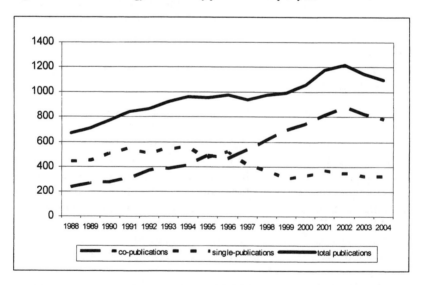

*Figure 6.2 Semiconductors: number of publications per year*

collaboration in publication is the dominant trend, both in biotechnology and in semiconductors.

Every co-occurrence of two organizations is counted as a collaboration. This means that a co-publication with $n$ organizations has $n(n-1)/2$ collaborations. As a result the focus lies on collaboration patterns as indicated by pairs of organizations. An individual publication can lead to multiple collaborations depending on the number of organizations involved. Alternatively, collaborations could be weighted by the number of pairs of collaborating organizations. The choice for a 'full count' of collaborations instead of a 'fractional' count is based on the fact that the main interest lies in patterns of collaborations.[3] With this in mind, it seems appropriate that an article with a higher number of collaborating organizations also has a higher weight (see also Mairesse and Turner (2005) for a similar way of reasoning).

For biotechnology a total number of 67 445 collaborations are counted and for semiconductors 15 498 collaborations are counted. The lower number of co-publications within semiconductors is due to the facts that the field (as measured with publications) is smaller and also the propensity to collaborate is lower. This can be seen in Table 6.2, which shows the share of co-publications (at the organizational level) in the total number of publications.

## 6.4 SPATIAL SCALES

The spatial scale of a collaboration was determined by analysing the addresses of the organizations involved. At the international level we distinguished between the collaboration at the EU level (a collaboration between an organization located in the Netherlands and an organization in one of the EU countries), the 'USA level' (all collaborations between Dutch and American organizations) and the international level (collaborations with other countries). Within the Netherlands we distinguished between the Nuts3, Nuts2, Nuts1 and national level. Nuts is the official EU classification of sub-national territories. In the Netherlands, a Nuts3 region is an urban region (most of the times consisting of a city and its surrounding municipalities), a Nuts2 region is a province (consisting of several Nuts3 regions) and the Nuts1 region corresponds to a 'country part', consisting of several Nuts2 regions.

There are 40 Nuts3 regions, 12 Nuts2 regions and 4 country parts. The spatial scale of each collaboration within the Netherlands is based on the co-location of both organizations in a region. So, a collaboration between

*Table 6.2  The share of co-publications in the total number of publications per year*

| | 1988 | 1989 | 1990 | 1991 | 1992 | 1993 | 1994 | 1995 | 1996 | 1997 | 1998 | 1999 | 2000 | 2001 | 2002 | 2003 | 2004 |
|---|---|---|---|---|---|---|---|---|---|---|---|---|---|---|---|---|---|
| Biotech-nology | 0.50 | 0.51 | 0.54 | 0.56 | 0.57 | 0.61 | 0.62 | 0.63 | 0.64 | 0.66 | 0.73 | 0.72 | 0.74 | 0.78 | 0.76 | 0.78 | 0.76 |
| Semicon-ductors | 0.35 | 0.37 | 0.35 | 0.36 | 0.42 | 0.42 | 0.43 | 0.51 | 0.47 | 0.57 | 0.63 | 0.70 | 0.70 | 0.69 | 0.72 | 0.72 | 0.71 |

organizations located in the same Nuts3 region is labelled as a Nuts3 collaboration and a collaboration between two organizations located in a different Nuts3 region but in same Nuts2 region is labelled as a Nuts2 collaboration and so on. Figure 6.3 shows the different regional classifications in a nested map. The four Nuts1 levels are indicated by the four different shades of gray; the Northern, Southern, Eastern and Western part of the Netherlands. Within each Nuts1 region, smaller regions can be identified by the thick black line. These are the 12 provinces or the Nuts2 regions. One can see that a Nuts1 region consists of several Nuts2 regions. A Nuts2 region consists of several Nuts3 regions indicated by the smaller gray lines.

*Figure 6.3 Regions in the Netherlands: biotechnology*

### 6.4.1 Biotechnology

Figures 6.4 and 6.5 show the development of the share of the different spatial scales in collaboration over time. Three-year moving averages are used in order to correct for some possible random variation, in order to reveal more clearly possible trends. Figure 6.4 shows the relative importance of different forms of international collaboration opposed to

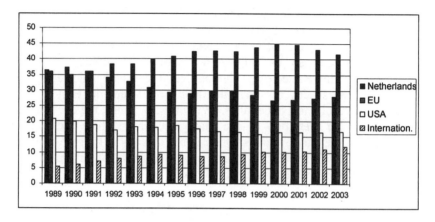

*Figure 6.4 Biotechnology: share of spatial scales of collaboration (3-year moving average)*

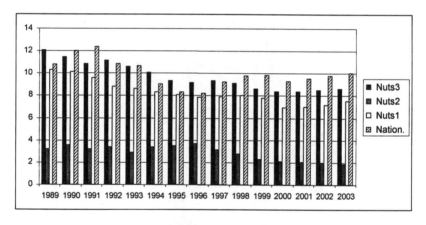

*Figure 6.5 Biotechnology: share of different spatial scales within the Netherlands (3-year moving average)*

collaboration within the Netherlands. The share of collaboration at the national level (between two organizations based in the Netherlands) is steadily declining, indicating that internationalization is a trend in the collaboration pattern of Dutch organizations.

A closer look at this trend shows that EU countries are becoming increasingly important until 2000 when their share is declining again. Yet the EU level remains the most important spatial scale. The share of collaboration with the USA is declining until 1999 but American organizations remain important partners for organizations in the Netherlands with a stable share around 16 percent from 2000 onwards. Collaboration with partners from other foreign countries is slowly increasing over time. In particular Canadian, Australian and Japanese organizations form important collaboration partners for organizations in the Netherlands.

Figure 6.5 shows the importance of different regional levels within the Netherlands; ranging from collaboration within the lowest regional level (Nuts3, n=40) to collaboration at the national level. On the y-axis the share of each spatial scale in the total number of collaborations (both at the national and international level) is indicated. Within the Netherlands, collaboration in biotechnology has a clear regional dimension. Collaboration at the Nuts3 level is more or less equally important as nationwide collaboration. Besides this, collaboration at the Nuts1 level is also frequently occurring. From Figure 6.4 it was already concluded that the share of collaboration within the Netherlands is declining compared to collaboration at the international level. Within the Netherlands the Nuts2 level seems not to be a relevant spatial scale for collaboration. Both the lower regional scale (Nuts3) and the higher scales (Nuts1 or the national scale) seem to be the relevant spatial scales for collaboration in science. The importance of the various spatial scales is relatively stable over time.

### 6.4.2 Semiconductors

Collaboration in the science base of semiconductors mainly takes place at the international level as can be seen in Figure 6.6. The share of collaborations at the national level is relatively low and slightly declining since 1989, as is the relative importance of the USA. Collaboration with countries from the EU is by far the most important, but the growing importance of collaboration with organizations from other countries is striking. This growth is mainly caused by the increasing collaboration with Japan, the new EU members in Eastern Europe and Russia.

Figure 6.7 shows the dominance of nationwide collaboration in semiconductors within the Netherlands. Over time, however, this is declining, whereas the share of collaboration at the lowest regional level (Nuts3) is increasing. So within an overall declining importance of collaboration within the Netherlands (as concluded from Figure 6.6), collaboration at the regional level is becoming more important.

Contrasting the spatial scales of collaboration from the semiconductor and the biotechnology science base, one can see that collaboration at the national level is less important in the former than in the latter. Besides

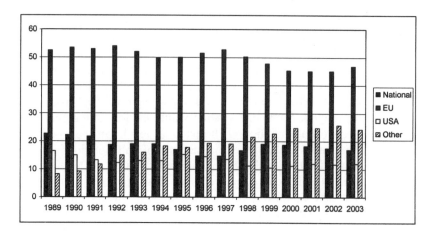

*Figure 6.6  Semiconductors: share of spatial scales in total collaboration
(3-year moving average)*

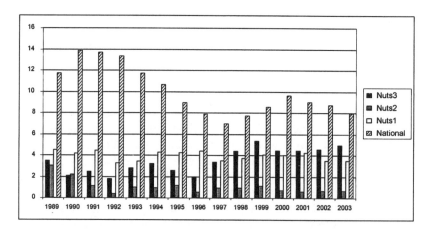

*Figure 6.7  Semiconductors: share of different spatial scales within the
Netherlands (3-year moving average)*

this, collaboration with organizations in other countries than the USA or
the EU countries seems to be more important in semiconductor
technology than in biotechnology. An important similarity is the
importance of collaboration with EU partners. Within the Netherlands,
collaboration in biotechnology is by far more regionalized than in
semiconductors.

## 6.5 ARE HETEROGENEOUS COLLABORATIONS MORE LOCALIZED THAN HOMOGENEOUS COLLABORATIONS?

Because we are especially interested in the spatial characteristics of collaborations between organizations with different institutional backgrounds, this part will focus on spatial patterns of different forms of collaborations in biotechnology and semiconductors technologies. We distinguished three different types of organizations: academic organizations, firms, governmental and non-profit organizations (including research organizions such as TNO in the Netherlands and the Fraunhofer institutes in Germany). Academic organizations are those organizations with the advance of science as primary goal – universities and other academic research organizations alike. Many governmental and non-profit organizations are additionally engaged in scientific research, but their main goal is often not the advance of science itself but lies merely in the use of the results of this research for society-broad goals. These types of organizations have been identified with the use of an algorithm that was based on a list of abbreviations and words. For example organizations with the abbreviation 'bv.', 'inc', 'GmbH' (and so on) were identified as firms. In the same manner academic organizations (based on abbreviations like 'univ' or 'academy') and governmental organizations were identified. This algorithm was then tested, manually inspected and improved several times on a changing subset of 2000 collaborations until more than 99 percent of the organizations were assigned correctly to one of the three types of organizations. Remaining collaborations with organizations that could not be identified have been deleted from the dataset.

Collaboration between the same types of organizations is labelled as a homogeneous collaboration and collaboration between different types of organizations is labelled heterogeneous.[4] Within biotechnology the share of homogeneous and heterogeneous collaborations is respectively around 55 and 45 percent and within semiconductors this is 65 and 35 percent. Over time there is little variation in the shares. Table 6.3 shows the absolute number of collaborations for each form of collaboration in both technologies. Not surprisingly, collaboration between academic organizations is the most important form of collaboration in science.

Collaboration between governmental organizations and academic organizations is also frequently occurring as is academic-industry collaboration. This is especially the case for semiconductors. The share

*Table 6.3  Number of collaborations and share in total of different forms of collaborations*

|  | Biotechnology | | Semiconductors | |
|---|---|---|---|---|
|  | Number of collaborations | Share in total (%) | Number of collaborations | Share in total (%) |
| Academic | 32 159 | 49.26 | 8 576 | 57.83 |
| Governmental | 5 049 | 7.73 | 365 | 2.46 |
| Industrial | 308 | 0.47 | 443 | 2.99 |
| Total homogeneous | 37 561 | 57.53 | 9 384 | 63.28 |
| Academic-industrial | 3 752 | 5.75 | 2 212 | 14.92 |
| Academic-governmental | 22 705 | 34.78 | 2 793 | 18.83 |
| Industrial-governmental | 1 313 | 2.01 | 441 | 2.97 |
| Total heterogeneous | 27 700 | 42.43 | 5 446 | 36.72 |

of collaborations between companies and between companies and governmental organizations is low. This is not because collaboration in fundamental research does not occur between firms, but this is unlikely to lead to a scientific co-publication.

Figures 6.8 to 6.11 show the spatial characteristics of heterogeneous and homogeneous collaborations within the Netherlands, for both biotechnology and semiconductor technology. The intensity of the interregional collaboration at the Nuts3 level is indicated by the thickness of the lines between two regions. The intensity of the intra-regional collaboration (collaboration between two organizations located in the same region) is indicated by the size of the dot in a region. For the sake of clarity, collaboration patterns are only shown if the intensity is higher than ten collaborations in the case of biotechnology and if the intensity is five collaborations or higher in the case of the semiconductor technology.

Some interesting things can be observed. Collaboration within biotechnology to a large extent takes place within the Western part of the country, within the economic centre called the Randstad. In Figures 6.8 and 6.9 the regions (including cities such as Amsterdam, Rotterdam, The Hague and Utrecht) belonging to the Randstad are indicated by a slightly lighter colour.

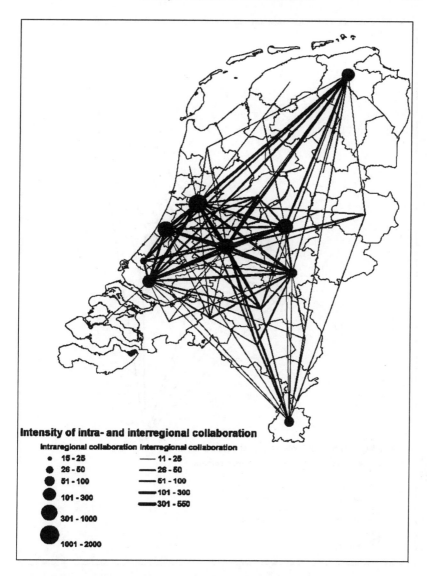

*Figure 6.8 Biotechnology: heterogeneous collaborations*

A closer look at the semiconductor technology reveals the presence of a concentration of (micro-)electronics firms and related organizations clustered around the Dutch electronics multinational Philips in one region. This region, South-East Brabant with the city of Eindhoven, is in-

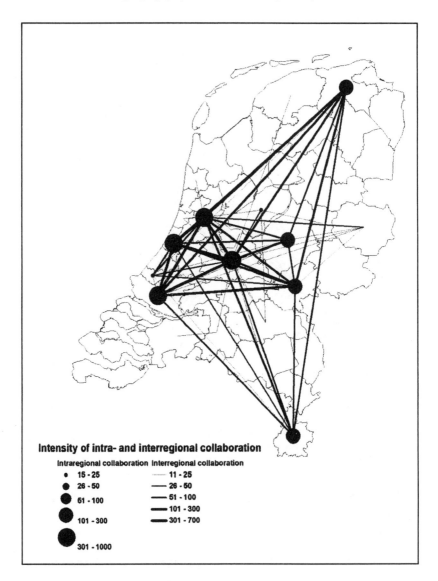

*Figure 6.9  Biotechnology: homogeneous collaborations*

dicated by a slightly lighter colour. Collaboration between organizations in this region is intensive and dominates the share of collaboration at the Nuts3 level.

Besides these observations, it can be seen that the spatial structures of

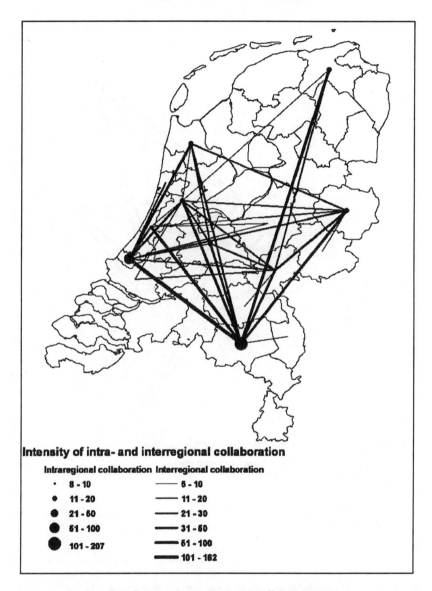

*Figure 6.10 Semiconductors: heterogeneous collaborations*

biotechnology and semiconductors are clearly different. Within each technology the homogeneous and heterogeneous patterns are more similar, although there are clear differences in intensities of collaboration both for intra- and interregional collaboration.

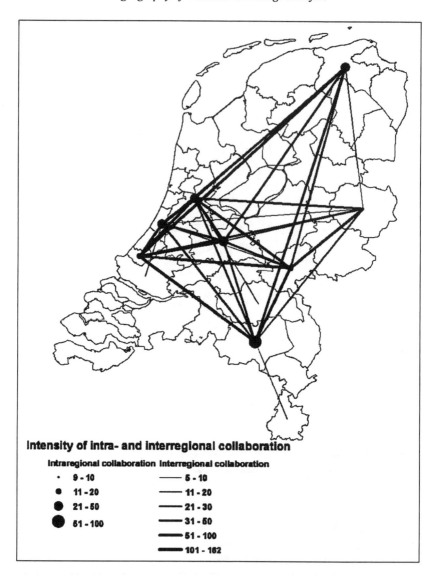

*Figure 6.11  Semiconductors: homogeneous collaborations*

We perform multiple significance tests on independency to see whether these different types of collaborations differ significantly in their spatial scale (Hair et al. 1998, p. 355). The Chi-square test of independence hypothesizes that spatial scale and form of collaboration are unrelated;

the column proportions are the same across columns and any observed discrepancies are due to chance variation. Because multiple tests are performed, the Bonferroni adjustment is applied. The Bonferroni adjustment ensures that the α-level of each individual test is adjusted downwards to ensure that overall value of α remains at chosen level.

First we tested whether or not heterogeneous collaborations are indeed more regionalized than homogeneous collaborations. Six ascending spatial scales were distinguished; ranging from Nuts3 level to countries outside the EU. Collaborations between organizations with the same institutional background were grouped under the label homogeneous and 'mixed' collaborations under the label heterogeneous. Tables 6.4 and 6.5 show the results for biotechnology and semiconductors.

The different spatial scales form the column categories and each column has a 'key' (A-F). The different forms of collaboration form the row categories. As said before, this test compares column proportions on significance differences for each row category. If there exists a significant difference between two column proportions, the key of the column with a significant smaller proportion appears under the column category with the larger proportion.

*Table 6.4   Biotechnology: the relative importance of spatial scales for heterogeneous and homogeneous collaborations*

|  | Nuts3 | Nuts2 | Nuts1 | National | EU | International |
|---|---|---|---|---|---|---|
|  | (A) | (B) | (C) | (D) | (E) | (F) |
| Homogeneous |  | A | A | A B C | A B C D | A B C D |
| Heterogeneous | B C D E F | D E F | D E F | E F |  |  |

*Note*: Results are based on two-sided tests with significance level 0.05.

Table 6.4 shows that heterogeneous collaborations in biotechnology are significantly more regionalized than collaborations between organizations with the same institutional background. Within the row 'heterogeneous', in the column 'Nuts3'or 'A', the letters B to F indicate that heterogeneous collaborations have a significant higher proportion in column A, than in all other columns. This shows that heterogeneous collaborations occur relatively more at the Nuts3 level than at all other distinguished spatial levels. A further look at Table 6.4 shows that at the other, higher, regional levels (Nuts2 and 1) heterogeneous collaborations also occur relatively more than at the national and the two international levels. The national level is more important than the EU and the international level.

In the case of homogeneous collaborations (collaboration between organizations with the same institutional background), almost the opposite can be seen. The letters A to D under column E (EU) and F (international) indicate that heterogeneous collaborations are significantly more occurring at the international and the EU level than at the lower spatial scales. Collaborations between organizations with the same institutional background are significantly more important at the national level than at the various regional levels. The regional levels at a higher spatial scale (Nuts1 and 2) are more important than the Nuts3 level, the lowest regional scale.

Figure 6.12 shows these results in a schematic way. The relative importance of the different spatial scale, as measured with a significant higher proportion than other spatial scales, is shown for both heterogeneous and homogeneous collaborations. The x-axis represents the different spatial scales, ranging from Nuts3 to international, and the y-axis represents the relative importance.

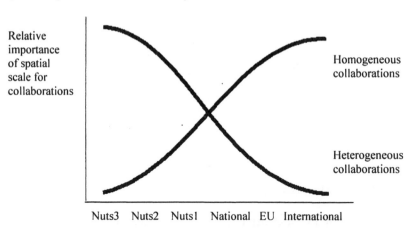

Relative importance of spatial scale for collaborations

Homogeneous collaborations

Heterogeneous collaborations

Nuts3   Nuts2   Nuts1   National   EU   International

*Figure 6.12  Schematic view on the relationship between geographical proximity and institutional proximity for collaboration in biotechnology in the Netherlands*

Note that the national level is the spatial scale which is 'equally important' for both forms of collaboration, indicating somewhat of a crossing point. This can also be concluded from Table 6.4 where the national level is more important than the regional levels for homogeneous collaborations and more important than the international levels for heterogeneous collaborations. Our hypothesis that heterogeneous collaborations are more regionalized than homogeneous collaborations can therefore be confirmed in the case of biotechnology.

Table 6.5 shows the results for semiconductors. To a certain extent, the results of collaboration of semiconductors resemble those of biotechnology. The Nuts3 and Nuts2 levels are significantly more impor-

*Table 6.5 Semiconductors: the relative importance of spatial scales for heterogeneous and homogeneous collaborations*

|  | Nuts3 | Nuts2 | Nuts1 | National | EU | International |
|---|---|---|---|---|---|---|
|  | (A) | (B) | (C) | (D) | (E) | (F) |
| Homogeneous |  |  | A B D | A B | A B D | A B D |
| Heterogeneous | C D E F | C D E F |  | C E F |  |  |

*Note*: Results are based on two-sided tests with significance level 0.05.

tant than other spatial scales for heterogeneous collaborations (indicated by the letters CDEF under A and B). As for biotechnology, heterogeneous collaboration also occurs relatively more at the national level than at the international. However, the national level is also more important than the Nuts1 level for heterogeneous collaborations, which is puzzling.

For homogeneous collaborations the EU and the international level are significantly more important than the national and regional levels. Surprisingly, homogeneous collaborations occur significantly more at the Nuts1 level than at the national level and no significant differences exist between the relative importance of the Nuts1 level and the international levels.

So the relationship between geographical and institutional proximity (as proxied by homogeneous relations) is not as clear for semiconductors as for biotechnology. However, with the notable exception of the Nuts1 level heterogeneous collaboration is more regionalized than homogeneous collaboration, which is significantly more internationalized. Our hypothesis can also be confirmed for semiconductors, although less convincingly than in the case of biotechnology.

However it is worth noting that there are a total number of 14 universities in the Netherlands. This means that in most Nuts3 regions there are no universities (and the region Amsterdam is the only region containing two universities). Universities are by far the most important academic organizations and therefore this can affect the results. For example, university-university collaboration is always (with the exception of Amsterdam) at least at the Nuts2 level, because there are no intra-Nuts3 opportunities for collaboration. Given the importance of universities in scientific knowledge production it seems common sense that homogeneous collaborations do not frequently occur at the lower regional levels. In order to control for this, the analysis is repeated with the Nuts1, the national, the EU and the international level. Collaborations at the Nuts3 and Nuts2 levels are grouped together with those at the Nuts1 level in order to see whether the results still hold.

Tables 6.6 and 6.7 show that, after taking into account the spatial distribution of universities, the conclusions hold. Comparing semiconductors and biotechnology, it is noticed that for semiconductors the national level (besides the regional Nuts1 level) is also significantly more important for heterogeneous collaborations than for homogeneous collaborations, whereas heterogeneous collaborations in biotechnology

Table 6.6  *Biotechnology: the relative importance of spatial scales for heterogeneous and homogeneous collaborations (regional scales aggregated at the Nuts1 level)*

|  | Nuts1 (A) | National (B) | EU (C) | International (D) |
|---|---|---|---|---|
| Homogeneous |  | A | A B | A B |
| Heterogeneous | B C D | C D |  |  |

*Note*: Results are based on two-sided tests with significance level 0.05.

Table 6.7  *Semiconductors: the relative importance of spatial scales for heterogeneous and homogeneous collaborations (regional scales aggregated at the Nuts1 level)*

|  | Nuts1 (A) | National (B) | EU (C) | International (D) |
|---|---|---|---|---|
| Homogeneous |  |  | A B | A B C |
| Heterogeneous | C D | C D | D |  |

*Note*: Results are based on two-sided tests with significance level 0.05.

only occur significantly more often at the subnational level.

The most important reason for this is the dominance of the multinational Philips (and the firms clustered around it) in the heterogeneous collaborations within the Netherlands. Located in the Eindhoven region in the south of the Netherlands, Philips collaborates with universities and other academic organizations all over the Netherlands. Philips is involved in 11.2 percent of all collaborations between 1988 and 2004 in the semiconductors technology. The dominance of Philips could be the reason for the sometimes puzzling results of earlier analysis. We repeated the analysis for semiconductors for both the aggregated regional levels and all regional levels. Tables 6.8 and 6.9 show the results.

Comparing Table 6.8 with Table 6.7, the influence of Philips becomes clear. Without collaborations where Philips is involved the national level becomes less important for heterogeneous collaborations than the Nuts1 level and homogeneous collaborations are also more occurring at the national level than at the regional level. Comparing Table 6.9 with

*Table 6.8 Semiconductors: the relative importance of spatial scales for heterogeneous and homogeneous collaborations (regional scales aggregated at the Nuts1 level, excluding collaborations with Philips)*

|  | Nuts1 | National | EU | International |
|---|---|---|---|---|
|  | (A) | (B) | (C) | (D) |
| Homogeneous |  | A | A B | A B |
| Heterogeneous | B C D | C D |  |  |

*Note*: Results are based on two-sided tests with significance level 0.05.

*Table 6.9 Semiconductors: the relative importance of spatial scales for heterogeneous and homogeneous collaborations (excluding collaborations with Philips)*

|  | Nuts3 | Nuts2 | Nuts1 | National | EU | International |
|---|---|---|---|---|---|---|
|  | (A) | (B) | (C) | (D) | (E) | (F) |
| Homogeneous |  |  | A B | A B | A B D | A B D |
| Heterogeneous | C D E F | C D E F |  | E F |  |  |

*Note*: Results are based on two-sided tests with significance level 0.05.

Table 6.5 similar observations can be made; no significant differences between the Nuts1 and the national level can be seen now. So Philips has certainly some impact in the earlier analyses. Philips collaborates heavily with academic organizations all over the Netherlands. As a result, many heterogeneous collaborations at the national level exist, thereby influencing the outcomes. The results of the analysis of the collaborations excluding those with Philips do not change our main conclusions. So after controlling for the spatial distributions of universities and the dominant position of Philips in semiconductor technology our hypothesis can be confirmed.

The outcomes of these analyses suggest that geographical proximity is more important for collaboration between different types of organizations (such as industry-university collaboration) than for collaboration between the same types of organizations. Although these outcomes are consistent with the theoretical considerations on the relative importance of geographical proximity for research collaboration in science, one should be aware of the fact that these outcomes could also be influenced by the spatial distribution of various organizations (especially universities). Although this is partly taking into account by the aggregation to the Nuts1 level, further research on this topic should try to include the spatial distribution of different types of organizations in the analysis.

## 6.6 CONCLUSION

In this study, an analysis of the spatial characteristics of collaboration in scientific knowledge production in the Netherlands is performed. Within science-based industries, collaboration between governmental, academic and private organizations in scientific knowledge production is an important and growing phenomenon. Based on theoretical insights from the literature of the geography of innovation it was hypothesized that collaborations between organizations with different institutional backgrounds were more regionalized than collaborations between organizations with the same background. Co-publications in scientific subfields that are relevant for technological innovation in semiconductors and biotechnology were used as a proxy for collaboration between organizations.

The main finding of this study is that heterogeneous collaborations are more regionalized than collaborations between the same kinds of organizations. Even after controlling for the spatial distribution of universities and the dominant position of Philips within the semiconductor technology in the Netherlands our hypothesis is confirmed. This can be seen as empirical evidence for the importance of geographical proximity for successful collaboration between organizations with different institutional backgrounds, like university-industry collaboration. Geographical proximity does not seem to play an important role for collaboration in science between organizations with the same institutional background. Homogeneous collaboration occurs relatively more at the international level, thereby indicating that institutional proximity is more important here than geographical proximity. Also these results illustrate the fact that science takes place in international epistemic communities centred around a global discourse.

## NOTES

1. We are very grateful to Stephaan Declerk of the Netherlands Institute for Spatial Research (RPB) for his research assistance and Hans van Amsterdam (RPB) for his help in data handling. Furthermore we would like to thank Attila Varga and an anonymous referee for helpful comments on an earlier version.
2. Correspondence address: Roderik Ponds, Netherlands Environmental Assesment Agency, P.O. Box 30314, 2500 GH, The Hague, the Netherlands, roderik.ponds@pbl.nl.
3. As a result a publication with for example authors from four different organisations leads to six collaborations.
4. Building on the example of footnote 3, a publication involving two universities, a firm and governmental research institute will lead to one homogeneous collaboration (two universities) and four heterogeneous collaborations.

## REFERENCES

Ács, Z.J. (2002), *Innovation and the Growth of Cities*, Cheltenham, UK and Northampton, MA, USA: Edward Elgar.

Anselin, L., A. Varga and Z. Ács (2000), 'Geographic and sectoral characteristics of academic knowledge externalities', *Papers in Regional Science*, 79, 435-445.

Audretsch, D.B. and M.P. Feldman (1996), 'R&D spillovers and the geography of innovation and production', *American Economic Review*, 86 (3), 630-640.

Boschma, R. (2005), 'Proximity and innovation: a critical assessment', *Regional Studies*, 39 (1), 61-74.

Cockburn, I. and R. Henderson (1998), 'Absorptive capacity, coauthoring behavior and the organization of research in drug discovery', *The Journal of Industrial Economics*, 46 (2), 157-182.

Cohen, W.M. and D.A. Levinthal (1989), 'Innovation and learning: the two faces of R&D', *Economic Journal*, 99, 569-596.

Cohen, W.M. and Levinthal, D.A. (1990), 'Absorptive capacity: a new perspective on learning and innovation', *Administrative Science Quarterly*, 35 (1), 128-152.

Cohendet, P. and F. Meyer-Krahmer (2001), 'The theoretical and policy implications of knowledge codification', *Research Policy*, 30 (9), 1563-1591.

Cooke, P., M.G. Uranga and G. Etxebarria (1997), 'Regional innovation systems: institutional and organisational dimensions', *Research Policy*, 26 (4-5), 475-491.

Cowan, R., D. Foray and P.A. David (2000), 'The explicit economics of codification and the diffusion of knowledge', *Industrial and Corporate Change*, 6, 595-622.

Dasgupta, P. and P.A. David (1994), 'Toward a new economics of science', *Research Policy*, 23, 487-521.

DeSolla Price, D. (1984), 'The science/technology relationship, the craft of experimental science, and policy for the improvement of high technology innovation', *Research Policy*, 13 (1), 3-20.

Etzkowitz, H. and L. Leydesdorff (2000), 'The dynamics of innovation: from National Systems and "Mode 2" to a triple helix of university-industry-government relations', *Research Policy*, 29, 109-123.

Feldman, M.P. (1999), 'The new economics of innovation, spillovers and agglomeration: a review of empirical studies', *Economics of Innovation and New Technology*, 8, 5-25.

Feser, E.J. (2002), 'Tracing the sources of local external economies', *Urban Studies*, 39, 2485-2506.

Gertler, M.S. (2003), 'Tacit knowledge and the economic geography of context, or the indefinable tacitness of being (there)', *Journal of Economic Geography*, 3, 75-99.

Gittelman, M. and B. Kogut (2003), 'Does good science lead to valuable knowledge? Biotechnology firms and the evolutionary logic of citation patterns', *Management Science*, 49 (4), 366-382.

Greunz, L. (2003), 'Geographically and technologically mediated knowledge spillovers between European regions', *The Annals of Regional Science*, 37 (4), 657-680.

Hair, J.F., R.E. Anderson, R.L. Tatham and W.C. Black (1998), *Multivariate Data Analysis*, Fifth edition, London: Prentice Hall.

Jaffe, A.B. (1989), 'Real effects of academic research', *American Economic Review*, 79 (5), 957-970.

Jaffe, A.B., M. Trajtenberg and R. Henderson (1993), 'Geographic localization of knowledge spillovers as evidenced by patent citations', *Quarterly Journal of Economics*, 108 (3), 577-598.

Katz, J.S. and B.R. Martin (1997), 'What is research collaboration?', *Research Policy*, 26 (1), 1-18.

Lundvall, B.A. (1992), *National Systems of Innovations. Towards a Theory of Innovation and Interactive Learning*, London: Pinter.

Mairesse, J. and L. Turner (2005), 'Measurement and explanation of the intensity of co-publication in scientific research: an analysis at the laboratory level', NBER working paper 11172, Cambridge: National Bureau of Economic Research.

Marsili, O. (2001), *The Anatomy and Evolution of Industries: Technological Change and Industrial Dynamics*, Cheltenham, UK and Northampton, MA, USA: Edward Elgar.

Pavitt, K. (1984), 'Sectoral patterns of technical change: towards a taxonomy and a theory', *Research Policy*, 13, 343-373.

Rosenberg, N. (1990), 'Why do firms do basic research (with their own money)?' *Research Policy*, 19, 165-174.

Schmoch, U. (1997), 'Indicators and the relations between science and technology', *Scientometrics*, 38 (1), 103-116.

Varga, A. (1998), *University Research and Regional Innovation: A Spatial Econometric Analysis*, Boston: Kluwer Academic Publishers.

Verbeek, A., J. Callaert, P. Andries, K. Debackere, M. Luwel and R. Veughelers (2002), 'Linking science to technology-bibliographic references in patents, Volume 1: Science and technology interplay – policy relevant findings and interpretations', Report to the EC DG Research, project ERBHPV2-CT-1993-03, CBSTII action, Brussels (EUR 20492/1).

Wagner, C.S. (2005), 'Six case studies of international collaboration in science', *Scientometrics*, 62 (1), 3-26.

# 7. Academic knowledge transfers and the structure of international research networks[1]

## Attila Varga and Andrea Parag

## 7.1 INTRODUCTION

Although the majority of the literature on academic knowledge transfers focuses on the geographical aspects, several recently published papers raise the issue that besides pure spatial proximity to an academic institution some additional factors (such as local culture determining the extent of collaboration and the level of entrepreneurship or the spatial concentration of the system of innovation) might also be instrumental. Understanding the significance of those factors in regional economic development is at least as important for designing effective regional policies as improving our knowledge on the spatial proximity issue. Among the factors influencing academic knowledge transfers the specific role of scientific networking has not been touched upon very extensively in the literature. Scientific networking that may take different forms such as collaborative projects, co-publications or less formal meetings in conferences, workshops or seminars is a common means of advancing science, mutual learning, information sharing and gaining and maintaining attention among fellow scientists. Increasing specialization and competition in research as well as the rapid development of technologies that ease sustaining and expanding linkages among scientists over large geographical distances make it both possible and inevitable that collaboration among researchers working in different institutions has become a key to high level research productivity.

Research networking strengthens not only scientific productivity but also academic knowledge transfers to the industry. It is emphasized in the survey of Franzoni and Lissoni (2009) that scientific excellence and success in academic knowledge transfers (in the forms of patenting or spin-off firm foundation) do not necessarily contradict each other as successful academic entrepreneurs come disproportionately from the class of researchers with a brilliant scientific record. It has also been suggested in the literature that universities may act as key nodes channeling scientific-technological knowledge accumulated in (national or international) research networks to the regional industry via different mechanisms of localized knowledge flows such as patenting, licensing, spin-off firm formation, consulting or participating in collaborative R&D projects (Goldstein, Maier and Luger 1995).

Thus embeddedness in (regional, interregional or international) research networks may make a difference across universities with respect to their success in transferring knowledge to innovations. Ceteris paribus the same amount of university research expenditures might result in different levels of knowledge flows from academic institutions depending on how well they are integrated in scientific networks. Therefore the research question about the extent to which scientific networking influences academic knowledge transfers is indeed a relevant one. A major reason why the impact of research networking on academic knowledge transfers had not been tested systematically until very recently is that econometric estimations suffered from a technical barrier. Spatial econometric models with specific weights matrices (such as inverse distance weights as were done, for example, in Anselin, Varga and Ács 1997) had been the only possibilities until the rapid diffusion of Social Network Analysis (SNA) methods in innovation research (see Coulon 2005 and Ozman 2006 for reviews on the SNA literature of innovation research). As such SNA applications pave the way for more precise analyses.

The issue of the effect of research networks on academic knowledge transfers has been investigated in some recently published studies with the application of SNA methodologies. Based on their analysis with data on 109 European regions at NUTS 2 level Maggioni, Nosvelli and Uberti (2006) argue that participation in the EU 5th framework projects has a positive impact on regional innovation activity while Ponds, Oort and Frenken (2007) report significant interregional research networking effects on patenting using regional data of the Netherlands.

None of these recent studies addresses the question of the role of network structure in academic knowledge transfers. Nevertheless, the particular configuration of networks could make a difference in innovation as reported in several papers on industry networks. For example: Valente (1995), Cowan and Jonard (1999) and Spencer (2003) point to the significance of network structure; Ouimet, Landry and Amara (2004), Morrison and Rabellotti (2005) and Giuliani (2007) emphasize the role of network position; Giuliani (2004) finds that network density, strength of ties and external openness matters in innovation, while Ahuya (2000) reports that structural holes decrease innovation output.

Isn't it a realistic hypothesis that in addition to the pure size of research networks other structural features (such as the extent to which the network is concentrated around some 'stars' of the scientific field or the intensity of research relationships within the network) are also instrumental in academic knowledge transfers? While the size effect has already been investigated in the literature a more detailed analysis of the impact of network structure is still missing. We address the role of international network configuration in academic technology transfers with the application of recently collected data on international publication networks of selected research units at the University of Pécs. The second section explains the data, develops indices for different network characteristics and designs a comprehensive measure of academic network quality. In the third section (based on an extended knowledge production function framework) the effect of international network structure on university patenting is tested. We conclude with a summary section.

## 7.2 THE STRUCTURE OF INTERNATIONAL PUBLICATION NETWORKS

It is hypothesized in this chapter that structural features of research networks of universities are significant factors in knowledge transfers. Thus, ceteris paribus, even with similar levels of research expenditures universities may generate different economic impacts through knowledge transfers depending on the structure of their (regional, interregional or international) scientific networks. How can we define those network characteristics that are instrumental in knowledge transfers and how can we measure them? Can we even summarize those features in one particular index? These issues are the focus of this section.

While determining important network features in academic knowledge transfers our starting point is the systems of innovation (SI) literature (e.g., Lundvall 1992; Nelson 1993). According to this literature production of economically useful new knowledge depends to a large extent on three system characteristics: the number of actors involved in the system, the knowledge those actors have accumulated and the intensity of knowledge-related interactions among the actors during knowledge creation. Thus the efficiency of research networks in producing new knowledge can be approached by three features: the size of the network, the professional knowledge of individual scientists involved in the network and the frequency of their interactions (e.g., research collaboration, mutual learning).

We argue in this chapter that the quality of research network connections influences the scientific productivity of individual network members and, as such, academic knowledge transfers. How can we define the quality of a network connection and what are the structural features of a research network that determine it? Quality of a network connection reflects the level of knowledge (both tacit and codified) and information to which the individual researcher gets access by being linked to the network. This depends on the knowledge accumulated in the network and the position of the individual scientist within that network. Thus the knowledge to be accessed is related to the size of the network, the knowledge the members of the network possess, the intensity of science-related interactions among the actors and the network position of the individual researcher. Larger size, higher levels of knowledge of network members and frequent interactions among them are essential to guarantee the continuous extension of knowledge within the network (as described in detail by the SI literature) whereas network position could be extremely important for accessing that knowledge.

Research network position is either related to the knowledge (and reputation) of the researcher or to the knowledge (and reputation) of the immediate network partner of the researcher. There is a simultaneous relation between individual knowledge of the researcher and the number of linkages the researcher possesses in the network. Higher knowledge levels increase reputation that (via increased visibility) opens the possibilities for researchers to further increase the number of connections within the network whereas increased number of linkages lets them access and produce even higher levels of knowledge. Moreover it is also assumed that a favorable position in the network positively affects the position of the researcher's immediate network partner as well, first by providing him or her access to a considerable portion of knowledge accumulated in the network (and concentrated by the researcher with high reputation) and second (through more visibility) by offering good opportunities for increasing the number of his or her own connections. Therefore a researcher even with a lower level of scientific output can get access to a high level of knowledge (which may lead to increased research productivity) if the immediate partner enjoys considerable reputation.

Table 7.1 Selected features of international co-publication networks of sample University of Pécs academic units, 2000

| | International publications | UP co-authors | International co-authors of UP faculty | International co-authors of the international co-authors of UP faculty |
|---|---|---|---|---|
| Clinic of Neurology | 4 | 2 | 19 | 152 |
| Department of Anatomy | 18 | 11 | 6 | 102 |
| Department of Biophysics | 7 | 6 | 7 | 54 |
| Department of Immunology and Biotechnology | 4 | 3 | 13 | 77 |
| Department of Medical Chemistry | 7 | 9 | 31 | 191 |
| Department of Medical Genetics and Child Development | 3 | 1 | 6 | 92 |
| Department of Medical Microbiology and Immunology | 5 | 5 | 15 | 251 |
| Department of Neurosurgery | 5 | 5 | 10 | 145 |
| Department of Orthopedics | 7 | 8 | 12 | 53 |
| Department of Pathology | 6 | 7 | 9 | 141 |
| Department of Pediatrics | 12 | 8 | 9 | 169 |
| Department of Pharmacology and Pharmacotherapy | 4 | 1 | 2 | 23 |
| Department of Surgery | 3 | 3 | 10 | 136 |
| Institute of Organic and Medical Chemistry | 3 | 2 | 4 | 57 |
| Institute of Physics Department of Experimental Physics | 10 | 3 | 17 | 104 |
| Institute of Physics Department and Theoretical Physics | 6 | 6 | 9 | 28 |

Thus the advantage of a better quality network connection is that it increases research productivity both directly (with higher probabilities of achieving truly relevant results in collaboration) and indirectly (through learning and building further connections). As such the size of the research network, the intensity of knowledge-related linkages and the knowledge level of researchers (especially the knowledge of the immediate network partner) characterize network connection quality.

To study empirically the effect of research networks on academic knowledge transfers we use co-publication data collected for selected academic units at the University of Pécs (UP). We assume that the quality of international research network connection of each scientific unit (represented by international co-publications) influences the knowledge transfer activities of those academic units. The selection criterion was international publication excellence in hard sciences relative to the usual level of university research units at UP in these fields. The chosen year is 2000. UP Library and the ScienceDirect and EBSCOhost publication databases were the sources of data. Our focus is on those networks to which UP researchers are connected hence we collected data on research networks of international co-authors of each UP researcher in the sample. Table 7.1 lists the main features of the co-publication networks aggregated to academic units.

The international co-publication network to which each UP academic unit is connected is described by the numbers of UP scientists and their immediate research partners and the number of co-authors of the immediate international research partners. The size of research networks exhibits a considerable variation as demonstrated in Table 7.1. The internal structure of each network shows an even higher variability. This can be studied in Figures 7.1-7.3. Black triangles stand for the immediate research partners of UP scientists whereas their network members are shown by black squares (Hungarian co-authors) and circles (international co-authors). According to the simultaneous relationship between academic excellence (i.e., the knowledge of individual researchers) and the extent to which a researcher is connected with others in the network, the number of links of an individual scientist represents academic reputation. Our data allow us to judge network positions of UP researchers as well as their immediate publication partners. Since UP researchers do not play a central role in any of the networks investigated our analysis concentrates on the network positions of UP co-authors, the size of the network and the level of interactions within the network.

Network connection quality of individual research units varies widely in the sample. Some of the network connections might be described as

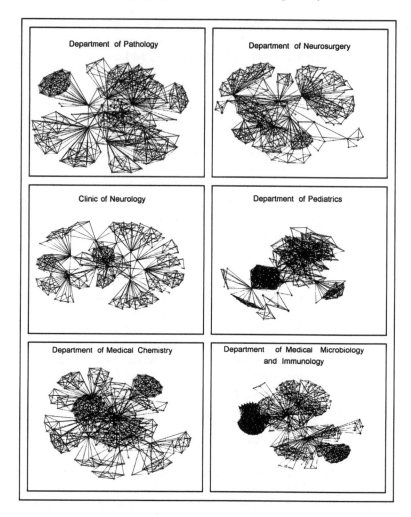

*Note*: Black triangles stand for the immediate research partner of UP scientists whereas their network members are shown by black squares (Hungarian co-authors) and circles (international co-authors).

*Figure 7.1  Large international networks*

'poor' such as the one of the Department of Pharmacology and Pharmacotherapy (Figure 7.3) where the two immediate international co-authors show modest levels of reputation (indicated by the number of their linkages) while the intensity of collaboration is also at a low level

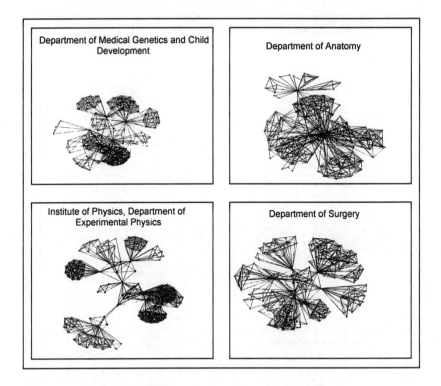

Department of Medical Genetics and Child Development

Department of Anatomy

Institute of Physics, Department of Experimental Physics

Department of Surgery

*Note*: Black triangles stand for the immediate research partner of UP scientists whereas their network members are shown by black squares (Hungarian co-authors) and circles (international co-authors).

*Figure 7.2  Medium size international networks*

in the network (indicated by the linkages connecting members of the network).

To take another example consider the Institute of Organic and Medical Chemistry (Figure 7.3) where one of the immediate international publication partners has several linkages but the small size of the network and also the rare occurrences of interactions among the partners (i.e., each paper is an 'island' with no 'bridges' among them) set the quality at a relatively low level.

On the other hand, the Department of Pediatrics (Figure 7.1) with the large size of the network it is connected to, the very intense collaboration among network members and the high concentration of linkages at some of the immediate research partners (who might even be 'star scientists' of their field) possesses a high quality network connection.

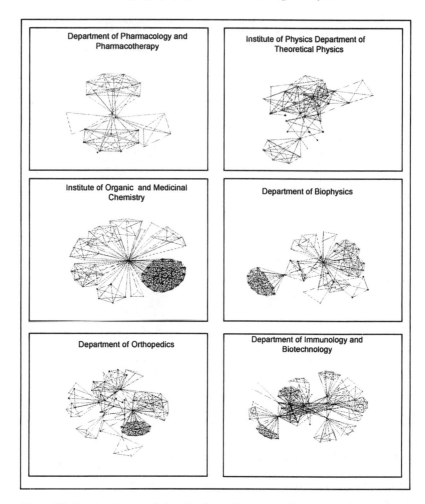

*Note*:  Black triangles stand for the immediate research partner of UP scientists whereas their network members are shown by black squares (Hungarian co-authors) and circles (international co-authors).

*Figure 7.3  Small international networks*

In order to study the effect of network connection quality we need to quantify those structural features that are instrumental in determining it. Since the measures to be used in the current analysis should be comparable across networks with different sizes, commonly applied indices such as centrality (which could be useful for measuring reputation) or density (for quantifying the intensity of network connections) are not suitable (Scott 2000). As such appropriate indices need to be developed before studying the network effect on knowledge transfers.

To measure the size of the network of academic unit $i$ we introduce the following index:

$$SIZE_i = \text{(Network members)}_i/\text{(Network members)}_{max}$$

Thus the values of *SIZE* are between 0 and 1 where the academic unit with the largest network gets the value of 1.

As shown in Figures 7.1-7.3 network position of UP international publication partners could be decisive while network connection quality is determined. How to measure this position? We start with the experience that the knowledge of a researcher determines his or her position in a scientific community and this position is reflected by the number of linkages the researcher possesses. Thus the better the network position of a scientist the more concentrated the network around him or her. The following formula calculates the index of knowledge concentration by immediate research partners of each academic unit:

$CONCi$ = (average number of international co-authors of immediate UP co-authors)$_i$/(average number of international co-authors of immediate UP co-authors)$_{max}$

The values of *CONC* range between 0 and 1. The higher the value of *CONC*, the better the average position of UP publication partners of a research unit.

The index INT measures the level of integratedness of the network. By integratedness we intend to quantify the intensity of linkages among network members.

$INT_i$ = [(Average number of linkages on a paper)/(average number of linkages among co-authors on a paper)]$_i$/[(Average number of linkages on a paper)/(average number of linkages among co-authors on a paper)]$_{max}^2$

The higher is the value of *INT*, the larger is the relative number of linkages bridging communities of co-authors of different papers. Hence *INT* measures the intensity of interactions among network members and its value ranges between 0 and 1.

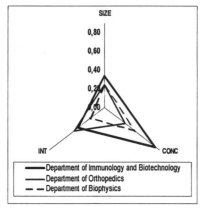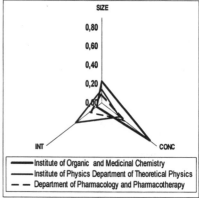

*Figure 7.4 SIZE, CONC, INT: large international networks*

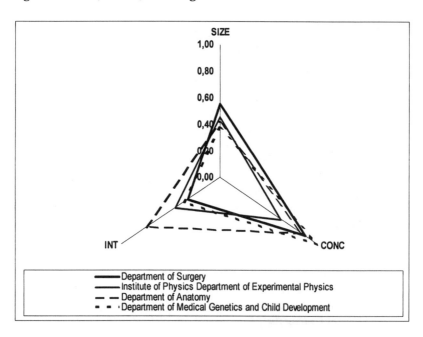

*Figure 7.5 SIZE, CONC, INT: medium size international network*

Figures 7.4-7.6 present the values of *SIZE, CONC* and *INT* for the studied publication networks classified into the three network size categories. A comparison of patterns in Figures 7.1-7.3 and 7.4-7.6

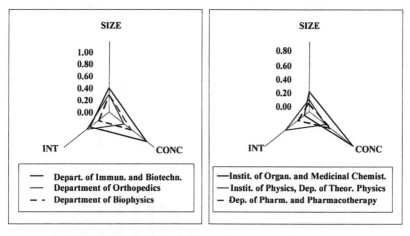

*Figure 7.6 SIZE, CONC, INT: small international networks*

suggests that the three measures follow the three network characteristics very closely.

The quality of a network connection reflects the three structural characteristics and is in a positive relationship with all of them. How could we integrate the three indices into one to measure network connection quality? The intuition behind the solution comes after studying the triangles of Figures 7.4-7.6: the composite quality index (*NETQUAL*) for each academic unit is the area of the respective triangle representing the unit divided by the maximum possible area of the triangles. Thus the closer the value of *NETQUAL* is to 1 the higher the quality of network connection of an academic unit resulting from a particular combination of *SIZE, CONC* and *INT*. Figure 7.7 exhibits *NETQUAL* values.

## 7.3 EMPIRICAL MODEL, DATA AND RESULTS

Expenditures on research and development are key determinants of scientific success. While modeling university knowledge transfers R&D expenditures are commonly applied input measures in empirical studies. However, even with expenditures at similar levels the impact of university research could be different depending on various factors such as the development of the innovation infrastructure, entrepreneurship or cultural factors like the openness to cooperate in innovation. We hypothesize in this chapter that academic technology transfers are also

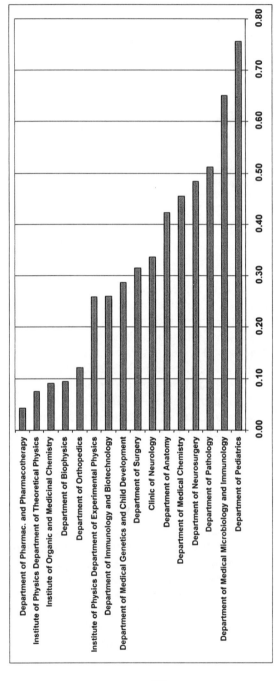

Figure 7.7 NETQUAL values by UP academic units

150

related to the quality of network connections and this effect alters the impact of R&D expenditures.

To empirically test our hypothesis we follow the approach of Varga (2000, 2001) and Ács and Varga (2005) to develop a hierarchical regression framework within the knowledge production function approach of Griliches (1986) and Jaffe (1989). The empirical model is written in the following form:

$$K_t = \alpha_0 + \alpha_1 RD_i + \alpha_2 Z_t + \varepsilon_t, \qquad (7.1)$$

where $K$ is economically useful scientific knowledge, $RD$ is research and development expenditures, $Z$ is for additional explanatory variables (such as a variable measuring the experience in industrial problem solving) and $\varepsilon$ is the error term. Observational units are groups of university researchers specialized in particular research fields.

We assume that $\alpha_1$ is not constant across observational units but depends on research network features. Thus the model in (7.1) is then extended by the following equation:

$$\alpha_{1,t} = \beta_0 NET_t, \qquad (7.2)$$

where $NET_t$ stands for a particular characteristic of the research network of observational unit $i$. Therefore to account for the impact of networking the estimated equation gets the following form:

$$K_t = \alpha_0 + \beta_0 NET_t RD_i + \alpha_2 Z_t + \varepsilon_t \qquad (7.3)$$

In the following empirical analysis we study the impact of research networking on university patenting a particular type of academic knowledge transfer. Data come from two sources. The first is the publication database of UP academic units that has already been explained. The second data source is a result of a survey of UP research groups conducted in 2006 (Szerb and Varga 2006). Table 7.2 explains the data in details.

Reflecting the fact that $K$ in (7.1) is measured by count data we run negative binomial count regressions. Estimation results are presented in Table 7.3. According to expectations R&D expenditures enter the equation with a positive and significant parameter (M1). The effect of experience in industrial problem solving (measured by the number of collaborating firms) matters for the case of Hungarian firms but not for

*Table 7.2 Description of the applied data*

| Variable | Explanation | Number of research groups** | Minimum | Maximum | Average | Standard deviation |
|---|---|---|---|---|---|---|
| PATANUM* | Number of university patents (2000-2005) | 23 | 0.00 | 5.00 | 0.39 | 1.16 |
| PROJBUD17 | The value of the seven most important projects in Euro (2000-2005) | 23 | 50 000 | 3 701 000 | 894 000 | 1 144 000 |
| CONC | Knowledge concentration (year 2000) | 23 | 0.29 | 1.00 | 0.66 | 0.25 |
| INT | Intensity of interactions (year 2000) | 23 | 0.09 | 1.00 | 0.47 | 0.21 |
| SIZE | Network size (year 2000) | 23 | 0.09 | 1.00 | 0.46 | 0.28 |
| NETQUAL | Network connection quality (year 2000) | 23 | 0.04 | 0.76 | 0.32 | 0.21 |
| COBHNUM* | Hungarian firms collaborating in innovation (2000-2005) | 23 | 0.00 | 5.00 | 1.74 | 1.36 |
| COBFNUM* | International firms collaborating in innovation (2000-2005) | 22 | 0.00 | 2.00 | 1.09 | 1.02 |

*Notes:* * Medium of the range.
  ** A particular academic unit might contain several research groups. This results in different observation numbers in Tables 7.1 and 7.2.

Table 7.3 Negative binomial count estimation results for the number of university patents, selected University of Pécs hard sciences research groups, 2000–2005

| | M1 | M2 | M3 | M4 | M5 | M6 | M7 | M8 | M9 | M10 | M11 |
|---|---|---|---|---|---|---|---|---|---|---|---|
| C | -2 866*** (0 914) | -6 025** (2 642) | -6 983** (3 548) | -5 916** (2 596) | -2 715*** (0 996) | -7 797** (4 048) | -4 062*** (1 403) | -8 209* (4 670) | -2 943*** (1 059) | -6 369** (0 027) | -8 695 (5 392) |
| PROJBUD | 1 01E-06*** (3 33E-07) | 1 30E-06*** (5 04E-7) | 1,17E-06** (4 93E0-7) | | | | | | | | |
| PROJBUD*SIZE | | | | 9 89E-07* (5 51E-07) | 1 02E-06 (6 65E-07) | | | | | | |
| PROJBUD*CONC | | | | | | 2 42E-06* (1 29E-06) | 1 52E-06** (6 70E-07) | | | | |
| PROJBUD*INT | | | | | | | | 1 73E-06* (9 75E-07) | 1 14E-06* (6 10E-07) | 1 70E-06* (8 99E-07) | |
| PROJBUD*NETQUAL | | | | | | | | | | | 3 27E-06* (1 96E-06) |
| COBHNUM | | 0 991* (0 564) | | 1 014* (0 549) | 0 581 (0 325) | 1 479* (0 833) | 0 752** (0 376) | 1 532* (0 931) | 0 575* (0 335) | 1 166* (0 620) | 1 685 (1 092) |
| COBHNUM*COBFNUM | | | 0 861 (0 537) | | | | | | | | |
| PHARMA1 | | | | 1 838 (1 520) | | 5 606** (2 346) | | 4 350** (2 030) | 4 429*** (1 675) | 4 429*** (1 675) | 6 426** (3 183) |
| LR-Index (Pseudo R²) | 0 34 | 0 45 | 0 45 | 0 50 | 0 14 | 0 56 | 0 31 | 0 57 | 0 18 | 0 52 | 0 58 |
| Log Likelihood | -12 766 | -10 41 | -10 19 | -9 604 | -16 285 | -8 409 | -13 047 | -8 174 | -15 556 | -9 198 | -7 999 |
| N | 23 | 23 | 22 | 23 | 23 | 23 | 23 | 23 | 23 | 23 | 23 |

Notes: Estimated standard errors are in parentheses; *** is significance at 0.01; ** is significance at 0.05; * is significance at 0.10.

153

international companies (M2 and M3). One of the pharmaceutical research groups (PHARMA1) shows exceptionally successful knowledge transfer activities (five accepted patents) during the time period under consideration. To account for potentially different mechanisms of knowledge production at this group we introduced the PHARMA1 dummy.

Our base model is M4. Compared to M2 model fit increased somewhat in M4 (the LR index[3] went up from 0.45 to 0.50) but the parameter of PHARMA1 is not yet significant. Models M5, M7 and M9 estimate the impacts of network characteristics in focus such as network size (*SIZE*), knowledge concentration (*CONC*) and intensity of research collaboration (*INT*). M6, M8 and M10 have the same setups but treating PHARMA1 separately. In general we found marginally significant effects of network features (P < 0.10). It is also evidenced that PHARMA1 follows a different pattern from the rest of the research groups. The PHARMA1 dummy enters the models with significant and positive parameters. Additionally, introduction of this dummy variable increases regression fit considerably. M11 shows that the effect of network connection quality on university patenting is also positive and marginally significant (P < 0.10). This model provides the best fit to the data which is an important further evidence for the network quality effect.

Regression results support the hypothesis that the impact of academic R&D expenditures on knowledge transfers varies according to the quality of research network connections. To what extent does this impact differ across research units? Which network characteristics have the strongest influence on the quality effect? The next step in the analysis is to answer these questions.

Substituting the estimated $\beta_0$ from M11 to (2) and calculating $\alpha_{1,i}$ for each research unit results in the Alpha *NETQUAL* values of Figure 7.8. The figure demonstrates that the impact of R&D expenditures on university patenting shows notable variations across academic units depending on their network connection quality. The straight line indicates the value of $\alpha_1$ as estimated in M4. This parameter shows the average effect of R&D on university patenting with no respect on the differences in network connection quality. On the other hand, Alpha *NETQUAL* varies widely: there is 18 times the difference between the minimum and the maximum estimated values of $\alpha_{1,i}$.

Can we weight the impacts of different network characteristics on the network connection quality effect? The regression output in Table 7.4 evidences that the position of the immediate co-author in the research

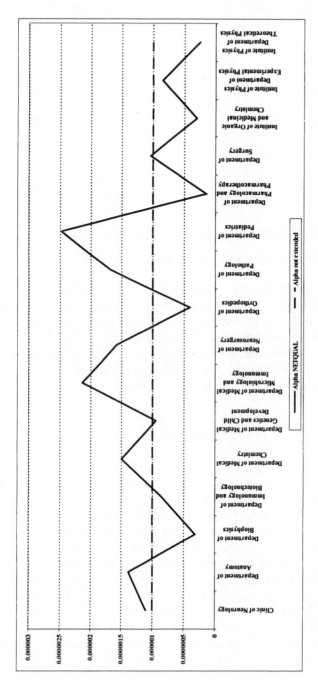

*Figure 7.8 The influence of network quality: varying R&D expenditure impacts on university patenting*

*Table 7.4  The effect of network characteristics on the network connection quality parameter (Log($\alpha_1$))*

| | |
|---|---|
| Constant | 12.512*** |
| | (0.392) |
| Log(*BETACONC*) | 0.709*** |
| | (0.033) |
| Log(*BETAINT*) | 0.568*** |
| | (0.023) |
| Log(*BETASIZE*) | 0.635*** |
| | (0.024) |
| $R^2$ | 0.998 |
| N | 23 |
| F-statistic | 3467.506*** |

*Notes*: Estimated standard errors are in parentheses; *** is $p < 0.001$.

network (measured by the concentration of knowledge by the international partner) is the most influential network characteristic to determine the differing effects of R&D on university patenting. The estimated parameter indicates that a 1 percent change in *CONC* results in a 0.71 percent average change in the estimated $\alpha 1$ values of the research units. This is followed by the size of the network and the intensity of collaborations among researchers in the network.

## 7.4 SUMMARY AND CONCLUSIONS

Transfers of economically useful scientific knowledge from universities to industry could generate substantial economic growth as the experiences of classical high technology regions (e.g., Silicon Valley) and emerging new technology centers around the world well demonstrate this effect. It is evidenced in the literature that the effectiveness of academic knowledge transfers is related to several factors. Our study focuses on the role of research network connection quality in this respect. Research network connection quality determines the stock of knowledge to which the individual researcher has access by being linked to other researchers. It is related to the knowledge accumulated in the network and also the position from which the researcher enters the network.

Applying recently collected data on international publication networks of selected hard sciences research units of the University of Pécs, this chapter analyses the effects of network size, concentration of knowledge at immediate international publication partners and intensity of interactions among network members on university patenting. The main achievements of this study can be summarized as follows:

1. the term 'network connection quality' is introduced to estimate the impact of research networks on academic knowledge transfers;
2. appropriate indices measuring size and concentration of networks and interaction intensity among network members are developed;
3. a composite index of network connection quality is introduced;
4. the effects of individual indices of network structure characteristics and the composite network connection quality index on university patenting are estimated within the knowledge production function tradition;
5. the importance of individual network characteristics for the impact of network connection quality is tested.

Our results indicate that the quality of international network connections matters for academic knowledge transfers. Thus not only is the distribution of public research expenditures across different research projects important but also the position from which researchers enter international networks and the level of knowledge accumulated in those networks. The main policy consequence of this study is that the set of tools of knowledge-based economic development should include not only R&D promotion but also clever ways of supporting academic research networking. For the University of Pécs it is found that promoting connections to higher position international scientists would be the most advantageous way of strengthening the network quality effect on university patenting.

We need to mention the limitations of the current study. These include first that we collected only one year of publication network data. More years would perhaps alter our results. Also we were not able to account for the scientific quality of publication partners of immediate international colleagues of UP researchers. Although this would not change our results with respect to the examined network structure characteristics, its impact on the overall quality of network connections might be interesting. Future research based on data collected for several universities will certainly extend our knowledge on the relationship between research networks and academic technology transfers even further.

## NOTES

1. This research is supported by the 'CrosboR&D' INTERREG (SL-HU-CR/05/4012-106/2004/01/HU-12), the IAREG FP7 and the 'VERINEKT' NKFP projects (KF-30-3372/2004). The authors wish to express their thanks for the helpful comments by two outside referees.
2. The average number of linkages per paper measures the linkages among authors. It is calculated in the following manner: $N*(N-1)/2$, where $N$ is the average number of co-authors on a paper. The average number of linkages on a paper is the ratio of the size of the network and the number of articles.
3. The LR index relates log-likelihood values of the estimated equation with constant term only to its actual version.

## REFERENCES

Ács, Z.J. and A. Varga (2005), 'Entrepreneurship, agglomeration and technological change', *Small Business Economics*, 24, 323-334.

Ahuya, G. (2000), 'Collaboration networks, structural holes, and innovation: A longitudinal study', *Administrative Science Quarterly*, 45, 425-455.

Anselin, L, A. Varga and Z.J. Ács (1997), 'Local geographic spillovers between university research and high technology innovations', *Journal of Urban Economics*, 42, 422-448.

Coulon, F. (2005), 'The use of social network analysis in innovation research: A literature review', Unpublished manuscript.

Cowan, R. and N. Jonard (1999), 'Network structure and the diffusion of knowledge', MERIT Working Papers.

Franzoni, C. and F. Lissoni (2008), 'Academic entrepreneurs: critical issues and lessons for Europe' (in this volume).

Goldstein, H., G. Maier and M. Luger (1995), 'The university as an instrument for economic and business development', in D. Dill and B. Sporn (eds), *Emerging Patterns of Social Demand and University Reform: Through a Glass Darkly*, Oxford, UK: Pergamon.

Griliches, Z. (1986), 'Productivity, R&D, and basic research at the firm level in the 1970s', *American Economic Review*, 76, 141-154.

Guiliani, E. (2004), 'Laggard clusters as slow learners, emerging clusters as locus of knowledge cohesion (and exclusion): a comparative study in the wine industry', LEM Working Papers, 9, Saint Anna School of Advanced Studies, Pisa.

Guiliani, E. (2007), 'The selective nature of knowledge networks in clusters: evidence from the wine industry', *Journal of Economic Geography*, 7, 139-168.

Jaffe, A. (1989), 'Real effects of academic research', *American Economic Review*, 79, 957-970.

Lundvall, B.A. (1992), *National Systems of Innovation*, London: Pinter Publishers.

Maggioni M., M. Nosvelli and E. Uberti (2006), 'Space vs. networks in the geography of innovation: a European analysis', Paper presented at the ADRES Conference on networks and innovation and spatial analysis of knowledge diffusion, Saint Etienne.

Morrison, A. and R. Rabellotti (2005), 'Knowledge and information networks: evidence from an Italian wine local system', CESPRI Working Papers.

Nelson, R. (1993), *National Innovation Systems. A Comparative Analysis*, Oxford: Oxford University Press.

Ouimet, M., R. Landry and N. Amara (2004), 'Network positions and radical innovation: a social network analysis of the Quebec optics and photonics cluster', Paper presented at the DRUID Summer Conference 2004 on Industrial dynamics, innovation and development, Elsinore, Denmark.

Ozman, M. (2006), 'Networks and innovation. Survey of empirical literature', BETA Working Paper, 7, University of Strasbourg.

Ponds R., F. van Oort and K. Frenken (2007), 'Interregional collaboration networks and regional innovation', Paper presented at the 47th Congress of the European Regional Science Association, Paris.

Scott, J. (2000), *Social Network Analysis. A Handbook*, Sage.

Spencer, J.W. (2003), 'Global gatekeeping, representation, and network structure: a longitudinal analysis of regional and global knowledge-diffusion networks', Working Paper, George Washington University.

Szerb, L. and A. Varga (2006), 'The innovation capacity of the small and medium sized enterprises in the South-Transdanubian region of Hungary and the research and innovation transfer potential of the university of Pécs', Final research report: Business potential of R&D activities in the university environment and their transfer to SMEs in the Cross-Border Region (CrosboR&D) project, SL-HU-CR/05/4012-106/2004/01/HU-12 (Slovenia, Croatia, Hungary).

Valente, W.V. (1995), *Network Models of the Diffusion of Innovations*, Cresskill, NJ: Hampton Press.

Varga, A. (2000), 'Local academic knowledge spillovers and the concentration of economic activity', *Journal of Regional Science*, 40, 289-309.

Varga, A. (2001), 'Universities and regional economic development: does agglomeration matter?', in B. Johansson, C. Karlsson and R. Stough (eds), *Theories of Endogenous Regional Growth – Lessons for Regional Policies*, Berlin: Springer, pp. 345-367.

PART THREE


Knowledge transfer mechanisms: academic
entrepreneurship and graduate mobility

# 8. Academic entrepreneurs: critical issues and lessons for Europe

## Chiara Franzoni and Francesco Lissoni

### 8.1 INTRODUCTION

This chapter surveys the notion of 'academic entrepreneur', as it emerges from a wide range of contributions to the economics and sociology of science. Insights from those contributions are then used to examine critically the most recent literature on academic spin-offs and university-industry technology transfer.

The chapter proceeds in a cumulative fashion. We start first with the rhetorical device of putting forward a 'straightforward definition' of academic entrepreneurship, one which is most intuitive and at the same time traceable in many recent policy initiatives, both in the US and in Europe (section 8.2).

We then move on to survey the socio-economic literature dealing with the notion of 'entrepreneurship' in academic research. We suggest that contemporary science is the result of an 'entrepreneurial' effort, undertaken both by individual scientists and by the academic institutions that host them. The intensity and specific features of the entrepreneurial effort depend very much on the institutional characteristics of national academic systems, which we outline by looking briefly at the history of the US and French systems, the latter taken as the extreme example of the European case (section 8.3).

In section 8.4 we examine the recent literature on spin-off firm creation, and briefly touch upon some related issues on intellectual property rights over academic research results. We suggest that both patenting and spin-off creation result from the broad entrepreneurial agendas described in section 8.3, and not merely from the individual scientists' profit-seeking attitudes.

In section 8.5 we propose several policy implications and directions for future research.

## 8.2 ACADEMIC ENTREPRENEURS:
## THE 'STRAIGHTFORWARD' DEFINITION

At first glance, the definition of academic entrepreneur (AE) looks straightforward: the AE is a university scientist, most often a professor, sometimes a PhD student or a post-doc researcher, who sets up a business company in order to commercialize the results of her research. It is the nearest possible definition to the classical one of entrepreneur,[1] enriched of the qualifying adjective 'academic', to stress that the innovations introduced by the entrepreneur originate from the research she conducted as part of her 'other job' as a university scientist.

This straightforward definition cannot but please the policymaker, as it attributes to this special breed of entrepreneur, who already benefits the society through innovation and job creation, an additional social function: the valorization of academic research, often funded by the public purse and targeted to fundamental objectives. And whenever the university administration supports the new company through equity participation, the successful AE will also deserve praise for having contributed to the financial health of her institution.

The straightforward definition is also easy to conceptualize and popularize, as it fits closely a linear model of university-industry and science-technology relationships, with the university being largely in charge of producing science, and the industry being responsible for processing it as an intermediate input to technology.[2]

According to this model, inventions of high commercial value follow inevitably from 'pure' or 'basic' research, the only problem being the uncertain timing. At worst, the flow from science to technology may be subject to interruption if nobody undertakes the applied research and development efforts necessary to turn the inventions into viable innovations.

In this view, any scientist whose shelves are full of prototypes and proofs of concepts awaiting commercial development is a potential AE, who simply lacks adequate economic incentives and/or is refrained from venturing into business by the 'Ivory Tower' culture of academia, one which condemns commercialization activities (and therefore makes economic incentives irrelevant). In order to straighten up the incentives, the policymaker is then called to establish a clear IPR regime over the results of public funded research (by assigning them either to the scientist or her university, or to any private partner joining the research project) and to promote a cultural change among scientists. A clear definition and attribution of IPRs will in turn create a market for university-based inventions, whose development will be paid for by private investors, in exchange for exclusive licensing rights.

The influence of this perspective on European policymakers, both at the national and at the Union level, is witnessed by the large number of IPR reforms concerning universities that have been introduced all over the Old Continent in the past ten years or so. Such reforms have been most often modelled upon earlier US pieces of legislation, such the Stevenson-Wydler Act and the Bayh-Dole Act of 1980, both of them inspired by the wish to create a market for inventions derived from public science.[3]

As for the creation of an entrepreneurial culture among university scientists, training courses in business and law for scientists and technology transfer officers have proliferated throughout Europe. In a similar vein, it is now common for public research funding agencies to require all applicants to outline a clear 'exploitation plan' for the promised results.[4]

More generally, the repeated statements about the existence of a supposed 'European paradox' (according to which many EU countries would be holding prominent worldwide positions in terms of scientific achievements, but wouldn't be able to 'translate' them into technological advantages) clearly remind us of the 'shelved-inventions' metaphor, and the straightforward definition of AE.[5]

Unfortunately, both the shelved-inventions metaphor and the straightforward definition of AE do not take into full account the complexity of the system of economic incentives that affects academic scientists' behaviour. The linear model of science-technology interaction that inspires both the shelved-inventions metaphor and the straightforward definition of AE has been heavily criticized by economists and sociologists alike. These critics suggest that scientific advancements which are susceptible of practical applications do not merely drop onto technology from above, but are elicited and made possible by the latter, whose autonomous progress poses challenging research questions, produces data, allows for new experimental settings, and creates scientific instruments.[6]

As for the incentives, the straightforward definition of AE represents scientists as individuals free of any contractual engagement (i.e. dedicated to 'pure research' for the sake of advancing knowledge), who must decide point-blank whether or not to activate a contract with a business company (either by licensing their inventions or by taking an equity position in a start-up). Alas, contemporary academic scientists are far from free from contractual obligations: they are employees subject to the control either of the State or of their universities, and linked to other faculty members and students by a number of formal and informal obligations. In addition, they have long-term career plans which have to be taken care of in order to be successful.

In synthesis, the straightforward definition of AE sweeps under the carpet too many details of the contemporary features of the economics of academic science. In order to take in full account the complexity of the phenomenon, academic research should rather be conceived as a 'scientific enterprise', in which career-motivated scientists act as research-oriented entrepreneurs, whose approach to commercial activities depends upon a broader career strategy.

## 8.3  ENTREPRENEURSHIP IN SCIENCE AND ACADEMIA: A BROADER VIEW

Entrepreneurship is quite a popular word in a number of studies dealing with the philosophy, sociology, economics, and history of science. Far from being occasional and inconsistent, its use points to well-defined and historical features of contemporary science.[7]

### 8.3.1  Entrepreneurship as an Individual Feature: the Sociology of Scientist-Entrepreneurs

The contemporary sociology and economics of science describe the organization of scientific research, especially in experimental sciences, as necessarily entrepreneurial. Scientists at the head of large laboratories perform a number of activities which are typical of the modern entrepreneur, such as setting up and managing increasingly complex organizations, and providing them with adequate funding and human capital. More generally, scientists with innovative research agendas have to broker relationship with agents outside the universities (especially policymakers and industrialists) looking for political and material support for that agenda. As for risk-taking, a typical trait of entrepreneurs as defined by many economic theories, innovative scientists venture outside the boundaries of established disciplines or research lines, looking for scientific breakthrough that could earn them fame, but also taking a high risk of not getting any result.

The starting point of many recent essays is Robert Merton's portrait of academic scientists as individuals engaged in careers based upon peers' recognition of their contributions to the advancement of knowledge

(Merton, 1973). In Merton's view, such recognition primarily takes the form of acknowledging the scientist's priority claims of having made an important discovery. Philosophers and economists of science have gone a long way in exploring how the quest for priority may shape social relationships in science, and have reinforced the notion that being credited with one or more 'discoveries' (through the mechanism of bibliographic citations and, possibly, eponymy) is essential to a scientist's career (Kitcher, 1993; Dasgupta and David, 1994).

A more complex view of how scientists manage their careers according to entrepreneurial criteria comes from the sociological tradition of 'science studies' (Callon, 2002), and a number of related contributions to the history of science and technology (Latour, 1988; Lenoir, 1997).[8]

This literature explores in greater depth the relational aspect of the scientific enterprise. Scientific facts are not merely 'discovered' by the first scientist who solves a theoretical puzzle or creates an innovative experimental routine (and thus wins the priority race). Rather, they are established laboriously by obtaining social consensus on the relevance of the topic, on the legitimacy of the theoretical assumptions, and on the solidity of experimental routines. Such consensus has to be gained both from fellow scientists (especially within one's own disciplinary field) and from other relevant actors, such as businessmen and policymakers.

Fellow scientists can validate the contents of a scientific paper or programme by citing it as a legitimate source of information, or they can condemn those contents by neglecting the paper as irrelevant or poorly conceived. Their consensus has to be elicited either by indirect means (e.g. by choosing the best publication outlet or through a perfunctory use of paper citations) or by more direct ones, by establishing social ties through research co-operation, conference invitations, and joint lobbying for economic resources from state and industry.

In this respect, businessmen and policymakers can be instrumental in providing funds, data, scientific materials and instruments, as well as ethical validation. Participation in science policy forums, policy and ethical committees, and scientific boards of large companies are all necessary activities for senior scientists to support the activity of their laboratories.

If seen within this context, IPR management, consulting, and equity participation to spin-off companies are not simply market and market-like activities which take time away from research, but indeed necessary steps, conditional to the scientific entrepreneur's chief goal of setting up or expanding her lab, and promoting her academic career (OECD, 1999, p. 37).

Faced with this rich analysis, the straightforward definition of AE proves to be inadequate. Individual scientists who engage in ambitious research programmes need resources to pursue their objectives, and nurture actively extra-academic contacts to that end. Ease of patenting and access to resources for setting up a company are welcome insofar as they are instrumental in widening and thickening the scientist's network. They will be disregarded if they do not fit in the research agenda. At the same time, more research funds, or more career opportunities, although totally unrelated to any technology transfer target or firm creation objective, may naturally push more scientists to pursue the ambition of setting up their own laboratory; and it may well be that, by doing so, those scientists will reach out of the academic walls anyway.

Lenoir's (1997) portrait of German physiologists, physicists, and chemical scientists in the nineteenth century confirms this view. In particular, Lenoir compares scientists engaged in academic careers within the boundaries of established disciplines with those whose research agenda foresaw the birth of a new discipline, or required disciplinary boundaries to be redrawn, either to allow for interdisciplinary work or to establish new hierarchies between disciplines.

Lenoir's scientist-entrepreneur first aims at acquiring superior skills and technical expertise in handling complex experimental procedures and equipment, so that other scientists will find it hard to disprove his experimental results, and will require his approval or help to validate their own findings. Then the scientist-entrepreneur will promote a wider agenda, which aims at proving the social benefits bestowed by the new disciplinary programme for society. Practical applications of the new scientific discipline (were it nineteenth century organic chemistry or twentieth century nuclear physics) are proved through patenting, licensing, consulting, and the encouragement of start-ups by young colleagues and students.[9]

Summing up, academic entrepreneurship, as part of the more general phenomenon of scientific entrepreneurship, proves to result from a more complex bundle of strategies and incentives than envisaged by the straightforward definition.

### 8.3.2 Academic Entrepreneurship as an Institutional Feature: US vs. European Universities

One of the best-known papers on entrepreneurship in academia is Henry Etzkowitz's (1983) essay on 'Entrepreneurial scientists and entrepreneurial universities in American academic science'. The 'entrepreneurial university' is there portrayed as the outcome of a revolutionary process started in the US with the Big Science programmes launched in the aftermath of World War II. In a later paper, Etzkowitz (2003) suggests that, slowly but inevitably, European research-oriented universities will leave room to profit motives for their research, and turn into entrepreneurial ones as their overseas counterparts.

At a closer look, however, the American model of 'entrepreneurial university' appears to be rooted much more deeply in the gradual evolution of US universities from teaching colleges of divinity and liberal arts to modern research institutions. By contrast, many contemporary efforts to promote entrepreneurial attitudes in universities outside the US are at odds with an institutional history of central planning and control.

The US university system has been, since its early boom in the first half of the nineteenth century, a heterogeneous collection of a large number of autonomous institutions cherished by their local communities or by religious groups and individual philanthropists (Rudolph, [1962] 1990). Their faculty members were neither subject to their students' control (as in Italian medieval universities) nor ever served as civil servants paid by the state, as in most European countries. Since their inception, the president and board of trustees of US colleges exercised a degree of local control which federal and state governments never managed to overcome. Attempts to centralize the university system have always been overthrown (both by the oldest private colleges and the more recent state universities), even at times when financial distress could have advised otherwise (Trow, 2003). Nowadays, autonomy is one of the greatest strengths of the US universities, and this is also the main background reason for their transformation into entrepreneurial organizations.

Since right after World War II, and well through the 1960s, the extraordinary success achieved by basic science applications to military technology legitimized the well-known exponential increase of federal funding of academic research. However, Vannevar Bush's famous report 'Science, the Endless Frontier', while convincingly making the case for large public funding of research universities, failed to persuade US lawmakers of the need to set up a centralized body for the administration of all funds (Graham and Diamond, 1997).

A number of concurrent institutions still provide research grants in various fields: the National Institute of Health, the National Science Foundation, the Ministry of Defence, and other governmental bodies run projects both in separate scientific fields and in a few overlapping ones. The possibility to be financed, in certain fields, by different agencies has helped to keep alive a healthy heterogeneity of research targets (also within the same scientific field) and administration models.

All of these programmes rely on the so-called principal investigator (PI) principle, by which individual scientists (not their departments or their institutions) are made entirely responsible for a project. A strong individual research record is instrumental for the PI to win the grant in order to set up or expand her own laboratory or research group. Individual scholarship, and not any political objective of equal distribution of resources, becomes the key allocation criterion for research funding. As a consequence, universities have always engaged in a race to recruit the most talented scientists, whose contribution is decisive to get public funds.

At the same time, individual scientists engage in self-promotion activities leading to winning and then managing the grants. They devote their effort not only to publishing, attending conferences and scientific meetings, but also to establishing relationships with one or more funding agencies, aiming at influencing the choice of the research topics to be funded, as well as networking at the academic level for recruiting brilliant young scientists to the ever-increasing needs of their laboratories. As employees of their universities, and not of the federal government or of the individual states, the academic scientists are let free and possibly encouraged to engage in these typical entrepreneurial activities, as long as this enhances their university's reputation and financial health. As a result, the US academic system has witnessed in recent years an overall tendency of research teams to increase in size and complexity in all scientific fields, albeit at different rates (Adams et al., 2005).

Etzkowitz (1983) describes this pattern as one of diffusion of 'quasi-firms' (laboratories and research groups), whose survival and expansions depend upon chasing and managing funds, recruiting skilled employees, delivering results, and moving up to higher level funding agencies.[10] PIs provide the necessary entrepreneurial efforts and skills to do the job, in exchange for a large bite of the credit for the success of the scientific enterprise. Stephan and Levin (2002) offer a similar view.

One cannot fail to see here a strong parallel with the redefinition of academic entrepreneurship we have proposed in section 8.3.1. The parallel extends from individual scientists to academic institutions, to the extent that the latter are also engaged in a competitive effort to establish new research lines and disciplines, to solicit funds from both industry and the central governments, and to attract the best scientists for those purposes.

This is a far cry from the way academic research is organized in Europe. The most striking differences in institutional settings between US and European academic systems are well exemplified by French universities, whose history is one of abrupt termination and slow re-creation under tight centralized control.

French medieval universities were abolished by the Revolution, as part of the wider effort to reduce the influence of the Catholic church in education. The task of educating the technical and administrative elites was then assigned to the so-called Grandes Ecoles modelled after the Ecole des Ponts Chaussées, founded in 1775 by the king.[11]

Later on, with the end of the Republic and the creation of the Empire, a brand new institution was set up in between 1806 and 1808, charged with the task of educating a new generation of teachers, lawyers, medical doctors, and the ranks and files of public administration: the Imperial University. Its lecturers were asked to act as civil servants, organized along rigid disciplinary lines and within regional faculties under the State's control. It was not until 1896 that the regional faculties were transformed into local universities, and not until the 1970s that they gained a substantial degree of organizational (but not yet financial) autonomy (Neave, 1993). Still nowadays, the entire process of recruitment occurs at the national level, and the mobility of academic staff across universities is very limited.

For a long time, French universities were devoted only to teaching. Research tasks were assigned to specialized institutes, often founded around a new discipline, such as the Institute Pasteur (1887), or under the direct control of a ministry, as in the case of various agricultural agencies. In 1939 the National Centre for Research (CNRS) was founded. Still well into the 1990s, CNRS employed over 14 000 full-time researchers, and even more were employed by the other PROs, as opposed to 45 000 university professors, whose research engagement was, at best, on a part-time basis.

Although since the late 1980s more and more CNRS labs have been moved within academic walls, a vertical hierarchy of labs exist, starting from those staffed exclusively by CNRS personnel (funded directly by CNRS and the Ministry of Education), followed by those staffed both by CNRS and university personnel, and down to those staffed entirely by university faculty, with no access to CNRS funds (Larédo and Mustar, 2001; Neave, 1993).[12]

In recent times, French policymakers have borrowed heavily from the straightforward definition of AE we outlined in section 8.2, but at the same time they have been unwilling to allow for more autonomy of both the universities (which still cannot manage freely their personnel, real estate, and finance) and their researchers (whose contacts with industry are regulated in great detail).[13] While US policies on intellectual property rights and academic spin-offs are often imitated, very little is retained of the lessons derived from the long US history of generous support to fundamental research, faculty mobility, and university autonomy, nor from the role these features play both in promoting technology transfer and in shaping scientific entrepreneurship.

Rigidities such as those described for France are common throughout continental Europe, starting with large countries such as Germany and Italy (Clark, 1993; Romano, 1998; Jong, 2007). Here, as in France, a stark contrast exists between policy measures undertaken to encourage the commercialization of academic research activities, and the widespread reluctance to give more autonomy to universities. In other words, while the straightforward definition of AE seems to be highly popular, no room of action is given to entrepreneurial scientists and universities.

## 8.4 ACADEMIC ENTREPRENEURSHIP AND FIRM CREATION

In this section we review the specific literature on AEs' contribution to firm creation, through the lenses of our broader definition of entrepreneurship in science. The literature we examine comes by and large from the US, where the debate on university-technology transfer has revolved around the evaluation of the effects of the Bayh-Dole Act and related increase in university patenting; as such, it focuses on the commercialization of patented research results, either through licensing or firm creation or both.

Space constraints force us to avoid discussing the phenomenon of university patenting in depth (for surveys, see Mowery et al., 2001; and OECD, 2003). As an introduction to this section, however, it is worth mentioning that the number of patents taken over academic research results has been growing incessantly over the last 20 years, both in the US and in Europe. Those 'academic patents' account for no less than 4 percent of total domestic patents in the US, and similar figures have been estimated for France, Italy, Sweden, Finland and Norway.[14] In science-based technologies such as Biotech, percentages can easily climb well over 15 percent.

However, the US and Europe differ in the attribution of property over academic patents. While more than 60 percent of such patents in the US are owned by universities, in Europe the same percentage is around 10 percent. Conversely, over 60 percent of European academic patents are owned by business companies, while the same percentage for the US is estimated at no more than 25 percent. For an explanation of these figures, which owe very much to the institutional features of academic systems as described in section 8.3, see Lissoni et al. (2008). Here it suffices to say that, at least until recently, the issue of commercializing academic patents was by and large felt only by US university administrators, their European counterparts having solved the problem by leaving all IPRs in their professors' hands, and from those hands into business companies' hands.

US universities' tradition of patent management is not recent, and certainly dates back to before the Bayh-Dole Act. However, until the 1980s, management practices essentially were reduced to patent licensing, either directly or via specialized technology brokers such as the Research Corporation or WARF, the Wisconsin Alumni Research Foundation (Apple, 1989; Mowery and Sampat, 2001).

But with the university patent explosion of the 1980s and 1990s, it soon became clear that in many cases the licensing of potentially valuable patents was not easily achievable for a number of reasons (Jensen et al., 2003; Thursby et al., 2001). First, since in many cases academic inventions were disclosed at a proof-of-concept stage, it was hard to convince a firm to take on the long and risky development work needed to bring a final product to the market. Second, in many cases, this work could not be effectively done by an external firm alone, because the tacit and know-how dimension of the knowledge involved was too high. Third, many of the most promising cutting-edge and disrupting technologies are of no interest to large incumbents and would make a good investment only for venture capital and high-risk equity markets.

At the same time, with the development of biotech companies in the US, several successful examples of superstar scientists that had raised huge amounts of capital in the market by selling the equity of their start-ups were impressing public opinion, and seemed to suggest that academia and industry could join their effort to leverage a new generation of high-tech companies, characterized by a strong research focus.

Business angels and venture capitalists started to knock on the universities' doors, in search not only of promising business ideas, but also of qualified consultants' and peers' opinions to evaluate and manage the strategic choices of their biotech portfolios.

These new opportunities were received favourably by university administrators, who soon adapted their regulations to allow giving equity capital and branding to start-ups, and to ensure job security and institutionalized temporary leave to professors on 'entrepreneurial duties'. Many technology managers saw academic spin-offs as a sort of advanced solution to technology transfer that would help in finding viable commercialization strategies to growing patent portfolios (Franklin et al., 2001).

### 8.4.1 Early Studies: Academic Knowledge as a Non-Tradeable Asset

Since the beginning of the 1990s, many scholars, especially in the US, have investigated the individual motivations and the rationale behind the claim for a proactive role of university-based scientists in the generation of new high-technology applications for nascent industries.

Early contributions tended to stress that university-based scientists own a specific set of knowledge and information, enabling them to spot valuable opportunities of investment, which would remain hidden to other people. Hence a scientist may have a comparative advantage *vis-à-vis* other potential entrepreneurs in the recognition of promising businesses, thanks to the idiosyncratic knowledge gained while working on a scientific discovery.

This view was supported by several pieces of empirical evidence, especially with regard to emerging high-tech industries. For instance, Zucker and Darby (1996) suggested that the most successful biotech companies were co-publishing with university professors and showed that their commercial success, in terms of the number of products developed and commercialized, was positively associated with the scientific eminence of researchers participating in the scientific board and holding equity stakes. In a later study, co-publications were also shown to explain a firm's patent citations rate, suggesting the idea that a stronger technological base would produce higher quality patent applications in fields characterized by a high strategic value of IPR assets (Zucker et al., 1998). Shane and Stuart (2002) studied the probability of success of 134 new ventures exploiting MIT inventions and found that both the academic rank of the inventor and the number of MIT patents in the company portfolio were likely to increase the probability of an IPO and decrease the failure rate.

The attention of early studies was especially focused upon the growing US biotechnology industry and on its innovative potential, as compared to more traditional drug industry and market incumbents. Certainly the idiosyncratic features of that industry, one wherein scientific results are often immediately suggestive of commercial applications (very much in the spirit of the linear model), make any generalization hard and warn against placing too much emphasis on early results from the literature.

With regard to the choices on the structure of ownership, those contributions stressed that cutting-edge science is naturally attached to individuals and, because of the poor absorptive capacity of the environment, transfer could not occur through simple licensing, but required aligning the professor's remuneration to the success of the venture (Audretsch, 1995; Audretsch and Stephan, 1999). This seemed particularly the case of newly created firms, which can be shaped around the emerging scientific culture and may be better suited to the exploitation of new and radical technologies (Henderson, 1993).

In such a context, because the intellectual capital was seen as the true key asset, the founding of a firm looked like a unique means for the scientist to extract private gains from her idiosyncratic knowledge. Additionally, since the diffusion of this knowledge is naturally bounded by face-to-face interactions, the literature foresaw a lesser need to engage in enforcement and protection of IPRs (Audretsch, 1995). Hence, the mantra went that the best scientists enjoy both a superior access to high-value knowledge and a stronger natural excludability; leading to higher-value entrepreneurial opportunities in the selection phase and sustainable competitive advantages later on (Zucker et al., 1998).

Besides, in highly incomplete informational contexts, the scientific reputation of the academic entrepreneur, or the rank of the related institution could have been used by the stakeholders as an indirect signal of the high prospective value of the venture (Stuart and Ding, 2004; Shane and Khurana, 2003). In the absence of more accurate information, a researcher's eminence could serve to proxy the strength of a start-up company's technological base whereas the star scientist's research specialties would signal the future technology strategies that the company would have undertaken (Audretsch and Stephan, 1996). In a study of biotechnology IPOs, Stephan and Everhart (1998) found that the amount of funds raised and the initial stock evaluation of firms were positively associated with the reputation of the university-based scientist associated with the firm. *Ceteris paribus*, Di Gregorio and Shane (2003) found that spin-off companies from top universities were more likely to attract venture capitals than those from less prestigious institutions, whereas Franklin et al. (2001), in a survey of key competitive factors conducted among UK technology managers, reported that the researcher's reputation was ranked immediately after their scientific preparation and that this was especially true for higher performing and more experienced universities.

### 8.4.2 Incentive Problems Rediscovered

Following this line of thought, at the beginning of the 1990s most academic administrations, technology managers and venture capitalists were especially stressing the technical content of university applications, which they expected to be more radical and broader in scope than innovations with purely industrial backgrounds. Nevertheless, the emphasis on the knowledge capital and on the alleged superior technological endowments eventually faded at the end of the 1990s, when broader studies reported mixed evidence. For instance, Nerkar and Shane (2003) found that the top technological level of MIT start-ups reduced failure rates only in low-concentration industries. The same study also re-established the importance of industry differences in terms of patent effectiveness and appropriability regimes in explaining venture success.

In the meantime, with the help of policymakers, an increasing number of universities had invested in (often unprofitable) technology transfer activities (Thursby and Thursby, 2002). As a consequence, doubts emerged on whether the importance of firm creation from academia had been possibly over-emphasized, possibly beyond any true economic advisability, both in terms of economic gains and of professors' intentions, which brought into play an entirely new set of problems.

As soon as the profit started to become a concern of universities at the institutional level, technology managers discovered that a good technological endowment or the expectations of business profits were, in many cases, not enough to justify or to convince a scientist to take part in a venture, as ultimately entrepreneurship also meant risk-taking, a strategic vision and possibly a life change. Indeed 'entrepreneurial-type' scientists, in the straightforward definition of AE we proposed above, were hard to find. A considerable mismatch of objectives between faculties, technology managers and investors was affecting transactions (Siegel et al., 2003).

Despite their technological strengths, newborn firms were frequently reported to be unsuccessful because of a failure in complying with the market needs. Field studies and extensive interviews to technology managers portray scientists as individuals with a good taste for science, but with relatively naive ideas about the pursuit of market goals (Thursby and Thursby, 2003c).

The knowledge-endowment argument and its related theory of entrepreneurship hence lost much of their appeal, as a stronger trade-off between scientific and market concerns was brought back to the forefront of analysis.

What falls down in the straightforward notion of AE applied to spin-off policies and strategies is not the capacity of scientists to offer a valuable pool of technological opportunities to market investors, or really the role of the 'knowledge entrepreneur' in chasing market opportunities. Rather, the focus is shifted towards the alignment of a scientist's objectives to the goals of a nascent firm, where the expected gains of a scientific entrepreneur are seen not only as profit in the case of firm success, but also come in the form of increased availability of funds for complementary research.

To scientists concerned with their academic careers, research funds made available through the firm's R&D activities may be particularly appealing in so far as they may serve to buy instruments and data, hire additional personnel, pay for travel to conferences, and generally enlarge the professor's budget for research. Hence, the decision to start up a company would depend in good part on the researcher's expectations of engaging in stimulating, fruitful and possibly generously funded development activities, rather than on expectations of profit and growth, especially when she is not required to put a big share of the equity upfront.[15]

In addition, because the gains from big research budgets vary with the different stages of a career, the propensity of faculty members to engage in interchanges with industry was seen to be also dependent on lifecycle effects and on the choices of investigative pathways (Thursby and Thursby, 2003a). Whenever the contiguity of scientific and industrial effort faded, monetary incentives should be raised to compensate for the time taken by purely commercial activities with an unclear effect on the academic career (Thursby and Thursby, 2003b).

The idea that, in many cases, market goals as such simply fail to produce a set of incentives compatible with the day-to-day life of the entrepreneurial scientists has been commented on in many surveys. Jensen et al. (2003) report that scientists may voluntarily retain disclosures of potentially marketable technologies and suggest that the opportunity cost of development activities was stronger for higher quality scientists, whose inventions arise typically at a very embryonic stage. Franklin et al. (2001) report that technology managers indeed regard the academic founders of their spin-off companies as entrepreneurial individuals with good commitment on the research projects, but they signal a stronger mismatch of perceived goals as the most common cause of venture failure.

The researcher's attitude towards pure scientific investigation, the privilege of having her own lab and enlarging her group of graduate students frequently clashes against reward schemes based upon commercialization. Not surprisingly, many scholars report that the problem arises most often when the development stage is nearly completed and the firm has to promote a general shift of goals towards the industrialization of the product and/or to cope with marketing and financial pressures (Shane, 2004; Vohora et al., 2004). It is at that stage that financial constraints challenge the availability of funds for further development and laboratory work and the appeal of having sponsored additional research fades.

In the follow-up of a survey conducted on 62 US universities in 1990s, Jensen et al. (2003) describe the relationship linking university administration, technology managers and individual scientists as an agent-principal game-theoretic model. Scientists are seen as positively reacting to both monetary incentives, and to the share of sponsored research they may obtain for their labs, but, because high quality faculties would disclose inventions at a more embryonic stage, willingness to disclose would depend more substantially on the latter than on the former.

Besides, the opportunity costs faced by scientists would not just depend on exogenous preferences and personal interests, but also on the availability of other funds, on other appointments and on purely life-cycle effects. In this respect, older scientists may be more willing to cash in the market gains of their knowledge assets than their younger colleagues because they have already achieved the highest academic ranks (Audretsch and Stephan, 1996). This could also be the case for professors of continental European countries, where the academic environment is characterized by lower competition and by job security. For instance, Audretsch (2000) found that the probability for an individual scientist to create a private venture is higher for older professors, suggesting the idea that academic entrepreneurship becomes a more viable option when career pressures have cooled down and the scientist has coped with the concern of establishing her scientific position in academia. This can be especially true within the contexts in which social rules discourage for-profit activities, in which case, only older and highly reputed scientists may dare to undergo non-traditional academic pathways (Stuart and Ding, 2004). For younger scientists, as newly qualified PhD students and research assistants, the founding of a venture may rather become appealing as a viable strategy to exit academia (Franklin et al., 2001; Roberts, 1991).

### 8.4.3 Business Creation vs. Patent Licensing: Do we Really Need Academic Spin-offs?

Although university patents, spin-off company creation, consulting and joint research agreements are often addressed as separate, alternative transfer mechanisms, in practice, commercializing a piece of university research may require a variable mix of all those instruments. For instance, in a recent survey on commercialization of US academic research, it emerged that licensing contracts made by technology transfer offices in the majority of cases involve royalties, annual fees, equity, milestones and consulting agreements (Thursby et al., 2005). The question of what instrument is best suited to transfer different pieces of knowledge has been the focus of many recent contributions. The central argument is that the market inefficiencies in the transfer of knowledge can be corrected by involving in the ownership structure (with some risk-taking positions) the party that possesses the most idiosyncratic assets, as suggested by the 'straightforward' notion of AE.

Because scientists' knowledge is characterized by natural excludability, it resists codification in a fully transmittable form and tends to stick to individuals, even after a patent has been filed or an article published. At the same time, many academic inventions are no more than a proof of concept at the frontier of knowledge. It follows that, in order to take up the nutshell technology and undertake the final development stage on their own, firms need to recruit the scientist as a partner or stakeholder: in the absence of her personal involvement, they would not be able to profit from the innovation (Jensen and Thursby, 2001). Therefore, as academic scientists face a stronger need of becoming entrepreneurs, the higher is the degree of sophistication of their technology compared to that of the outside business world (Shane, 2004).

Besides, the decision of whether or not the exploitation of a technology is best achieved by patent licensing or by a start-up depends on the technological regime and on the appropriability of the innovation. In low-appropriability patent regimes, licensing may be hard and innovations may not be commercialized because of a lack of incentives, but if the knowledge is also characterized by natural excludability, the creation of a company exploiting a scientist's idiosyncratic knowledge may become the only viable transfer option (Shane, 2004).

Some empirical evidence in support of this thesis has been provided both in case studies and empirical analyses. Shane (2001b, 2002) found that the probability of an MIT invention resulting in a patent application was higher in strong appropriability regimes. In a related study he also found that the spin-off rate increased with the novelty and importance of the technology behind it (Shane, 2001a).

In a study of the technology transfer activities at University of California, Lowe (2002) found that patents characterized by a stronger scientific base and a higher degree of tacitness were significantly more likely to be licensed to their original inventors, thus supporting the idea that spin-off creation is necessary when the scientist's knowledge is highly uncodified and idiosyncratic.

Finally, Feldman et al. (2002) report that the willingness of US universities to take up equity in a new venture was generally higher among longer-experienced technology offices, which suggests that the equity positions of university administrations may offer a second-best solution to the problem of achieving higher transfer of knowledge to the market, one that perhaps involves a lower risk of diverting good scientists from their original tasks.

### 8.4.4 (Intended and Unintended) Consequences of Academic Entrepreneurship

The argument that academic entrepreneurship may do a non-replaceable job in fostering the emergence of new generations of high-tech firms and the renovation of local economic systems has been widely popularized by policymakers in many European countries. In addition, common arguments in favour of academic spin-off creation normally emphasize that the core attitude of a spin-off company for experimentation would resist the start-up phase and result in a superior propensity of the firm to deliver continuous innovation later on.

However, if one considers the widespread consensus on those claims, it comes as a surprise that little assessment has been undertaken so far on the actual performances and contributions of academic venturing to technological change and local development. As we look at the empirical literature, even notwithstanding the problem of the reliability of field analyses in the absence of a clear-cut definition of academic spin-off (Pirnay et al., 2002), we have little more than anecdotes on success stories of university-based inventions that were incorporated into a firm, developed a successful application, grew big and eventually clustered other firms (see Roberts, 1991). Research on the biotechnology sector, which we have mentioned in section 8.4.2, suggests that the presence of a scientist has a positive effect over start-up success. Nevertheless, those results have hardly been extended to different industries (Nerkar and Shane, 2003) and to institutional frameworks other than the US.

When it comes to appreciating the actual contribution of academic ventures, only some very preliminary evidence is available that proves the supposed higher performances of spin-off companies either in terms of innovativeness, or in terms of employment created and new product developed and sold. Mustar (1997) reports that the R&D intensity of French academic spin-offs was higher than that of other new-technology-based start-ups. Similar results were found for samples of UK firms.[16]

Perhaps some stronger, though highly industry-specific, evidence has been provided in support of the claim that companies founded by academic personnel were likely to locate around universities (Audretsch and Stephan, 1996; Zucker et al., 1998). This can somehow be a desirable feature from the point of view of policymakers, concerned with fostering economic development locally, and for university administrators alike, to the extent that spin-off companies may serve as good partners for joint research and technology licensing later on. The clustering choices observed in many research spin-offs may reflect the initial need for part-time scientists to locate close to their academic jobs and to a hard-science environment, in order to comply with their multi-task careers (Audretsch and Stephan, 1996). However, it is dangerous to push this observation further and take it as a confirmation of a higher-than-average focus on innovation and high-technology content of spin-off firms (see Shane, 2004), as ultimately location entails a strong path-dependent component.

Overall, as we look at successful case studies, it is worth asking how these really benefited from their origin in an academic environment: Did they enjoy access to the unique knowledge offered by universities? Or did they merely benefit from the support given in terms of credibility and networking in a context of jeopardized information? Or, finally, did they simply gain from being close to good training sites and qualified scientific consulting markets?

This is a question worth asking because spin-off activities also bring several downsides and costs, even beyond the general costs and risk of the investments.

Major opportunity costs faced by university administrators, irrespective of their civil service mission, are at least of two kinds. First, universities may lose good scientists or may simply divert them from high quality publications and teaching. Second, at a more fundamental level, they may be afraid of losing their long-lasting reputation of reliable and non-opportunistic agents, which is fundamental to their ability to act as a broker for the market of technology, as well as for their more traditional goals. This concern seems to have been understated more in the literature than in practice. For instance, some institutions, such as the University of Cambridge (UK), although proactive in business creation, refuse to commit their commercialization activities to a purely profit-oriented mission and describe their role as one of facilitators in the diffusion of knowledge for the benefit of society. In practice, concerns have been expressed that professors may use students as low-paid employees and indiscriminately re-sell the effort of collective commitments. Shane (2004) reported that, in order to cope with the problem of moral hazard, many US faculties have also introduced a general prohibition for scientists to work at the same research project both in their internal unit and in their external private ventures, after a person died at the University of Pennsylvania Medical School during the test of a therapy developed by an academic spin-off.

## 8.5 CONCLUSIONS

Academic entrepreneurs who are active in patenting, firm-founding, and more generally in technology transfer, come disproportionably from the ranks of scientific entrepreneurs with a brilliant scientific record, possibly oriented to fundamental research. These scientists' economic agenda is centred upon entrepreneurial efforts within the university, aimed at gaining reputation through discipline building, creation and management of laboratories and research teams, and an appetite for the economic resources necessary to pursue those goals.

To those scientists, patent licensing and spin-off creation are appealing not just because of the expectation of profits, but also because they offer valuable opportunities to enlarge their sphere of influence, to empower their internal and external consensus, and inflate the budgets available for their research. Hence, any wise policy of technology transfer in academia should move from a broad consideration of the overall personal incentives faced by scientists and framed within the context of academic careers.

The complexity of academic scientists' incentives to commercialize their discoveries suggests an immediate policy conclusion, albeit a speculative one (at this stage of research): the two objectives of promoting academic entrepreneurship and restraining public expenditures for academic science (which are often found to go hand in hand in

Europe) are largely incompatible. Starving academic science does not push 'unruly' scientists to apply their knowledge more thoroughly to technologically relevant issues; it merely stifles the entrepreneurial spirits of the younger and more dedicated researchers, from whose ranks we expect the most active producers of patents, companies, and any other form of technology transfer effort to emerge. Additionally, when the goals of science and market diverge, the cost of convincing good scientists to take part in commercial activities increases and technology managers may end up with only untalented scientists.

In this view, the much larger success of the US academic system in fostering academic entrepreneurship, compared to Europe, can be explained as a mere reflection of the US's large success in fostering scientific entrepreneurship as such. In turn, this success depends on the long-standing institutional features of the various national university systems. These institutional features do not simply affect the intensity of patenting and firm creation activities. More generally, they explain to what extent commercial activities may or may not help scientific entrepreneurs to progress in their careers. Among those institutional features, university autonomy, personnel mobility, and the principal investigator principle stand out as the most prominent. Patent-based and spin-off-based technology transfer is by and large the product of a specific institutional history, that of the US research universities, where these features have been prominent. Every introduction of those issues within the various European university systems should require first and foremost strong reflections and adjustments that take into account institutional, organizational and environmental characteristics of academic research at the national level.

The main limitation of the analysis we conducted in this chapter is the absence of considerations on the demand side of the market for academic inventions. By and large, however, this is not our choice, but a reflection of the strong supply-side orientation of the literature we chose to review. Future theoretical efforts to conceptualize AE properly will have to take demand into proper account.

As for empirical research, this will have to be directed towards a better measurement of entrepreneurial activities taking place in universities, without drawing any preconceived distinction between the industrial exploitation of research results, and more traditional efforts to build academic careers within the university via breakthroughs into new research fields and the creation of new research groups, labs, and departments.

## NOTES

1.   'The entrepreneur is the head of the firm and coordinates the factors of production; introduces new methods, products, and processes and creates opportunities for growth; bears the risks connected with his or her activities; and enjoys power and high status in capitalist market societies' (Martinelli, 2001; p. 4545).
2.   For a critical synthesis of the linear model of science-technology interaction, see Kline and Rosenberg (1986). On the influence retained by the linear view among policymakers, scientists, and the popular press see David (1997, especially pp. 8-9) and Martin (2003, p. 9).
3.   The Bayh-Dole Act, originally intended to promote the exploitation of public-funded research by small companies, assigns to academia all the intellectual property rights on the results of federally funded research (in doing so, it imitated similar provisions taken by the National Science Foundation in the 1970s). The Stevenson-Wydler Act lays out similar provisions for federal laboratories. On the relationship between the linear view of science-technology interaction and the Bayh-Dole Act, see Colyvas et al. (2002) and Mowery (2001, p. 28). On the wave of Bayh-Dole-like pieces of European legislation see OECD (2003) and, for a critique, Mowery (2001; pp. 31-40) and Mowery and Sampat (2005). Pavitt (2001) offers a more general critique of the European policymakers' tendency to learn the wrong lesson for the US science policy experience.
4.   See the example we put forward in section 8.4.3.
5.   On the European paradox see Caracostas and Muldur (2001). For a recent critique of the argument, see Dosi, Llerena and Sylos-Labini (2005). It is worth pointing out that the paradox argument has been applied to other countries and regions before Europe, and surfaces cyclically in the history of science and technology policies. For example, the Bayh-Dole Act and the Stevenson-Wydler Act we mentioned above followed an intense debate on the apparent failure of the US national innovation system to translate its undisputed scientific leadership into an equivalent technological dominance, at a time when the latter was disputed by Japan and, to a lesser extent, Germany. The European Commission itself, based the European paradox argument very much upon pre-existing literature on the 'Swedish paradox', whose proponents also lamented the Scandinavian country's inability to get enough technological advancements from a world class scientific research system (Edquist and McKelvey, 1998; Jacobbson and Rickne, 2004).
6.   For a classic treatment, besides Kline and Rosenberg (1986), see Rosenberg and Nelson (1994).
7.   For a survey on the use of the entrepreneurship concept in universities, see also Keast (1995).
8.   A major point of contention between the new economics of science and the science studies approach relates the public good nature of scientific knowledge (Callon, 1994; Cowan, David and Foray, 2000). This dissent leads the two schools to judge differently the systemic outcome of strengthening the IPR

regime over academic research results. This point is not of pre-eminent interest here.

9.  We find other striking accounts of this breed of scientists in Latour's (1988) portrait of Louis Pasteur, and in Mowery and Sampat's (2001) and Apple's (1989) biographical notes on Frederick Cottrell and Harry Steenbock (see section 4.1).

10. According to Crow and Bozeman (1998) laboratories represent the core 'production unit' of science; similar findings are reported, for a few case studies, by Slaughter and Leslie (1997, chapter 5). About Europe, see Carayol and Matt (2004).

11. The most prominent Ecoles are still nowadays the Ecole Polytechnique (1793), the Ecole des Mines (1793), and the Ecole Pratique des Hautes Etudes (1868).

12. Similar arrangements exist also for other public research centres, such as INSERM, the national institute for health research.

13. Llerena et al. (2003) describe how a new innovation law was introduced in 1999, in order to provide new incentives for researchers to engage in collaboration with industry, by taking leave up to six years to set up a new company, or by holding equity positions in hi-tech start-ups. Applications, however, ended up not being examined by individual universities, but by an overly cautious national commission. A similar fate occurred, in 2001, to a set of recommendations concerning IPRs in PROs, issued by the Ministry of Research (Gallochat, 2003).

14. On France, Italy and Sweden, see Lissoni et al. (2007), who also provide a comparison with the US. On Finland, see Meyer et al. (2003). On Norway, see Iversen et al. (2007).

15. This point is well illustrated, for a small sample of French academic start-ups, by Shinn and Lamy (2006).

16. On this issue, see Shane (2004, chapter 2).

## REFERENCES

Adams, J.D., G.C. Black, J.R. Clemmons and P.S. Stephan (2005), 'Scientific teams and institutional collaborations: evidence from US universities, 1981-1999', *Research Policy*, 34, 259-285.

Apple, R.D. (1989), 'Patenting university research. Harry Steenbock and the Wisconsin Alumni Research Foundation', *Isis*, 80, 375-394.

Audretsch, D.B. (1995), *Innovation and Industry Evolution*, Cambridge, MA: MIT Press.

Audretsch, D.B. (2000), 'Is university entrepreneurship different?', mimeo.

Audretsch, D.B. and P.E. Stephan (1996), 'Company-scientists locational links: the case of biotechnology', *American Economic Review*, 86 (3), 641-652.

Audretsch, D.B. and P.E. Stephan (1999), 'Knowledge spillovers in biotechnology: sources and incentives', *Journal of Evolutionary Economics*, 9, 97-107.

Callon, M. (1994), 'Is science a public good?', *Science, Technology, and Human Values*, 19 (4), 395-424.

Callon, M. (2002), 'From science as an economic activity to socioeconomics of scientific research', in Mirowski and Sent (eds), *Science Bought and Sold. Essays in the Economics of Science*, University of Chicago Press.

Caracostas, P. and U. Muldur (2001), 'The Emergence of the New European Union Research and Innovation Policy', in P. Larédo and P. Mustar (eds), *Research and Innovation Policies in the New Global Economy*, Cheltenham, UK and Northampton, MA, USA: Edward Elgar.

Carayol, N. and M. Matt (2004), 'Does research organization influence academic production? Laboratory level evidence from a large European university', *Research Policy*, 3, 1081-1102.

Clark, B.R. (ed.) (1993), *The Research Foundations of Graduate Education: Germany, Britain, France, United States, Japan*, University of California Press.

Colyvas, J., M. Crow, A. Gelijns, R. Mazzoleni, R.R. Nelson, N. Rosenberg and B.N. Sampat (2002), 'How do university inventions get into practice?', *Management Science*, 48 (1), 61-72.

Cowan, R., P. David and D. Foray (2000), 'The explicit economics of knowledge codification and tacitness', *Industrial and Corporate Change*, 9, 212-253.

Crow, M. and B. Bozeman (1998), *Limited by Design*, Columbia University Press.

Dasgupta, P. and P.A. David (1994), 'Toward a new economics of science', *Research Policy*, 23, 487-521.

David, P.A. (1997), 'From market magic to calypso science policy. A review of Terence Kealey's *The Economic Laws of Scientific Research*', *Research Policy*, 26, 229-255.

Di Gregorio, D. and S. Shane (2003), 'Why do some universities generate more start-ups than others?', *Research Policy*, 32, 209-227.

Dosi, G., P. Llerena and M. Sylos-Labini M. (2005), 'Science-Technology-Industry Links and the "European Paradox": Some notes on the dynamics of scientific and technological research in Europe', *LEM Working Paper* 2005/02, Scuola Superiore S. Anna, Pisa.

Edquist, C. and M. McKelvey (1998), 'High R&D intensity without high tech products: a Swedish paradox', in K. Nielsen and B. Johnson (eds), *Institutions and Economic Change: New Perspectives on Markets, Firms and Technology*, Cheltenham, UK and Lyme USA: Edward Elgar

Etzkowitz, H. (1983), 'Entrepreneurial scientists and entrepreneurial universities in American Academic Science', *Minerva*, 21, 198-233.

Etzkowitz, H. (2003), 'Research groups as "quasi-firms": the invention of the entrepreneurial university', *Research Policy*, 32, 109-121.

Feldman, M., I. Feller, J. Bercovitz and R. Burton (2002), 'Equity and the technology transfer strategies of American research universities', *Management Science*, 48 (1), 105-121.

Franklin, S.J., M. Wright and A. Lockett (2001), 'Academic and surrogate entrepreneurs in university spin-out companies', *Journal of Technology Transfer*, 26, 127-141.

Gallochat, A. (2003), 'French Technology Transfer and IP Policies', in *Turning Science into Business. Patenting and Licensing at Public Research Organizations*, Paris: Organisation for Economic Co-operation and Development.

Graham, H.D. and N.A. Diamond (1997), *The Rise of American Research Universities: Elites and Challengers in the Postwar Era*, Johns Hopkins University Press.

Henderson, R. (1993), 'Underinvestment and incompetence as a response to radical innovation: evidence from the photolithographic alignment equipment industry', *Rand Journal of Economics*, 24 (2), 248-270.

Iversen, E.J., M. Gulbrandsen and A. Klitkou (2007), 'A baseline for the impact of academic patenting legislation in Norway', *Scientometrics*, 70, 393-414.

Jacobsson, S. and Rickne A. (2004), 'How large is the Swedish "academic" sector really? A critical analysis of the use of science and technology indicators', *Research Policy*, 33, 1355-1372.

Jensen, R. and M.C. Thursby (2001), 'Proofs and prototypes for sale: the tale of university licensing', *American Economic Review*, 91, 240-259.

Jensen, R., J.G. Thursby and M.C. Thursby (2003), 'Disclosure and licensing of university inventions: "The best we can do with the s**t we get to work with"', *International Journal of Industrial Organization*, 21, 1271-1300.

Jong, S. (2007), *Scientific Communities and the Birth of New Industries. How Academic Institutions Supported the Formation of New Biotech Industries in Three Regions*, unpublished PhD dissertation, Florence: European University Institute.

Keast, D.A. (1995), 'Entrepreneurship in universities: definitions, practices, and implications', *Higher Education Quarterly*, 49, 249-266.

Kitcher, P. (1993), *The Advancement of Science*, Oxford University Press.

Kline, S. and N. Rosenberg (1986), 'An overview of innovation', in R. Landau and N. Rosenberg (eds), *The Positive Sum Strategy*, Washington DC: National Academy Press, pp. 275-305.

Larédo, P. and P. Mustar (eds) (2001), *Research and Innovation Policies in the New Global Economy*', Cheltenham, UK and Northampton, MA, USA: Edward Elgar.

Latour, B. (1988), *The Pasteurization of France*, Harvard University Press.

Lenoir, T. (1997), *Instituting Science. The Cultural Productivity of Scientific Disciplines*, Stanford University Press.

Lissoni, F., P. Llerena, M. McKelvey and B. Sanditov (2008), 'Academic patenting in Europe: new evidence from the KEINS Database', *Research Evaluation*, 16, 87-102.

Llerena, P., M. Matt and V. Schaeffer (2003), 'Evolutions of the French innovation policies and the impacts on the universities', in A. Geuna, A. Salter and E. Steinmuller (eds), *Science and Innovation: Rethinking the Rationales for Funding and Governance*, Cheltenham, UK and Northampton, MA, USA: Edward Elgar.

Lowe, R.A. (2002), 'Entrepreneurship and information asymmetry: theory and evidence from the university of California', mimeo.

Martin, B.R. (2003), 'The changing social contract for science and the evolution of the university', in A. Geuna, A.M. Salter and E.D. Steinmueller (eds), *Science and Innovation: Rethinking the Rationales for Funding and Governance*, Cheltenham, UK and Northampton, MA, USA: Edward Elgar.

Martinelli, A. (2001), 'Entrepreneurship', in N.J. Smelser and P.B. Baltes (eds), *International Encyclopedia of the Social and Behavioural Sciences*, Elsevier.

Merton, R.K. (1973), *The sociology of science: theoretical and empirical investigations*, N.W. Storer (ed.), University of Chicago Press.

Meyer, M., T. Sinilainen and J.T. Utecht (2003), 'Toward hybrid triple helix indicators: a study of university-related patents and a survey of academic inventors', *Scientometrics*, 58, 321-350.

Mowery, D.C. (2001), 'The United States national innovation system after the cold war', in P. Larédo and P. Mustar (eds), *Research and Innovation Policies in the New Global Economy*, Cheltenham, UK and Northampton, MA, USA: Edward Elgar.

Mowery, D.C. and B.N. Sampat (2001), 'Patenting and licensing university inventions: lessons from the history of the research corporations', *Industrial and Corporate Change*, 10/2, 317-355.

Mowery, D.C. and B.N. Sampat (2005), 'The Bayh-Dole Act of 1980 and university-industry technology transfer: a model for other OECD governments', *Journal of Technology Transfer*, 30, 115-127.

Mowery, D.C., R.R. Nelson, B.N. Sampat and A.A. Ziedonis (2001), 'The growth of patenting and licensing by US universities: an assessment of the effects of the Bayh-Dole act of 1980', *Research Policy*, 30 (2), 99-119.

Mustar, P. (1997), 'Spin-off enterprises. How French academics creates high tech companies: conditions for success or failure', *Science and Public Policy*, 24 (1), 37-43.

Neave, G. (1993), 'Séparation de corps. The training of advanced students and the organization of research in France', in B.R. Clark (ed.), *The Research Foundations of Graduate Education: Germany, Britain, France, United States, Japan*, University of California Press.

Nerkar, A. and S. Shane (2003), 'When do start-ups that exploit patented academic knowledge survive?', *International Journal of Industrial Organization*, 21, 1391-1410.

OECD (1999), *University Research in Transition*, Paris: Organisation for Economic Co-operation and Development.

OECD (2003), *Turning Science into Business. Patenting and Licensing at Public Research Organizations*, Paris: Organisation for Economic Co-operation and Development.

Pavitt, K. (2001), 'Public policies to support basic research: what can the rest of world learn from US theory and practice? (and what they should not learn)', *Research Policy*, 26, 317-30.

Pirnay, F., B. Surlemont and F. Nlemvo (2002), 'Toward a typology of university spin-offs', *Small Business Economics*, 21, 355-369.

Roberts, E.B. (1991), *Entrepreneurs in High Technology*, Oxford University Press.

Romano, A. (1998), 'A trent'anni dal '68. "Questione universitaria" e "riforma universitaria" ', *Annali di Storia delle Università italiane*, 2.

Rosenberg, N. and R. Nelson (1994), 'American universities and technical advance in industry', *Research Policy*, 23, 323-348.

Rudolph, F. [1962] (1990), *The American College and University: A History*, University of Georgia Press.

Shane, S. (2001a), 'Technological opportunities and new firm formation', *Management Science*, 47 (2), 205-220.

Shane, S. (2001b), 'Technological regimes and new firm formation', *Management Science*, 47 (2), 1173-1190.

Shane, S. (2002), 'Selling university technology: patterns from MIT', *Management Science*, 48 (1), 122-137.

Shane, S. (2004), *Academic Entrepreneurship. University Spin-offs and Wealth Creation*, Cheltenham, UK and Northampton, MA, USA: Edward Elgar.

Shane, S. and R. Khurana (2003), 'Bringing individuals back in: the effect of career experience on new firm founding', *Industrial and Corporate Change*, 12 (3), 519-543.

Shane, S. and T. Stuart (2002), 'Organizational endowments and the performance of university start-ups', *Management Science*, 48 (1), 154-170.

Shinn, T. and E. Lamy (2006), 'Paths of commercial knowledge: forms and consequences of university-enterprise synergy in scientist-sponsored firms', *Research Policy*, 35, 1465-1476.

Siegel, D., P. Westhead and M. Wright (2003), 'Assessing the impact of university science parks on research productivity: explanatory form-level evidence from the United Kingdom', *International Journal of Industrial Organization*, 21, 1357-1369.

Slaughter, S. and L. Leslie (1997), *Academic Capitalism: Politics, Policies, and the Entrepreneurial University*, Johns Hopkins University Press.

Stephan, P.E. and S. Everhart (1998), 'The changing rewards to science: the case of biotechnology', *Small Business Economics*, 10, 141-151.

Stephan, P.E. and S.G. Levin (2002), 'The importance of implicit contracts in collaborative scientific research', in P. Mirowski and E.-M. Sent (eds), *Science Bought and Sold. Essays in the Economics of Science*, University of Chicago Press.

Stuart, T.E. and W.W. Ding (2004), 'When do scientists become entrepreneurs? The social structural antecedents of commercial activity in the academic life sciences', mimeo.

Thursby, J.G. and M.C. Thursby (2002), 'Who is selling the ivory tower? Sources of growth in university licensing', *Management Science*, 48 (1), 90-104.

Thursby, J.C. and M.C. Thursby (2003a), 'Are faculty critical? Their role in university-industry licensing', *NBER WP* no. 9991, National Bureau of Economic Research.

Thursby, J.C. and M.C. Thursby (2003b), 'Patterns of research and licensing activity of Science and Engineering Faculty', mimeo.

Thursby, J.C. and M.C. Thursby (2003c), 'Industry/university licensing: characteristics, concerns and issues from the perspective of the buyer', *Journal of Technology Transfer*, 28, 207-213.

Thursby, J.G., R. Jensen and M.C. Thursby (2001), 'Objectives, characteristics and outcomes: a survey of major US universities', *Journal of Technology Transfer*, 26, 59-72.

Thursby, M.C., J.G. Thursby and E. Dechenaux (2005), 'Shirking, shelving, and sharing risks: the role of university license contracts', mimeo.

Trow, M. (2003), 'In praise of weakness: chartering, the university of the United States, and Dartmouth College', Center for Studies in Higher Education Research and Occasional Paper CHSE. 2.03, Berkeley: University of California.

Vohora, A., M. Wright and A. Lockett (2004), 'Critical junctures in the development of university high-tech spin-out companies', *Research Policy*, 33, 147-175.

Zucker, L.G. and M.R. Darby (1996), 'Star scientists and institutional transformation: patterns of invention and innovation in the formation of biotechnology industry', Proceedings of the National Academy of Sciences of the United States of America, Colloquium paper, 93, 12709-12716.

Zucker, L.G., M.R. Darby and M.B. Brewer (1998), 'Intellectual human capital and the birth of US biotechnology enterprises', *American Economic Review*, 88, 290-306.

# 9. Firm formation and economic development: what drives academic spin-offs to success or failure?

## Knut Koschatzky and Joachim Hemer

## 9.1 INTRODUCTION

The knowledge economy represents an essential challenge for regional innovation systems. The main focus in knowledge-driven innovation systems is on the organizational change in innovation activities, as well as on general structural shifts within the system. New forms of knowledge generation, knowledge transfer and the activation of new, respectively unexploited, technology and knowledge potentials are closely linked to these structural shifts (Miles, 2003). In this context, newly founded firms with high knowledge content can and should play a significant role in regional modernization. Thus in the last decades a dynamic development can be observed in knowledge-intensive firm formation (cf. Kerst, 1997). The dynamics in this enterprise population are often generated via the demand of existing enterprises for new, advanced, knowledge-intensive and specialized products and services (Fritsch and Müller, 2006; Fritsch and Schmude, 2006).

From a regional point of view, policymakers and managers of research organizations have discovered that the formation of technology-oriented and knowledge-intensive firms can function as an essential and system-bridging actor group which could actively contribute to regional development and change (Almus et al., 2001; Meyer-Krahmer and Lay, 2001; Wood, 2002). Assuming that the more the public sector is able to support firm formation through the promotion of spin-offs from public research organizations or through the provision of supportive framework conditions like public funding, the more strongly it can govern the structural ties and thus the regional integration of these firms and their contribution to the regional economy (Lockett et al., 2005). In this respect, academic organizations, i.e. universities and non-university research institutes, can act as breeding grounds for knowledge-intensive start-ups. Academic spin-offs are knowledge-based firms newly founded by graduates or by academic staff members of universities and research institutes. Academic spin-offs deserve special attention because they are expected to strengthen future-oriented sectors of the economy, to grow more rapidly than 'normal' company start-ups and therefore to contribute more than these to the industrial and economic structural change in regions and entire economies (Egeln et al., 2004, p. 207). The popular understanding is that the idea to start academic spin-off is mainly triggered by the economic exploitation of a scientific idea or a research result.

On the basis of five theory-led assumptions formulated in the next section, it is the objective of this chapter to analyse the success and critical factors during the start-up and early development process of these firms. Using case study data from 59 academic spin-offs in Germany, of which 26 come from the new federal states (plus Berlin)[1] and 33 from the old federal states,[2] it will be shown to what extent academic spin-offs are successful in a general business management sense, which regional and non-regional factors contribute to this success and which future promotional measures could contribute to the success of academic spin-offs and thus to economic and regional development.

## 9.2  KNOWLEDGE-INTENSIVE FIRM FORMATION AND ECONOMIC DEVELOPMENT

Economic competitiveness depends increasingly on the extent to which endogenous knowledge and technology potentials are successfully activated and converted into value added (Cooke, 2002). This activation and conversion activity can take place within already existing enterprises, but – increasingly more important in knowledge-based economies – in connection with start-ups. Knowledge-intensive start-ups are knowledge-intensive in the sense that their fundamental key functions are to be seen in knowledge transfer, diffusion and utilization, as well as the integration of various knowledge stocks and competences (Strambach, 2001, pp. 62-63; Antonelli, 1999, p. 254). They act as transmission agents in knowledge processes and are characterized particularly by a knowledge- and human-capital-intensive performance (Lambooy, 2004). This first short reflection results in

**Assumption (1)**: Academic spin-offs play a highly important role in the process of knowledge transfer.

Besides corresponding impacts for the economy as a whole, innovative and knowledge-intensive new firms possess all important functions with regard to technological and knowledge modernization and to regional structural change. High start-up rates and the technical advances at least partly driven forward by innovative new firms are often connected with high innovation rates and efficiency increases. The latter can also be of an individual nature, for instance, the market entry of new enterprises can stimulate the existing enterprises to intensify innovation activities in order to catch up with competitiveness (Fritsch and Müller, 2004).

**Assumption (2):** High-tech spin-offs and start-ups play a significant role in structural change and in the regional technological and knowledge modernization process.

The quantitative and qualitative dimension of start-ups depends strongly on the interrelationships within a regional innovation system (Cooke, 2004). Almus et al. (2001) demonstrate that regional supply and demand conditions exercise a decisive influence on the regional start-up intensity. Besides the size of the region in question (number of inhabitants, number of employees), variations in the endowment with firm formation relevant characteristics and their systemic interactions affect the number and development of new enterprises (Nerlinger, 1998, p. 57). Other recent studies stress the importance of the presence of a university as an additional locational factor. Audretsch et al. (2005) found out, however, that not only is the mere presence of a university important, but also the type of knowledge and the spillover mechanisms. In natural sciences, for example, new firms tend to choose a location closer to universities with high outputs of students (specificity of human capital), while this is not the case for social sciences. On the other hand, for social sciences it is not the student output which is important, but the type of scientific knowledge which is less codified and more tacit compared to the natural sciences. This forces new firms with natural science backgrounds to locate closer to universities where they are able to profit from knowledge spillovers (ibid., p. 1119). Egeln et al. (2004, p. 221) arrived at a similar conclusion. According to their empirical study on the location decisions of spin-offs from public research institutions, spin-offs with a natural science background tend to locate rather close to their parent institute[3] because they rely strongly on technology pushes, while service-oriented spin-offs rely more on demand factors and thus locate closer to their markets. Since according to Egeln et al. (2004) the majority of spin-offs is service-oriented, the overall picture of spin-off locations reveals a pattern of a rather distant location to the parent institute. In this respect, spin-offs contribute to a higher extent to interregional and not only to intraregional knowledge transfer. In general, if urbanization economies are less pronounced, spin-offs are likely to locate at different locations.

**Assumption (3):** Start-up intensity is strongly influenced by regional supply and demand conditions.

This makes it clear that the level of the regional spin-off formation rate is influenced by different locational factors, such as a favorable combination of technological, industrial, social, institutional and other 'soft' factors. Regions offering these attributes are called 'entrepreneurial regions'. Such regions do not only exercise a positive influence on the quantitative start-up activities as a whole (intensity and dynamics), but may also have an important stabilizing function for enterprises in their early life phase (cf. also Reynolds et al., 1994). The most important factors which influence the regional entrepreneurial activities are (according to Birch, 1987; Malecki, 1994; Sternberg, 2000; Niese, 2003; Bergmann, 2004):

1. the number, types, and mix of incubator organizations[4] (e.g. private companies, R&D organizations, universities etc.),
2. the entrepreneurial 'atmosphere'; for instance, previous spin-off founders may serve as role models or examples for new entrepreneurs,
3. the technological regimes (entrepreneurial vs. routinized regimes),
4. the social mix of places, especially regarding educational level,
5. the region's ability to attract and retain educated people and also entrepreneurs,
6. the quality of government and public support measures for entrepreneurship,
7. the existence of 'intermediary institutions' and the provision of financing (private financing institutions and public banks).

**Assumption (4):** Regional framework conditions as defined by the criteria of an entrepreneurial region exercise a positive stabilizing effect upon start-ups during their early phases.

Typically, it is the urban agglomerations which dispose of characteristic features for stimulating start-ups and which are favored spaces for public promotion measures with regard to fostering entrepreneurship. Although Shaver and Flyer (2000) argue that knowledge- and technology-intensive firms do not benefit from agglomeration externalities because their technology and knowledge base spills over to competitors, and thus they have little motivation to geographically cluster, these firms, be they start-ups or established companies, are a major characteristic of urban agglomerations. For Malecki (1994), agglomeration advantages make urban spaces the preferred sites for technology- and knowledge-intensive start-ups (for a recent empirically-based analysis of new firm formation in cities, see Fritsch et al., 2004). Nevertheless, semi-urban agglomerations or even peripheral regions are also locations of research organizations. In these regions, however, these organizations fulfill a different, in most cases a more pronounced, role in stimulating knowledge exchange and innovation than in urban agglomerations, where they are only players among many other important knowledge-generating and -diffusing actors. Besides their role as knowledge producers, universities and non-university research organizations can actively contribute to transferring implicit knowledge into economic value added.

According to Paul Romer's model of endogenous technological change, the economy consists of three sectors: the research sector, which uses human capital and the existing stock of knowledge to produce new knowledge; the intermediate goods sector, which uses the designs from the research sector together with existing output to produce inputs for the final-goods sector; and the final-goods sector, which uses labor, human capital intensive labour and a set of producer durables that are available to produce final output (Romer, 1990, p. S79). Since knowledge and human capital can be accumulated, a dynamic research sector can contribute to the economic growth of a national (or regional) economy. Universities and research organizations as major elements of the research sector should not only act as a passive knowledge supplier, but its members (mainly researchers and scientists) can actively contribute to the knowledge transfer process between the sectors through blueprint developments (patents) and their commercialization, that is licensing and spin-off activity (cf. Etzkowitz and Leydesdorff, 2000; Etzkowitz, 2002). Since the endogenous growth theory stresses the importance of knowledge as a key source of economic growth and thus the role of the small business sector in knowledge generation (Audretsch, 2004), the public promotion of new firm formation, e.g. via the support for academic spin-offs, can enhance the system as such and contribute to regional development.

**Assumption (5):** Public funding of academic spin-offs supports their regional integration and their contribution to the regional economy.

In general, it is assumed that universities and different kinds of research organizations can contribute to a better exploitation of the commercial use of their creativity and research activities by acting as incubators for new knowledge-based firms and can thus fulfill different functions in the start-up process of these firms and in their region, depending on its organizational fabric.

Taking these assumptions as basic guidelines of this chapter, the following sections are organized in a way which aims to answer the following research questions:

1. What role do universities and non-university research organizations play in the support of academic spin-offs?
2. What effects does this support have on the development of the new firms?
3. Regarding the supportive role of the different academic organizations and their effects on the spin-off firms, what conclusions can be derived for regional innovation policy addressing different kinds of organizational structures in regions?
4. Can the abovementioned assumptions be verified by empirical evidence?

## 9.3  BRIEF DESCRIPTION OF THE GERMAN RESEARCH SYSTEM

This section serves to illustrate the complex structure of the German research system. It is characterized by (a) several hundreds of universities which all are obliged to conduct higher education, research, and technology transfer, (b) the existence of many non-university research organizations and, last but not least, (c) the strong role of industrial R&D. Whilst the federal states ("Länder") are responsible for most of the universities,[5] the group of non-university research institutes comprises several different types of organizations with different missions and roles in the research system, legal statutes and funding schemes. The following figure provides a brief overview of the German research system.

The group of non-university research organizations can briefly be characterized as follows:

1. Max Planck Society (MPG) which carries out basic research with about 24,000 employees of which are 11,000 young scientists in natural, medical and social sciences in approximately

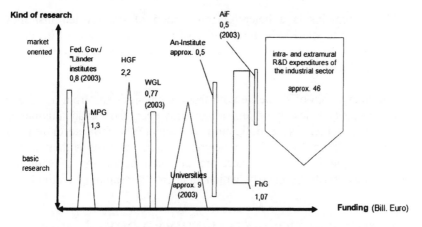

*Source*: Koschatzky et al. (2008).

*Figure 9.1  The German research landscape in 2004*

80 establishments (funded by the federal government and federal states).

2. Helmholtz Association of National Research Centers (HGF) which carries out applied R&D in 15 large technology-oriented research centers with 28,000 employees, primarily state (funded by the federal government and federal states).

3. Fraunhofer Society (FhG) which carries out contract research and technological development with approximately 15,000 employees in various fields of technology, natural sciences and economics in 56 institutes, funded mainly through contracts both from industry and public (including EU and foreign governments).

4. Leibniz Association (WGL) which conducts basic and applied research with a staff of approximately 14,000 persons in about 80 institutes, funded from various public and private sources according to individual schemes.

5. Some 600 or more so-called 'An-Institute', i.e. institutes attached to universities by means of co-operation agreements, but legally independent of them. They cover the whole range from basic to applied research or even product development, consultancy, technology transfer services etc., thereby meeting many needs of society and industry. They are either run or funded by public institutions or by the private sector (mostly via legal forms like associations or limited companies including universities and companies as shareholders).

6. What remains is a heterogeneous group of state-owned institutes dedicated to specific research fields of public interest and of institutes of collaborative industrial R&D run or partly financed by industry (so called AiF Institutes).

In Germany, spin-offs emerge both from universities and from the above non-university research organizations. However, their exact number is unknown, as there is no obligation for founders to indicate or register their spin-off plans to their parent organizations, as long as they do not apply for public support. Therefore the total population of spin-offs cannot be described precisely, which hampered our sampling method.

## 9.4 STUDY DESIGN AND METHODOLOGY

For data collection we chose a case study approach. Although this methodology limits the transferability of conclusions, it offers the chance to collect many details about each single case (i.e. the spin-off firm) which otherwise could not be obtained. In our analysis we therefore draw on detailed explorative case studies of spin-off enterprises, consisting of document research and in-depth interviews, not only with the company founders and present-day CEOs, but also with persons who accompanied the firm formation processes as consultants or mentors, financiers, sponsors, colleagues or former principals of the founders (Hemer et al., 2007).

The case studies were selected from five types of German scientific organizations from the range described above: universities including universities of applied science ('Fachhochschulen'), plus the four most important non-university research organizations in Germany – HGF, FhG, MPG and WGL.

Fifty-nine case studies with spin-offs were conducted in total, 33 in East Germany including Berlin and 26 in West Germany: 20 case studies were carried out during a study on behalf of the Federal Research Ministry (BMBF) finished in summer 2005 (Hemer et al., 2006); 39 further cases are part of a study done for the Office of Technology Assessment at the German Parliament (TAB) in 2006 (Hemer et al., 2007). Selection criteria were to have good coverage by

1. type of parent organization,
2. spatial distribution of parent institute (northern, eastern, southern and western Germany),

3. age of spin-off (two groups were formed: older foundations until 2000 and younger ones founded after 2000).

The cases were drawn from a list of about 800 spin-offs[6] which had been compiled by means of thorough internet research around major universities and from lists the non-university holding associations provided. In some cases they also helped in approaching the founders or CEOs of the candidate case study companies, but generally the study team tried to address the representatives of the firms personally and directly.

No more detailed selection criteria was deployed since the population of academic spin-offs in Germany is virtually unknown and the sampling was not meant to be representative, as the approach was a purely explorative one.

Due to the case study approach and the limited number of cases, a selection bias among the analysed firms cannot be ruled out. This is especially the case for the exclusion of unsuccessful spin-offs in our sample. The distribution by types of scientific organizations, technology fields and locations is shown in Table 9.1.

The founding dates of the 59 case study firms ranged between 1990 and 2004. This means that the foundations encompassed widely differing founding conditions.

## 9.5 EMPIRICAL EVIDENCE: ROLE AND IMPACTS OF ACADEMIC SPIN-OFFS IN GERMANY

### 9.5.1 Role of Universities and Non-university Research Organizations in the Support of Academic Spin-offs

**Support strategies of parent organizations**
Regarding the type and degree of support the parent organizations provided to their spin-offs and to the respective founders, the following distinctions must be made. There are official support policies by the authorities responsible for the scientific institutions. In Germany, different federal ministries are responsible for the four groups of non-university research organizations and the Länder ministries for education and cultural affairs for the universities. These responsibilities are mostly laid down in the form of regulations, directives or recommendations. After some years of experimenting and learning, these policies now tend to converge, although differences still exist, especially as far as the direct

*Table 9.1   Distribution of case studies by their founding and current
              locations*

| West German Länder | No. of founding locations/current locations | East German Länder and Berlin | No. of founding locations/current locations |
|---|---|---|---|
| Baden-Württemberg | 10/10 | Berlin | 6/6 |
| Bavaria | 6/6* | Brandenburg | 1/2* |
| Bremen | 0/0 | Mecklenburg-West Pommerania | 3/3 |
| Hamburg | 0/1* | Saxony | 9/9 |
| Hesse | 3/2** | Saxony-Anhalt | 2/2 |
| Lower Saxony | 4/4 | Thuringia | 4/4 |
| Northrhine-Westphalia | 7/7 | | |
| Rhineland-Palatinate | 1/1 | | |
| Saarland | 2/2 | | |
| Schleswig-Holstein | 1/1 | | |
| Abroad | 0/1** | | |
| Total | 34/35 | | 25/26 |

*Notes*: * One further location was established in this Land after foundation in
         another.
       ** After foundation in Hesse the company moved abroad.

*Source*: Hemer et al. (2006, 2007).

financial support or the equity participation in spin-offs is concerned.
However, according to their legal and statutory structures and depending
on their degree of autonomy, the research entities interpret official
policies regarding spin-off activities vary individually and practice their
degree of freedom in a specific manner. This results in some remarkable
differences in the policies 'on the spot', i.e. by each individual professor
or institute director. Depending on the institute's mission, its research
field (basic or applied research; natural, engineering or social sciences)
and research tradition (industry co-operation, market orientation), the
spin-off tradition of the institute, its links to industry, etc., the director
and his team may be open or even proactive towards entrepreneurial
activities, or reluctant or even negative. This has strong impacts upon the
form, quality and intensity of support the potential founder is actually
being offered. And above all, the full spectrum of support is not offered
to all founders in all cases. Especially those institutes that follow a

selective support policy seem to deliberately differentiate their offers according to the nature and quality of the spin-out project (criteria being e.g. strategic relevance for the institute itself, scientific esteem, market maturity of the product or service envisaged, qualification and personal characteristics of the founding team etc.).

In our case study sample, specific strategy patterns can be recognized. We differentiated between activities of either the parent organizations (PO) or the parent institutes (PI) respectively in the preparatory/founding phase and during further development phases of the spin-offs. Four different strategic patterns could be elaborated for the parents represented in the sample, as shown in Table 9.2.

*Table 9.2   Four strategy patterns practiced by the parent organization (PO)or parent institute (PI) and their distribution by 59 case studies*

| Pat-tern no. | Characteristics of support provided by PO or PI | Frequencies | | |
|---|---|---|---|---|
| | | West Germ. | East Germ. | Total in % |
| 1 | Minimal support during foundation phase, more reliance on founder's individual initiatives | 12 | 8 | 34 |
| 2 | Mainly personal support, information, sensitization, 'instigation' of researchers to found, advice and coaching, use of rooms and equipment, continued payment of salaries, IPR[7] management<br>Low selectivity | 7 | 8 | 25 |
| 3 | Information, coaching, brokering of contacts, systematic evaluation of business concept, IPR management, use of rooms and equipment, continued payment of salaries, equity participation, financial management<br>Concentration on specific technology<br>Medium selectivity | 9 | 5 | 24 |
| 4 | Preparation of spin-off's products and markets by near-to-market R&D in PI, close links between R&D, manufacturing and sales<br>Concentration on specific technology<br>High selectivity | 5 | 5 | 17 |

*Source*: Hemer et al. (2006, 2007).

In order to arrive at an appropriate evaluation of the strategies and their relevance for a re-adjustment of the promotion of start-ups, the patterns must be assessed in their specific contexts. The constitutional difference is important: non-university organizations work in specialized fields; universities, on the other hand, work in a wide field of disciplines and technology areas according to their remit. This fundamental difference also has consequences for spin-off behavior. Non-university research organizations – in particular those of pattern 4 – support founders primarily in the areas of competence in which the institute excels. The constant contact with industry, e.g. in the framework of co-operative projects, and the clearly defined resource specificity of the institutes of pattern 4 make possible a thorough preparation of products and markets for start-ups. In this respect, such spin-offs have a shorter 'lean period' to tide over between foundation and product development and market introduction. However, spin-off activities are only undertaken very selectively by the institutes following this pattern.

Promotion strategies according to strategy pattern 3 are more open in this regard and can react more flexibly to spin-off plans which were not directly developed on the basis of an institute-specific research project. The strength of this type is based on a broad knowledge of certain markets and, above all, of the behavior of banks and private equity investors. Spin-off plans can also be promoted in cases where not a technology, but 'only' a knowledge transfer is taking place. Universities, on the other hand, must distribute their scarce funds and possibilities in many competence fields and therefore fall behind, almost inevitably, with regard to their specific resources profile. Their support must therefore take other forms. They can invest less in the preparation of products and markets, but more in personal key competences of graduates, doctoral students and staff interested in spin-offs. Their educational mission qualifies them to impart social and business administration or managerial competences and to 'instigate' spin-offs in all phases of the study course. However, with regard to networking with industry or the trade and finance sector, university spin-offs have inferior starting conditions to spin-offs from non-university research institutes. This is because German universities traditionally have less strong links to industry as they are more inclined towards basic research and in addition, as staff fluctuation is so high it is – much too often – very difficult to establish and maintain stable competences and client relations with industry. Thus, faculties can provide their spin-offs with less robust and market-related contacts and the spin-offs can also profit less from a specific, clearly defined research path and a correspondingly accumulated knowledge base.

The strategies of the investigated universities obviously lead to an externalization of the founding risks to the founder teams to a large extent. According to the universities' apparent strategies, it is the sole responsibility of the founders and of market forces to make a start-up a success – in contrast to the non-university research organizations of pattern 4, which prepare the founders well and release them from the institution with only a low founding risk.

## Research result exploitation and transfer process

The possession of, or access to, individual intellectual property rights (IPR) is important for technology start-ups to rapidly overcome market entry barriers and to gain market shares. However, policies differ on the part of the enterprises, parent organizations and investors regarding how to handle IPR. Whether a parent organization agrees to transfer property rights to the spin-off depends on its general IPR or commercial exploitation policy, and on the (subjective) valuation of the patent in each individual case. Table 9.3 displays seven different forms of relations spin-offs maintained with their parent organizations in respect of IPR matters as observed in the 20 cases of the first study.[8]

The above distribution of variants of IPR agreements shows how frequently the founders are also (co-)inventors of the products or processes which form the basis for the spin-off foundation. This illustrates how strong the founders' personal engagement or involvement in the spin-off projects are (spin-off projects often being their 'brainchild'). On the other hand, it is not absolutely necessary to dispose of patents in all cases; depending on the nature of the invention or innovation, it might even be advisable to market it immediately rather than wait for a patent to be awarded and spend a fortune on the process.

The parent organizations often pursue very traditional patenting policies which, however, seem to be slowly becoming more flexible. Traditionally, they regard and treat their patent portfolio as a valuable asset that can be marketed and that may have a positive effect on their balance sheets and may raise their valuation. More flexible behavior would mean that in the case of each individual patent a trade-off is made between its role as an asset (i.e. it can be traded or can yield royalties) and a role as a strategic variable that can be deployed to build a network of related industry partnerships, to prepare new research fields and position the institution with leading and patented developments, and to generate an unbeatable scientific position in the global innovation cycle.

Research results in the form of new knowledge or newly developed technologies formed the basis in all east German cases of the sub-sample

*Table 9.3  Spectrum of IPR relations between spin-off and parent organization (PO)9*

| Category no. | Observed characteristics of IPR relations | Frequencies | | |
|---|---|---|---|---|
| | | West Germ. | East Germ. | Total in % |
| 1 | No agreement with PO since patents are not relevant for the spin-off | 2 | 1 | 15 |
| 2 | No agreement with PO since innovative product or process is not patented | 2 | | 10 |
| 3 | Founder(s) was/were (co-)inventor(s), PO owns the patent and grants license to spin-off | 2 | 6 | 40 |
| 4 | Founder(s) was/were (co-)inventor(s), PO grants patent to founder(s) or spin-off, completely or partially | 1 | 2 | 15 |
| 5 | No agreement, founder(s) was/were inventor(s) and own(s) the patent | 1* | 1 | 10 |
| 6 | Novel product or process uses PO's know-how and IP, spin-off receives several licenses or one comprehensive license | 1 | | 5 |
| 7 | PO owns patent, no founder is (co-)inventor, but spin-off gets license | 1 | | 5 |

*Note*:  * Invention was done outside PO.

*Source*:  Hemer et al. (2007).

of 20 spin-offs. In west Germany, on the other hand, innovations originate rather due to the founders' own R&D efforts in the new enterprise or through external input, that is, not exclusively from within the research system. Only in 40 percent of the west German cases were specific R&D results or even inventions the cause of the spin-off foundation; 60 percent entered the market with self-developed technologies or technologies purchased from a third party. The parent institutes in these cases were not R&D partners of the enterprises at all. Our sample of case studies shows that the classical, linear transfer model only applies to a limited extent. The form of co-operation between the spin-off and the parent institute varies widely. In our first study we attempted to classify them and found five types of working relations in the innovation process in which both entities are involved (cf. Table 9.4).[10]

The first type corresponds to the traditional linear transfer model and

*Table 9.4  Types of working relations between spin-offs and parent institutes (PI)\**

| Type[11] | Observed characteristics of working relations | Frequencies | | |
|---|---|---|---|---|
| | | West Germ. | East Germ. | Total in %\* |
| 1 | R&D within PI until prototype, transfer to spin-off, further product/process development until marketability, marketing | 5 | 4 | 45 |
| 2 | Repeatedly R&D projects in PI, spin-off takes over for further development until marketability, close symbiotic collaboration, often mutual contracting | 1 | 3 | 20 |
| 3 | Product/process development solely by potential founder(s) within PO as scientist(s), no further involvement of PI, PI's support limited to provision of resources and facilities | | 2 | 10 |
| 4 | Product/process development solely within spin-off applying know-how acquired during founder team's former work in PI | 2 | 1 | 15 |
| 5 | No knowledge or technology transfer at all, no use of PI's know-how, product/process development solely within spin-off, totally decoupled from PI | 2 | | 10 |

*Note*: \* Basis is the sub-sample of 20 cases of the first study.

*Source*:  Hemer et al. (2007).

is – in the entire sample – by far the most frequent form of labor division. The most successful enterprises or those promising to be successful came from this group.[12] Category 3 appears to be less successful, because there is no connection between the R&D activities of the parent organization and those of the later founder. The first two classes demonstrate clearly, however, how important a sensible division of labor is in the sense of two equal R&D partners. In our case studies, not one of the spin-offs is merely an 'extended workshop' or only a service provider for the parent institutes; they all play an independent role in the R&D process.

On the whole, it can be stated that the parent institutes do not provide the founders with near-to-market prototypes for commercial exploitation as a rule; this finding was confirmed during the second study as well. This shows how necessary the extension of the innovation process or the transfer chain through independent spin-offs is; obviously the institutes are either not able or not willing to perform this last step in the innovation chain, for whatever reason.

Often the products, processes or services which are finally introduced to the market were not those the company was founded for. This can be a consequence of the deficits concerning market maturity of the projects taken over from the parent institute, forcing the company to change its business model, but it may also be due to the late recognition that the business model was commercially not at all viable.

### 9.5.2 Effects of Spin-off Support on the Development of the New Firms

**Performance assessment**
We developed a simple set of six success indicators to judge whether or not the companies investigated had performed positively so far or not. Otherwise it would have been impossible to draw conclusions about the specific effect of certain influential factors. The following set of success indicators seemed reasonable/viable:

1. survival for over five years,
2. stable or growing positive cash flow or profits over the last three years,
3. simultaneous job and sales growth over the last three years,
4. high market share with at the same time great market potential,
5. stable development with a positive trend over the last three years,
6. crisis-proven development (successfully managed turn around) with a final positive trend.

**Company performance**
Based on the available knowledge about the companies and the impressions obtained in the interviews of the 59 case studies, every firm was marked qualitatively with the help of the above six criteria on a four-step scale (from ++, +, 0 to −).[13] The following result was arrived at for the success of the 59 case study enterprises (cf. Table 9.5).

The support activities of the research organizations (cf. Table 9.2) were then mirrored against the success scores of this table. Looking, for instance, at the types of support strategies (Table 9.6) and working relations between spin-off and institute (Table 9.7), the effects on company performance are shown.

*Table 9.5 Success appraisal of 59 case study companies (no. of companies per score)*

| | |
|---|---|
| Companies that could already be regarded successful (score ++) | 10 |
| Companies which are on a potentially successful or at least promising route (score +) | 31 |
| Companies which still have to provide positive performance; their situation is still very unstable (score 0) | 14 |
| Companies that could already be regarded unsuccessful (score −) | 4 |

*Source*: Hemer et al. (2006, 2007).

*Table 9.6 Relation between support strategy and company performance[14]*

| Pat- tern no. | Characteristics of support provided by PO or PI | Performance of associated cases (absolute and relative frequencies)* |
|---|---|---|
| 1 | Minimal support during foundation phase, more reliance on founder's individual initiatives | 3 + (15%) 11 + (55%) 5 0 (25%) 1 − (5%) |
| 2 | Mainly personal support, information, sensitization, 'instigation' of researchers to found, advice and coaching, use of rooms and equipment, continued payment of salaries, IPR management  Low selectivity | 3 ++ (2%) 6 + (4%) 5 0 (33%) 1 − (7%) |
| 3 | Information, coaching, brokering of contacts, systematic evaluation of business concept, IPR-management, use of rooms and equipment, continued payment of salaries, equity participation, financial management  Concentration on specific technology  Medium selectivity | 2 ++ (14%) 9 + (64%) 3 0 (21%) |
| 4 | Preparation of spin-off's products and markets by near-to-market R&D in PI, close links between R&D, manufacturing and sales  Concentration on specific technology  High selectivity | 2 ++ (20%) 5 + (50%) 1 0 (10%) 2 − (20%) |

*Note*: *Frequencies refer to no. of cases in each category.

*Source*: Hemer et al. (2006, 2007).

*Table 9.7 Relation of forms of work division and company performance*[15]

| Type | Observed characteristics of work relations | Performance of associated cases (absolute and relative frequencies)* |
|---|---|---|
| 1 | R&D within parent institute (PI) until prototype, transfer to spin-off, further product/process development until marketability, marketing | 2 ++ (22%)<br>4 + (44%)<br>1 0 (11%)<br>2 – (22%) |
| 2 | Repeatedly R&D projects in PI, spin-off takes over for further development until marketability, close symbiotic collaboration, often mutual contracting | 1 ++ (25%)<br>3 + (75%) |
| 3 | Product/process development solely by potential founder(s) within PI as scientist(s), no further involvement of PI, PI's support limited to provision of resources and facilities | 1 0 (50%)<br>1 – (50%) |
| 4 | Product/process development solely within spin-off applying know-how acquired during founder team's former work in PI | 2 + (66%)<br>1 0 (33%) |
| 5 | No knowledge or technology transfer at all, no use of PI's know-how, product/process development solely within spin-off, totally uncoupled from PI | 1 + (50%)<br>1 0 (50%) |

*Note*: * Frequencies refer to no. of cases in each category.

*Source*: Hemer et al. (2007).

The share of successful or promising companies (scores ++ and +) per category is of the order of 60-70 percent in the strategy patterns 1, 2 and 4. Category 3 alone has a substantially higher rate with 78.5 percent. This might be a first indication that this strategy pattern has a specific comparative advantage as regards company performance, although a final conclusion cannot yet be drawn from this small sample. What can be stated, however, is that any of these four strategy types may contribute strongly to the creation of promising spin-offs. This must lead to the conclusion that success and performance are most probably driven more by the very individual firm characteristics and founding conditions that must come together to form favorable firm development constellations, a result which cannot really be satisfactory in the search for success factors. Further investigation in this subject field must concentrate on the question to what extent the different parent institutes have strong relations with firms dominant in the very sector or technology field and which conditions lead to the establishment of such strong relations. Also, the role further relations to another supporting institution (e.g. a corporate sponsor), parallel to the parent institute, can play on the path to good performance is worth investigating.

In a similar manner, we looked at the impact of work division practiced between the parent institute and the potential foundation or spin-off respectively on the firm's development route. Table 9.7 shows (albeit limited to the 20 cases of the first study only)[16] the correlation of performance and work division, here differentiated into five types.

This picture shows a clearer differentiation than the one above. Working relations of type 2 most often yield promising start-ups, followed by types 1 and 4. Supposing this pattern can be supported by more robust data, it would favor the prominent notion of very close and enduring collaboration between the spin-off and its parent institute. On the other hand, there are no signs that the other forms of working relations would be clearly disadvantageous.

### 9.5.3 The Impact of Locational Factors Related to the Parent Institute

According to the case studies, the regional environment defined by the classical infrastructure endowment had little impact on the development of the new firms, at least during the start-up phase. This is, however, slightly different with regard to the scientific environment. The founders stressed the dynamism or fertility of the parent institute's scientific network, both on the local or regional and on the transregional scale. As far as the parent institute maintained contacts with either the regional or local scientific community or with regional company networks or clusters, the founders had the opportunity to deploy them. However, in our sample, most of the spin-offs had been founded around high-tech niche products with a high degree of innovation (due to the sampling criteria), for which a regional market could hardly exist yet. Therefore it is comprehensible that local or regional economic integration was weak at the beginning. This result is similar to the findings of Egeln et al. (2004), although they attributed low intraregional knowledge transfer effects to service-oriented types of spin-off.

Nevertheless, soft location factors played a significant role, since most of the founders tend to locate in spatial proximity to their parent institute, often even in the direct neighborhood. This was due to the benefits they received through direct knowledge exchange (bi-directional) with their former colleagues and principals, through easier recruitment of specified staff, through mutual utilization of resources and personnel, and through easier procedures to arrange joint projects or other forms of collaboration. After some years of existence, when the company develops positively, it tends to emancipate itself from the parent institute and develops its own target markets, networks and a proprietary knowledge base. The relations with the parent organization may become weaker; the opportunities for contacts decrease. According to our findings, this will be the normal process of disembeddedness/independence which means that spin-offs operating successfully in the market cannot be bound to the parent institute forever.

### 9.5.4 Implications for Regional Development

**Financing and growth**
Although we have not found proof that scarcity of capital is a major reason for poor performance of the firms, it undoubtedly can hinder ambitious founders from growing faster and thus limits the possible contribution of the firm to the growth of the regional economy. The empirical findings have, again, shown the well-known phenomenon of the early stage private financing gap.[17] The companies start with scarce resources, small staff and modest ambitions. This cannot only be traced back to a distinct skepticism on the part of the founders regarding bank loans, equity capital or on the part of private VC investors and banks being risk averse and reluctant. Founders also maintain quite subjective, and much too often justified prejudices towards bank financing and private equity. Support for spin-offs via public promotion during the pre-seed, seed, foundation and build-up phase appears in this situation to be a necessary subsidiary consequence.

Our case studies show frequently that founders – in the presence of an 'adequate' total financing concept in their eyes – believed that it was unnecessary to tap further capital sources than those they had. Therefore the growth potential of the founding firms is often not fully exploited. (Regional) public support schemes should insist on a financing mix meaningful for the company, which is composed from various sources. Public financing should strengthen existing potentials instead of completely compensating financing gaps. The government (as promoter) can establish conditions for promotion so that an adequate mixture of promotional funds, own and outside capital as well as cash flow financing is achieved in the companies' development phase.

In the few cases where a firm was founded at another location than that of the parent institute, or when it relocates afterwards, we observed that it had been lured away by attractive public promotion. This was mainly the case when firms moved from west German *Länder* to eastern ones. Other locational factors could not compete with such offers in these cases.

### Improved strategic management and commercial qualifications for the founders

Many academic spin-offs lack, from the very beginning, a professional, structured enterprise strategy. Sometimes even clearly defined targets are missing. Remarkable too is the frequent lack of knowledge about the situation in the target markets. Elementary key information such as market potential, competitive situation, own market share etc. are often not known or are wrongly estimated. Actually, these are central requirements of a business plan which is presented to a sponsor or investor. A start-up should not receive public funds if the business plan does not qualitatively and quantitatively fulfill the minimum standard of corporate planning.

One support activity in that direction is qualification programs by parent organizations (e.g. in the EXIST framework[18]) or by public bodies. From our case studies we can conclude that such qualification offers, which should be local, can be of decisive significance for the development of the firm. It should also be considered whether – for promoted start-ups – they should be made obligatory. However, a careful selection of the training schools and subject matter must be made as this market is confused and there is no efficient quality control.

### Starting points for the parent organization's spin-off policy

The scope of support actually provided by the parent organizations is very broad. For the strategy patterns of firm formation support by the parent institutes presented in Table 9.2, the possibilities of stronger links between science and industry should be examined. A solution which would be uniformly applicable does not appear practicable against the background of specific strengths, preferences and development potentials. The strengths of pattern 2 lie in the 'instigation', the early sensitization for the funding and orientation of the subject or project towards application-oriented and near-to-market activities in research and teaching. Against this background, specific other approaches to the targeted extension of founding promotion also emerge. In strategy pattern 3 (support in qualification and organization of knowledge transfer) and 4 (maximum support and high selectiveness), which are rather to be encountered in non-university organizations, the advantage lies essentially in the high specificity and selectivity. Institutes already active for years in the relevant research markets can prepare the markets for the spin-offs, just as they can also help the founders to prepare in a more targeted manner for their entrepreneurial tasks. Highly specialized, technical expert and excellent enterprise units can be created which from the very beginning can take a top place and earn a high market share.

The better this succeeds, the sooner the founders will be able to emancipate themselves from the parent institute. Spin-offs from institutes with a classical scientific (i.e. not exploitation-oriented) set-up, strive to draw a clear demarcation line with regard to work style and organization to the parent institute, even relocating from its former regional environment. Nevertheless, our case studies show that not only a rapid disembeddedness, but also an enduring, close co-operation with the parent institute can be ways to success.

## 9.6 SUMMARY OF THE FINDINGS

As already pointed out in section 9.4, the case study approach provided us with detailed insights into the foundation and development process of the analysed spin-offs, but implied limitations to deriving general conclusions from our findings. Nevertheless, with regard to the assumptions formulated in section 9.2, we came to the following conclusions for the sample of spin-offs studied in this chapter:

1. *Academic spin-offs play a highly important role in the process of knowledge transfer*: This is supported by our observation; many spin-offs hold either licenses granted by their parent organizations or even own patents. However, it must be emphasized that not all spin-offs build their business concept upon developments made in the research establishment. But in any case, knowledge that was acquired during the former employment in the scientific environment is deployed to execute the business concept, and vice versa. Knowledge the entrepreneurs gain in the markets may also be fed back in cases where they maintain a close relationship with their former colleagues. What can also be concluded from the empirical analysis is that the

assumed linearity in technology transfer processes, e.g. that foundation ideas are triggered by the exploitation of research results, is a popular relic of formerly prominent perceptions. Only in a few cases are the spin-offs built upon some already exploitable research results of the parent institutes. What can be observed is that the founders more often have a strong personal affinity and motivation to proceed further with their current research subjects or to bring about something useful for society, while the institute itself has little interest in commercializing the research results. An even more important finding is that, if technological developments prepare the ground for the spin-off project, these are rarely developed in the parent organizations to a marketable state, mostly not even ready for prototyping. The founders have, in these cases, to invest substantial extra efforts to get the products ready for market, which usually exhausts the resources they have acquired as start-up funding.

2. *High-tech spin-offs and start-ups and play a significant role in structural change and in the regional technological and knowledge modernization process*: To play this role requires a good integration into the regional economy and innovation system. The majority of the case study spin-offs missed this integration, at least in their early stage of existence. Good linkages exist within the scientific community in which academic founders feel familiar, but they are poorly networked with the commercial and business sectors of the regional economy. Such links tend to develop step-by-step, but rather on a transregional, often international scale, than on a regional scale (Koschatzky and Stahlecker, 2006). In theory, high-tech spin-offs should possess these capabilities, but in practice they do not seem to have them. Another issue is that the spin-offs analysed by our case studies lack sufficient financial sources to be able to grow fast enough. Additionally, the founders dramatically lack the necessary growth perspective and the management skills to steer their companies to the growth path needed to meet the expected impact on regional structural change or on the technological and knowledge modernization process.

3. *The start-up intensity is strongly influenced by regional supply and demand conditions*: Although this assumption was often supported by other studies (e.g. Almus et al., 2001), we cannot find much evidence in our data. There was nearly no regional demand for the products and services of the spin-off firms in terms of technology input from other than the parent institute. Only in cases where larger companies of related sectors are located in the same region might they become clients or R&D partners of the new firms. This was more the exception

than the rule. The issue of regional supply structures, however, has not been investigated in detail, and supply structures have not been voiced as important for the development of the spin-off firms.

4. *Regional framework conditions as defined by the criteria of an entrepreneurial region exercise a positive stabilizing effect upon start-ups during their early phases*: This effect could be supported for soft locational factors. Founders in such regions reported the encouraging and supportive effect of a dynamic and pro-entrepreneurial atmosphere, as far as they are linked to some local networks at all. They benefit from role stories of former firm formation processes, from assertive, encouraging attitudes of university professors, institute directors or well-known entrepreneurs. Knowledge transfer in the sense of 'embedded dynamic relations' (Lambooy, 2004) plays an important role in this respect. On the other hand, in cases where the physical, institutional and service infrastructure is fairly similar in every place, even in peripheral locations, this infrastructure endowment was reported to be not decisive. It can thus be concluded that spin-offs located in less entrepreneurial or peripheral regions apparently were affected by the lack of a supportive entrepreneurial environment, especially in the early phase of their development.

5. *The public funding of academic spin-offs supports their regional integration and their contribution to the regional economy*: This assumption sounds reasonable, but cannot be supported in a clear-cut manner by our data. The correlation strongly depends on the appropriateness, complementarity and diversity of the support measures offered. There are clear indications that, for instance, mere financial aid, particularly direct grants for start-ups, tempts new companies to operate in a rather uncommercial, namely more science-oriented manner which, as a rule, diminishes the chances to become established in the markets and accepted by industrial clients. We would rather reformulate the assumption, so that the more public funding is involved, the greater the academic spin-offs' distance to the markets.

Especially from assumption (4) it can be concluded that the different research organizations could perform specific functions in stimulating academic spin-off behavior and in the support of these firms in their early development stages. This holds especially true for regions which miss a well developed entrepreneurial climate and in which academic organizations (universities, technical colleges/universities of applied science, and research institutes) play a special 'lighthouse' role in the generation and distribution of knowledge. In Germany, much progress has been made in entrepreneurship education by universities and technical colleges after the mid 1990s (Koschatzky, 2003). Since academic firm founders still lack the necessary managerial and marketing skills to run a company successfully, entrepreneurship education and the essential networking between different partners for the training of entrepreneurs might serve as a fundament for learning processes between the parent organizations and other partners, e.g. from the venture capital market, in order to establish sustainable conditions for a further stimulation of academic firm formation and thus for economic growth, both in agglomerations and in peripheral regions.

## NOTES

1. The five new federal states (*Länder*) emerged from the former German Democratic Republic (GDR) after German reunification in 1990. They form, together with the state of Berlin, the six East German *Länder*.
2. The old federal states comprise the 10 *Länder* which constituted the 'old' FRG before reunification (West Berlin, then, was treated as the 11th quasi state but for political reasons was not part of the FRG).
3. Throughout this chapter, a difference is deliberately made between 'parent institute' and 'parent organization', the latter indicating the holding institution or the association to which the individual institutes belong, e.g. Fraunhofer institutes to the Fraunhofer Society FhG or Leibniz institutes to the Leibniz association WGL. This distinction is made to demonstrate the different levels of support and policies that play a role in assessing the support the spin-offs receive in practice (see also section 9.3 for details).
4. Incubator organizations are understood here as scientific breeding or parent organizations rather than as mere facilities housing start-ups and offering financing and office services.
5. There are also a number of privately run universities, like the International University Bremen.
6. It is obvious that this number is by no means the total population of academic spin-offs.
7. IPR = Intellectual Property Rights like copyrights, patents, licenses.
8. This aspect was not investigated in the second study.
9. Basis is the sub-sample of 20 cases of the first study.
10. This aspect was not investigated in the second study.
11. This is not meant as a ranking.
12. For the notion of success, see the next chapter.
13. In order to derive these performance marks, a simple scoring method was deliberately chosen in which each company was assessed for any of the six criteria, whereas both subjective judgments of the interviewers and available company data were used. The six individual scores of each company were

counted and the final mark per firm resulted from the balance of positive and negative scores.
14. Covering all 59 case studies of both studies.
15. Only including 20 cases of the first study.
16. The second study did not analyze this aspect.
17. Detailed data on this are not available.
18. EXIST is a federal program that supports the development of favorable conditions to generate spin-offs from German universities (cf. Meyer-Krahmer and Kulicke, 2002).

# REFERENCES

Almus, M., J. Egeln and D. Engel (2001), 'Determinanten regionaler Unterschiede in der Gründungshäufigkeit wissensintensiver Dienstleister' (Determinants of the regional differences in firm formation rates of knowledge-intensive services), *Jahrbuch für Regionalwissenschaft*, 21, 25-51.
Antonelli, C. (1999), 'The evolution of the industrial organisation of the production of knowledge', *Cambridge Journal of Economics*, 23, 243-260.
Audretsch, D.B. (2004), 'Sustaining innovation and growth: public policy support for entrepreneurship', *Industry and Innovation*, 11, 167-191.
Audretsch, D.B., E.E. Lehmann and S. Warning (2005), 'University spillovers and new firm location', *Research Policy*, 34, 1113-1122.
Bergmann, H. (2004), 'Gründungsaktivitäten im regionalen Kontext. Gründer, Gründungseinstellungen und Rahmenbedingungen in zehn deutschen Regionen' (Firm formation activities in the regional context. Founders, founder attitudes and framework conditions in ten German regions), Köln, Selbstverlag im Wirtschafts- und Sozialgeographischen Institut der Universität zu Köln.
Birch, D.L. (1987), *Job Creation in America. How our Smallest Companies Put the Most People to Work*, New York, Free Press.
Cooke, P. (2002), *Knowledge Economies. Clusters, Learning and Cooperative Advantage*, London, Routledge.
Cooke, P. (2004), 'Introduction: Regional innovation systems – an evolutionary approach', in P. Cooke, M. Heidenreich and H.-J. Braczyk (eds), *Regional Innovation Systems. The Role of Governance in a Globalized World*, 2nd edition, London, Routledge, pp. 1-18.
Egeln, J., S. Gottschalk and C. Rammer (2004), 'Location decisions of spin-offs from public research institutions', *Industry and Innovation*, 11, 207-223.
Etzkowitz, H. (2002), 'Incubation of incubators: innovation as a triple helix of university-industry-government networks', *Science and Public Policy*, 29, 115-128.
Etzkowitz, H. and L. Leydesdorff (2000), 'The dynamics of innovation: from national systems and "Mode 2" to a triple helix of university-industry-government relations', *Research Policy*, 29, 109-123.
Fritsch, M. and P. Müller (2004), 'Effects of new business formation on regional development over time', *Regional Studies*, 38, 961-975.
Fritsch, M. and P. Müller (2006), 'The evolution of regional entrepreneurship and

growth regimes', in Michael Fritsch and Jürgen Schmude (eds), *Entrepreneurship in the Region*, New York, Springer.

Fritsch, M. and J. Schmude (eds) (2006), *Entrepreneurship in the Region*, New York, Springer.

Fritsch, M., U. Brixy, M. Niese and A. Otto (2004), 'Gründungen in Städten' (Firm formation in cities), *Zeitschrift für Wirtschaftsgeographie*, 48, 182-195.

Hemer, J., H. Berteit, G. Walter and M. Göthner (2006), *Erfolgsfaktoren für Unternehmensausgründungen aus der Wissenschaft (Success Factors for Academic Spin-offs)*, Stuttgart, Fraunhofer IRB Verlag.

Hemer, J., M. Schleinkofer and M. Göthner (2007), 'Akademische Spin-offs – Erfolgsbedingungen für Ausgründungen aus Forschungseinrichtungen'(Academic spin-offs – Success factors for spin-off activities of research organisations, Studies for the Bureau of Technology Assessment at the German Parliament), Studien des Büros für Technikfolgen-Abschätzung beim Deutschen Bundestag, Berlin, edition sigma.

Kerst, C. (1997), *Unternehmensbezogene Dienstleistungen als Elemente einer innovativen Regionalökonomie* (Business-oriented services as elements of an innovative regional economy), Stuttgart, Akademie für Technikfolgenabschätzung in Baden-Württemberg.

Koschatzky, K. (2003), 'Entrepreneurship stimulation in regional innovation systems – public promotion of university-based start-ups in Germany', in D. Fornahl and T. Brenner (eds), *Cooperation, Networks and Institutions in Regional Innovation Systems*, Cheltenham, UK and Northampton, MA, USA: Edward Elgar, pp. 277-302.

Koschatzky, K. and T. Stahlecker (2006), 'Structural couplings of young knowledge-intensive business service firms in a public-driven regional innovation system. The case of Bremen, Germany', in M. Fritsch and J. Schmude (eds), *Entrepreneurship in the Region*, New York, Springer, pp. 171-193.

Koschatzky, K., S. Bührer, J. Hemer, T. Stahlecker and B. Wolf (2008), *Institute und neue strategische Forschungspartnerschaften im deutschen Innovationssystem*, (Institutes associated with universities and new strategic research partnerships in the German innovation system), Stuttgart, IRB-Verlag.

Lambooy, J.G. (2004), 'The transmission of knowledge, emerging networks, and the role of universities: an evolutionary approach', *European Planning Studies*, 12, 643-657.

Lockett, A., D. Siegel, M. Wright and M.D. Ensley (2005), 'The creation of spin-off firms at public research institutions: managerial and policy implications', *Research Policy*, 34, 981-993.

Malecki, E.J. (1994), 'Entrepreneurship in regional and local development', *International Regional Science Reviews*, 16, 119-153.

Meyer-Krahmer, F. and G. Lay (2001), 'Der Stellenwert innovativer Dienstleistungen in der Modernisierungsdebatte', (The significance of innovation services in the modernization debate), *WSI-Mitteilungen*, 6, 396-400.

Meyer-Krahmer, F. and M. Kulicke (2002), 'Gründungen an der Schnittstelle zwischen Wissenschaft und Wirtschaft – die Rolle der Hochschulen' (Firm formation at the interface between science and industry – the role of the universities), *Perspektiven der Wirtschaftspolitik*, 3, 257-277.

Miles, I. (2003), 'Services and the knowledge-based economy', in J. Tidd and F.M.

Hull (eds), *Service Innovation, Organizational Responses to Technological Opportunities & Market Imperatives*, London, Imperial College Press, pp. 81-112.

Nerlinger, E. (1998), *Standorte und Entwicklung junger innovativer Unternehmen. Empirische Ergebnisse für West-Deutschland* (Locations and development of young innovative firms. Empirical results for West-Germany), Baden-Baden, Nomos.

Niese, M. (2003), *Ursachen von Betriebsschließungen. Eine mikroökonomische Analyse von Probezeiten und Todesschatten im verarbeitenden Gewerbe* (Causes of firm closures. A microeconomic analysis of probation periods and death shadow in manufacturing industry), Münster, LIT Verlag.

Reynolds, P., D.J. Storey and P. Westhead (1994), 'Cross-national comparison of the variation in new firm formation rates', *Regional Studies*, 28, 443-456.

Romer, P.M. (1990), 'Endogenous technological change', *Journal of Political Economy*, 98, S71–S102.

Shaver, J.M. and F. Flyer (2000), 'Aggomeration economies, firm heterogeneity, and foreign direct investment in the United States', *Strategic Management Journal*, 21, 1175-1193.

Sternberg, R. (2000), 'Gründungsforschung – Relevanz des Raumes und Aufgaben der Wirtschaftsgeographie' (Research on firm formation – relevance of space and tasks of economic geography), *Geographische Zeitschrift*, 88, 199-219.

Strambach, S. (2001), 'Innovation processes and the role of knowledge-intensive business services (KIBS)', in K. Koschatzky, M. Kulicke and A. Zenker (eds), *Innovation Networks – Concepts and Challenges in the European Perspective. Technology, Innovation and Policy*, Heidelberg, Physica-Verlag, pp. 53-68.

Wood, P. (2002), 'Knowledge-intensive services and urban innovativeness', *Urban Studies*, 39, 993-1002.

# 10. On the economics of university ranking lists: intuitive remarks on intuitive comparisons[1]

## Ádám Török[2]

## 10.1 Introduction

It is a widely held belief that the superiority of the American R&D and innovation system over its counterparts in Europe (Rodrigues, 2003) is based, to a large extent, on the outstanding educational, research and also fundraising performance of American universities. The American system of higher education is often referred to as a benchmark for European efforts for making higher education and research in Europe more effective (European Commission, 2003a, 2003b). The mid-term reviews of the Lisbon Agenda stress Europe's lag in R&D with less emphasis on the condition of its education system (see Rodrigues, 2005).

Both intuitive thinking and widespread personal experience confirm the view that the American higher education system is the best in the world, but really convincing analytical evidence is not available yet in this respect. This might sound surprising since an entire industry ('a cottage industry', Thursby, 2000, 383) has recently developed in order to prepare quantitative rankings of universities. While the underlying idea is that universities compete with each other, institutions producing university ranking lists might also be in competition for wider readerships and a stronger influence on the choices of universities by future students. The question we have to ask at this point is twofold: (1) Is it really true that there is competition between universities analogous with the one in business known from a huge body of literature and everyday experience? (2) If competition is a fact in the university world, how can it be measured? A related question arises in case of an affirmative answer: are often published university ranking lists a good measure of that competition?

This chapter assesses the literature of university ranking lists first, putting it in the context of competitiveness analysis. A critical evaluation of two widely quoted international university ranking lists follows with an assessment of the ranking criteria used by them. The chapter concludes with remarks on the validity of university ranking lists.


## 10.2  COMPETITION BETWEEN UNIVERSITIES

Universities are expected to supply labour markets with qualified workforce. The globalisation of labour markets made universities institutions operating increasingly in competitive environments. An American, a German or a French university in the 1950s or 1960s served students mainly from the university's local environment and perhaps also from other regions of the same country, together with a limited number of students coming from Third World countries financed from development aid. Students wishing to study abroad may have encountered financial difficulties. Obtaining student visas was not quite as easy as in most industrial countries in the early 2000s. Information on foreign universities was only scarcely available.

Many if not most countries had state-owned higher education systems. This meant in many cases that government financing was assured independently from the amount of students admitted and degrees issued. The strengthening of budget constraints for state-owned higher education institutions resulted in a variety of solutions for linking university financing to university output (see Vossensteyn, 2004).

Such is, for example, the Hungarian system of unit based financing in which each additional student enrolled in a given university or college ensures a fixed additional amount of annual government financing for that institution. This system is considered to increase the financial efficiency of universities and constrain the waste of public money, but it has some shortcomings related to its being an incentive towards the overutilisation of the human and physical capital stock of universities and also towards underinvestment. (Morgan, 2000, 1)

Competition between universities has also been put in the focus of interest by the increasing popularity of the concept of competitiveness and the mushrooming on comparative international analyses of competitiveness (see Laursen, 2000; UNCTAD, 2002; Török et al., 2005). The idea of applying the concept of competitiveness in a macroeconomic approach and in cross-country comparisons has raised strong criticism (Krugman, 1994; Krugman and Obstfeld, 2000). Such negative assessments were, however, limited only to cross-country studies in which competitiveness levels of national economies were compared. The strategic management related literature (for a good survey of this literature see Moore, 2001) of competitiveness focuses on inter-firm or intra-industry analyses. This study is based on the hypothesis that higher education carries some characteristics of industrial sectors, and therefore it could be subjected to an inter-country competitiveness analysis. Our focus is primarily on whether international university ranking lists are in line with some important methodological requirements of competitiveness analysis.

University ranking lists may influence decisions of the supply and the demand sides of the market of higher education services regarding choices of financing, job applications in universities (supply side) as well as enrolment by students and choices of graduates by employers (demand side). R&D at universities do not belong to the classical education function of universities, but high-level R&D is key both to university financing and the maintaining of professional standards for higher education institutions operating in competitive environments. Therefore R&D should also be considered part of the competitive performance of a university.[3]

Competitiveness analysis has to cover both the inputs and the outputs of market players (Török et al., 2005, 21). Universities compete with each other for students, human and financial resources as inputs, and market shares measured on the output side. Regarding market shares as output-based measures of university competitiveness, the application of this measure may be controversial in practice because it depends on a number of non-quantifiable qualitative elements. With regard to methodological issues of competitiveness measurement, it also raises further questions about how 'market shares' could be compared between different universities. Market shares can be applied only to the comparisons of sales or exports of goods fully substituting each other. A student earning an MA degree in physics from university A may not be regarded as a complete substitute of a student of the same discipline graduating from university B.

## 10.3 UNIVERSITY RANKINGS IN THE LITERATURE

University ranking lists have been provided in an increasing supply since the late 1980s, but most of these lists appeared in non-scientific publications. This might be one of the reasons why no scientific standards have been developed in order to assure the methodological consistency of these lists. One can speak of a proliferation of such lists, even if many of them are only national. The business rather than scientific orientation of these lists seems to be reflected by the fact that two of the three kinds of university ranking lists contain data on education in economics or business. Such lists have been published on

1. universities (including comparisons of university departments);
2. economics departments; and
3. MBA programmes or schools of business/management.

These different kinds of ranking lists were published depending largely on their target audiences. General university ranking lists usually appear in widely read daily newspapers or political weeklies. Rankings of economics departments were published in journals of economics, and lists of MBA programmes or business schools can be usually found in the business press. Some of the most prestigious business journals regularly publish their own ranking lists of business schools.

The officially communicated purposes of such comparisons included

1. ranking scientific excellence between universities;
2. valuating investments in degrees with respect to future chances on labour markets; or
3. estimating the competitive positions of universities for fundraising and recruiting.

The three purposes are interrelated, but it seems obvious that one could generate three different ranking lists based on each of them. Scientific excellence is a criterion with a strong research orientation which means a university excelling in science may show mediocrity with respect to graduate or undergraduate education. The value of an investment in a degree issued by a given university may greatly depend on the weight of the business, the medical or the law school within that university. The reason is these are the degrees which promise above average incomes to their holders. The fundraising and recruiting potential of a university reflects its scientific excellence and its reputation on the labour market. There is an additional factor influencing the fundraising potentials of universities which is strongly linked to past reputation and performance, first of all in North America. This is the fundraising capacity of alumni and their associations. The alumni of the most prestigious universities are the best placed for raising additional financing for their old institutions of education. This is why there seems to be a strong link between past and present reputation and the financial potential of universities.

A wide analytical literature exists on comparisons and rankings of economics departments, units of higher education with comparable outputs and also with quite good perceptions of the competitive requirements of the labour market. An excellent survey of this literature is presented in Thursby (2000), although his article constrains itself only to a comparison of the research outputs of economics departments of American universities. Thursby's survey does not establish an explicit connection between the field of science taught in such departments and the methodology used in such comparisons. Still, it finds that most comparisons between economics departments[4] use similar methodologies and operate with similar variables. This is why the results of most of them are also similar.

Such rankings are based on elements of output such as publications and PhD degrees granted, but they do not cover the qualitative aspect of research and teaching, and neglect the importance of the innovative element in the activities of economics departments (Thursby, 2000). It would be an exaggeration to use innovation as a benchmark for economics departments as is usual in other, natural or technical, fields of science, but opening up new fields of research is also key to outstanding scientific production in social sciences.

Producing PhDs and publications may be a good output indicator for a higher education institution not related directly to business.[5] Economics departments grant degrees offering job prospects for example in the university world, at think tanks, government institutions and in strategic or analytical units of major firms. Their graduates and PhDs do not usually enter the labour market in the broader segments open to earners of business degrees.

MA programmes in economics and MBA programmes have common origins. Some parts of their curricula overlap, but their strategic orientations are significantly different. Knowing this, the analysts comparing MBA programmes with each other consider studying for an MBA at least as much an investment and a network building effort as a learning process. This is why these comparisons use indicators showing different aspects of return on such investment, not only the tangible ones.

The parameters used for ranking MBA programmes include such components which may make a given MBA degree especially attractive for its holder preparing for a career in business and interested in building a wide international network of professional relationships. MBA programmes are therefore usually ranked on the basis of

1. Post-degree earnings;
2. The international character of the faculty and the student body at the given business school;
3. The strength of the alumni network of the school; and
4. The business relations opened up to students by a given MBA programme and degree.

The rankings of MBA programmes and business schools can be considered as lists based on criteria of a certain understanding of competitiveness, but their methodological background is very special. MBA ranking lists are rather comparisons of firms operating in a special kind of market than of institutions producing and disseminating knowledge.

The most widely used and most often cited international ranking lists cover universities. Comparisons of university faculties or departments operating in the same field of science (e.g. engineering, law, economics, medicine or natural sciences) are common in national frameworks.

A feature shared by all university rankings (including those of economics departments and MBA programmes) is a lack of theoretical background in all of them, with one interesting exception to be explained below. The purpose of the rankings is not exactly declared in the introductions of most such analyses. Some comparisons simply say they want to show which universities are best, others produce rankings based on 'quality' (*The Times*, 2004). Methodologically demanding analyses of university or related rankings are sceptical regarding the scientific value of these ranking lists: 'Ranks, however produced, certainly provide entertainment, probably affect the hunt for graduate students and new faculty, and might serve to increase a department's resources' (Thursby, 2000, 383).

### 10.3.1  A Parallel with Chess

A meticulous and theoretically interesting ranking list of American colleges and universities was produced by Avery et al. (2004). This analysis attempts to create a ranking list based on the revealed preferences of students seeking admission to higher education. Revealed preferences are, in this case, the choices by students from the universities to which they applied and which admitted them to their BA programmes.

The choices of universities by students are represented as personal lists of preferences. The authors aggregate individual student choices in a model which is based on the logic of Elo ranking lists known from chess.

This means that university A may earn or lose ranking points if it ranks higher or lower than university B in the aggregation of student choices. Its actual gain or loss of points depends on the difference between its pre-comparison ranking points and those of B. If A is ranked much higher than B then B's better performance in getting students earns him relatively many points since the expected 'winner' would have been A (due to its higher former ranking position).

The logic of this kind of comparison is based exclusively on how universities are seen by students seeking admission to them (and also on how they prepare their strategies of application and immatriculation trying to optimize their payoffs), and not at all on what the current performance of a university is in education and research. Student strategies depend on factors such as the location of a university, financial aid a university can offer, and also on non-quantifiable cultural factors (e.g. the religious background of a higher education institution).

Students reveal their preferences based on a number of tactical considerations. They usually do not set up lists of the really best, but of the best they could still reckon to reach. Therefore the probably outstanding universities figure rather on the lists of the best students which may further improve their standing. Furthermore, they may see services offered by universities as packages of educational, financial aid and placement services. The value of a degree from a university depends on the combination of these factors by students with their individual consideration of these services. Some student choices may give priority to universities competing with price (including the terms of financial aid) rather than with quality.

This list creates a quite distorted picture of the American university world which can be explained with examples taken from chess. The leader of the ranking is Harvard University with an Elo number of 2800.[6] The chess world is structured in such a way that the best able to challenge the world champion have Elo numbers above 2600, those players with no less than 2400 are still considered professionals, the category above 2200 is that of the competent amateurs, but anybody in the category below 2000 has absolutely no chance to reach a draw with any member of the narrow international elite – even from a series of 50 games played in normal circumstances.[7]

The Avery list is based on the idea that students' revealed preferences cannot be manipulated by universities as can some other statistics, but it sets up a popularity list rather than a performance- and capacity-based list. It uses an analogy with chess which creates the impression that most colleges and universities in the United States have no chances in the competition with the best American universities.

## 10.4  UNIVERSITY RANKINGS AND COMPETITIVENESS

The university rankings do not explicitly refer themselves to the idea or concept of competitiveness, but it seems most of them have some vague understanding of competitiveness in their background. Still, no such reference is explicit in any of the ranking lists we have surveyed. The lack of theoretical basis for university ranking lists is all the more surprising given that their tradition goes back more than 20 years. The first such list was published by the US News and World Report in 1983 (*The Times*, 2004), covering only colleges and universities in the United States. University or related ranking lists became regular from the early 1990s in the industrial countries.[8]

The most widely known university or related ranking lists (excluding rankings of MBA programmes and business schools) are published by the political and business press able to reach the widest possible audiences. *The Times* (Top 100 Universities in the UK), *The Guardian* and *The Financial Times* regularly produce such lists in the United Kingdom, and the best known German lists come from the political weeklies *Der Spiegel* and *Die Zeit*.

We leave national lists aside in our further analysis. National lists often use country-specific parameters with little relevance for other countries.[9] We are mainly interested in the global picture of the world of higher education in general. Our main intent is looking for evidence on the assessment found in a number of Lisbon Agenda-related documents (European Commission, 2003a, 2003b; Rodrigues, 2003, 2005) that higher education in Europe lags behind the higher education system of the United States.

It would be interesting to see how this assessment is supported by two widely popular international ranking lists of universities. National ranking lists of universities and other higher education institutions are, however, much more popular than international ranking lists.

The techniques chosen for preparing international and national ranking lists of universities are not too different. The huge differences in their popularity seem to show that people read such lists in order to help their own or others' decisions on choosing a university from a given country, rather than to choose a country first and a university there afterwards. Choices of application or enrolment seem to be made more between the universities of a country already selected than between all universities of the world matching some criteria set by the potential applicant.

### 10.4.1 General Remarks on International Lists

The international university ranking lists do not have strict institutional criteria. This means their authors accept the national definitions of 'university' from all over the world. This flexibility has no alternative, since graduate degrees granted by any university of the world are accepted by any other university's PhD programme as a condition of application.

The term 'university' covers a wide range of institutions. For example, schools specialising only in agriculture, technology or medicine are called universities in many countries (e.g. Russia) even if the term itself means access to the 'universe' of sciences and not only to one restricted field. In other cases, some universities can boast of impressive indicators of output because they emerged from a series of university mergers as the leading national university of a country (Russia, India, Singapore), but therefore they are not completely comparable with smaller universities of other countries.

A factor distorting the picture offered by university comparisons is that several such ranking lists do not take institution size into account, although competitiveness analysis has two distinct approaches completing each other (see Török et al., 2005). The 'absolute' approach focuses on competitive performance regardless of efficiency (and thereby gives an implicit priority to larger players), whereas the 'per capita' approach allows for good competitive performance of efficient competitors of smaller sizes. University ranking lists use the absolute approach only, and this approach does not consider the size of inputs used for producing a unit of research or educational output. This is why the national 'superuniversities' of certain mainly non-OECD countries can fare quite well on the ranking lists.

Some comparisons use size as a variable irrespective of its relationship to output. Such lists give extra points to schools with high employment levels even if their large sizes are due only to their respective profiles. For example, many medical or agricultural schools maintain service and production facilities generating massive income from their core, market-oriented activities besides supporting education and research within the university.

Some comparisons (e.g. Thursby, 2000) consider research as the main output of a higher education institution. This approach makes so-called research universities[10] perform well in the rankings, even if they do only limited undergraduate and graduate education. The measurement and comparison of research outputs raises a number of methodological concerns (see Godin, 2003; Török et al., 2005), primarily with respect to the measurement and comparison of quality.

It would be simplified to say that a university can have only two kinds of output, namely research and education (i.e. students with degrees). Research output is linked to innovation output, but the two are markedly different from each other with innovation output more frequently produced by universities of technology than classical universities of science. Educational output has two major components with different parameters of quality: mass (graduate) degree output and quality (PhD) degree output. PhD output helps create synergies between university performance in research and in education, but the number of PhDs produced (see on this indicator Thursby, 2000) is not an appropriate measure of both outputs combined.[11]

We begin with the analysis of two international ranking lists of entire universities, the increasingly popular Jiao Tong list (Top 500) and the widely accepted *Times Higher Education Supplement* (THES) list (Top 200). Our purpose is less a detailed assessment of two lists selected from many than rather an inquiry into the logic and the economic ideas underlying university ranking lists in general, and the validity of this approach for international comparisons. Some elements of other lists are also considered while the two lists are being evaluated.

The structure of assessing the two lists will be as follows: (1) presentation of list; (2) presentation of ranking criteria; (3) comments on list; (4) comments on ranking criteria; (5) conclusions on the list in question.

## 10.5 THE JIAO TONG LIST ASSESSED

The international ranking list of universities published by a team of the Shanghai Jiao Tong University in early 2005 was fast to gain international reputation. This is a rather simple list with a certain lack of methodological sophistication. Its simplicity made it possible for its authors to compile data on many universities. The list may have become popular because the number of the world's 'top' universities is quite high here (the list ranks 500 universities altogether as opposed to only 200 in the THES list), thus many university leaders can claim, based on this list, that their institution belongs to the world elite.

This is probably the only worldwide ranking list of universities in which OECD countries are not overrepresented, and a number of Third World countries can be found on it. Russian, Polish, Czech and Hungarian universities can also be found in this club of apparent prominence.[12]

The list confirms widely held opinions considering American universities dominant in the world. The only European country with some universities matching American competitive standards seems to be the UK. The first 20 universities on the Jiao Tong list include only two British universities (Cambridge and Oxford) and one institution from Japan (Tokyo University). The United States is represented with 17 universities, including many great names in the world of higher education.

The validity and the methodological foundations of the Jiao Tong list can be questioned if positions between 100 and 500 are considered. The lack of sophistication of the method of measurement results in a lack of distinction among several dozens of universities. For example: no less than 100 universities are tied for position 202, and about the same thing can be observed for the subsequent ranking positions. Thus no less than 300 universities could claim that they belong to the first 200 in the world according to this list, and the first 300 could also mean about 400.

### 10.5.1 The Jiao Tong Criteria

The survey of the quantitative criteria of the Jiao Tong list confirms previously pronounced ideas about its simplicity which, however, does not necessarily mean clarity. Altogether six indicators are used, structured in four groups. The names of some groups speak of things different from the content of the indicators.

The first group of criteria ('Quality of Education') has a weight of 10 per cent. Under this heading, the alumni of the institution winning Nobel Prizes and Fields Medals[13] are counted. This criterion is a proxy of the past quality of education in the institution, but it doesn't have much to do with its current qualitative parameters.

The group 'Quality of Faculty' has two 20 per cent weights. The first proxy is the number of staff winning Nobel Prizes and Fields Medals while being employed by the given university (with more points given for more recent awards). This is a questionable measure due to, among other things, the sometimes weak relationships between scientific awards and real academic performance (on the mechanism of awarding Nobel Prizes see Hargittai, 2002, especially Ch. 3), and also to the fact that the competitive positions of universities tied for lower positions in the list could be strongly influenced by the presence of an award winner there even decades ago.[14] The strong American dominance in winning Nobel Prizes, mainly from the early 1980s on, may partly explain the excellent showing of US universities on the Jiao Tong list.

The other qualitative indicator of faculty strength is the citations performance of staff in 21 fields of science. There is a wide literature on the validity of citation indexes in showing scientific quality and performance (see a survey of this literature in Török et al., 2005).[15] Citation indexes are quite good indicators of research quality. They do not reflect, however, educational quality and performance which should also be key to evaluating universities in an international comparison.

The next group of indicators is called 'Research Output'. It reflects publications performance. Articles published in *Nature* and *Science* represent 20 per cent of the total score of the universities of the authors, and another 20 per cent may be obtained for publications in the Science Citation Index and the Social Sciences Citation Index. The Jiao Tong methodology is quite appropriate for evaluating co-authors' contributions (only the first or the corresponding author gets full credit for the article while other co-authors obtain lower points depending on the total number of authors, Shanghai Jiao Tong University, 2005). Still, university output embodied in publications does not say much about real performance in education. Using this indicator distorts the comparison of international higher education because it seems to reward universities for sending their best professors into research with some teaching load in PhD programs while undergraduate and graduate classes are increasingly taught by lower-level teaching staff. This shift of quality output from education to research is not well tackled by the Jiao Tong and similar lists.

The final group of Jiao Tong indicators is called 'Size of Institution' with a 10 per cent weight. It is used to achieve a balance between absolute and per capita (or efficiency-based) indicators of university quality and performance. The previously calculated indicators are size-adjusted here on the basis of the employment level (full time equivalent (FTE) academic staff) of each university in such a way that their total score is divided by the FTE number of academic employment. This size-adjusted calculation has, given its 10 per cent weight, only a minor influence on establishing the ranking position of a university.

### 10.5.2 Comments on the Jiao Tong Criteria and List

The Jiao Tong list has some shortcomings. To begin with, it lacks methodological sophistication which makes the list hardly useable for comparing universities outside the top 100 of the international field. The second important problem is that universities are regarded as merely research institutions and their output is not evaluated in terms of knowledge production and dissemination through education. This approach gives priority to research universities.

The Jiao Tong list also tries to consider institution size which is an element largely neglected by most other university ranking lists. This effort towards measuring institutional efficiency is not really successful in its present form, since it disregards the fact that many universities in the world are state-owned. Therefore their size often depends on exogenous, not necessarily efficiency related, policy decisions.

The Jiao Tong authors seem to see the university world as a set of individual researchers organised in educational institutions. They put a great emphasis on individual scientific excellence, but do not consider a number of other competitive factors such as, for instance, the ownership and financial background of the institutions ranked. While the list's approach is partly retrospective, it seems to establish too strong a link between past research achievements and current scientific performance. It gives priority to the human factor as the input to research performance, but it leaves aside the influence of public opinion on establishing university rankings. The opinion of the scientific/higher education community is, however, also an important factor in creating ranking lists reflecting competitive positions as the competitors themselves see them.

## 10.6 THE THES LIST ASSESSED

The THES list ranks only the first 200 universities of the world, but differentiates among them in a way more sensitive than the Jiao Tong list. The top 20 of the world is more international in this list, since only half of these are American universities. Besides five British universities and one from Japan, Singapore, Switzerland, Australia and China are also represented (with one university each) among the first 20. While 85 per cent of the top 20 are American universities in the Jiao Tong list, the THES list also gives an 85 per cent relative share to universities from the English-speaking world in the top 20. The continental European Union is not represented in any of the two ranking lists at all.

The very top is similar on both lists with the most prestigious American universities (Harvard, Berkeley, MIT and Stanford) in the leading positions. The THES list, however, uses a methodology which reflects the structure of the top and the distances between the leaders. If the leader's (Harvard's) score is 100 per cent, then 50 per cent belongs to position 11 (London School of Economics), 25 per cent to position 39 (Hong Kong University), and positions 197 to 200 (University of Bremen, City University of Hong Kong, Virginia Polytechnic and Rensselaer Polytechnic) have about 10 per cent.[16] These distances can be correctly interpreted only when the quantitative criteria of the list are known. Still, these distances seem to be increasingly biased when moving downwards on the list.

### 10.6.1 The THES Criteria

The ranking criteria in THES have a more sophisticated methodological background than the Jiao Tong ones. Moreover, the THES criteria try to establish a balance between the research and teaching performance of each university assessed. A balance between quantitative and qualitative parameters is also sought by the authors of the list.

Great weight (50 per cent) is given to a 'peer review' of universities based on a questionnaire sent to 1300 academics in 88 countries. This qualitative component was missing from the Jiao Tong list. Its inclusion in the comparison is important, because any selection of 'objective' quantitative criteria has a biased character, and the opinion of the international scientific community is key to creating a list synthetising quantifiable and non-quantifiable parameters. This qualitative factor may be overvalued among the THES criteria, since it may be influenced by university marketing and the strength of alumni networks as well. A university having a poor former track record and with less established connections in the academic world has no real chances of getting a high position on the THES list, even if it performs well with regard to the other criteria.

The too strong reliance on this qualitative parameter of university comparison may become counterproductive. If this criterion is used for the same purpose for several years in a row, it may become increasingly 'self-fulfilling'. It could add to the reputation of the universities already placed high on the list, and may detract from the reputation of one outside the list but wishing to appear on it. In an optimal case, it could be recommended that the expert panel of 1300 should not have anyone connected with the previous THES list on board, but fulfilling this requirement would exclude a very important and prestigious part of the international academic community from the panel.

The second criterion is called 'Research impact', and has a 20 per cent weight. It is calculated as the number of citations divided by the number of faculty members. This variable may be autocorrelated with the qualitative one since both are influenced by the prestige of a university as well as of its teachers. The idea of measuring the impact of university research as the sum of the individual impacts of its staff members is also controversial.[17]

The quality of teaching is estimated by the next parameter called 'Committment to teaching' with a 20 per cent weight again. This is a faculty-to-student ratio for each university, but a faculty member is not necessarily doing teaching if she is mainly paid for research. Being a faculty member is not equivalent to being committed to teaching. Moreover, too high faculty-to-student ratios may mean something other than the strong commitment of a higher education institution to teaching: namely (mainly for state-owned universities) such rigid systems of employment in which positions for civil servants are maintained in spite of poor student interest.

The last two criteria have only a 5 per cent weight each. This is why they can only colour the picture. Both reflect the international attractiveness and prestige of a university. The first one is called 'International student body' and is expressed by the percentage of foreign students. The second one is the 'Ability to bring in best academics from abroad' with the percentage share of foreign faculty. Such indicators can be often found in ranking lists of business schools where the creation of international personal networks is an important motivation for many students. University rankings use these indicators for slightly different purposes: the network building potential offered to possible students also matters, but the more important thing is the likelihood a foreign student or professor would choose the given university in order to pursue a career abroad.

Both these indicators show the attractiveness and prestige of a country and a university. High country prestige can offset the poorer reputation of a university to some extent, but this seems less true the other way round. Countries with good reputations regarding science and education could become overrepresented in university ranking lists giving too much weight to such kinds of indicators.

The international indicators used in the THES list are sensitive to changes in political climates and immigration rules, factors independent from university performance. From the early 2000s on, it is increasingly obvious that student visas and work permits are not easy to obtain to some countries with the most quality-sensitive and best performing education systems.

### 10.6.2 Comments on the THES Criteria and List

The peer review (questionnaire) method helps include universities from non-OECD countries which gives this list a really global character. Still, the strong weight of this qualitative indicator makes it that the list relies much upon worldwide university reputation.[18]

The THES list seemingly does not attach any special importance to scientific awards as the Jiao Tong list does. It is still obvious that a university's reputation may increase greatly if it has Nobel Prize winners on its payroll (Hargittai, 2002).

The THES list has a number of derivative lists in order to give a truly global picture of the quality of higher education. These sub-lists present rankings of American, European and 'Rest of the World' universities (almost all of them from Asia and the Pacific Rim[19]). The core list tells more about the European ranking than the derivative list of the top 50 schools in Europe.[20]

The Europe-50 list has no university from the new member countries of the European Union (in line with the Jiao Tong list where the best university of the new member countries, the University of Szeged, is tied for 202-301), but they are also missing from the Top 200. This may be explained by the poor international reputation of these universities and their not spectacular scientometric performance, related to their delayed integration in the international scientific community.

Other EU countries including Portugal and Greece (tiny Luxembourg should not be counted here) are missing not only from the Top 200 list, but also from the detailed list for Europe. Italy is represented only with two universities (La Sapienza, Rome and Bologna University) on the Top 200 list (positions 162 and 186 respectively) but these schools are missing from the narrower European elite shown by the Europe-50 list. So is Spain's Autonomous University of Madrid, placed 159th among the Top 200.

The poor showing of Southern Europe's universities on the THES lists may potentially become 'ominous' (*The Times*, 2004, 9) for the prospects of the knowledge-based society placed high on the Lisbon Agenda. But the underperformance of South and East Central European countries on the THES lists is in interesting contrast with another list not quoted here yet, one of the top 25 business schools in Europe (*Financial Times*, 2005) prepared with a different methodology not detailed here.[21]

The very poor showing of the Eastern and the Mediterranean region of the European Union in the ranking lists of universities in general may not be only a quality related issue, but could be linked also to differences in policy priorities. These countries have GERD[22]/GDP levels significantly below the EU average (see Rodrigues, 2003; Török et al., 2005). This reflects policies of higher education without a strong focus on research universities.

In spite of the deep methodological differences between the Jiao Tong and the THES lists of the top universities of the world, the picture of absolute excellence is about the same on both lists. None of them challenges the absolute leading role of US universities in the world with a supporting cast mainly from the United Kingdom and Japan. Striking differences can be observed between the two lists rather in their midfields, but the Jiao Tong list draws a very rough picture of this midfield. The THES list has more solid methodological foundations, but it still cannot be linked to theoretical approaches to competitiveness analysis.

The THES list may be considered as a 'middle approach' between Jiao Tong focusing on science and MBA lists focusing on business. It also makes a quite successful attempt at combining the R&D and teaching output orientations of universities in its quantitative assessment.

## 10.7  SOME ECONOMIC PROBLEMS OF THE UNIVERSITY RANKING LISTS

The lists surveyed left open one fundamental question: what is the measure of competitiveness of a university if any such measure could be found at all? The problem is that universities cannot be entirely compared to firms. Some universities (and mainly business schools) are for-profit organisations wishing to increase market shares or improve their profitability in a number of other ways,[23] but they also fulfil a cultural mission. Many universities owned by governments or endowments operate in non-business environments. Matching the expectations of politicians or trustees may require skills different from the ones appreciated by the business world.

The picture is made more complicated by the fact that all universities, regardless of their forms of ownership, need some kind of competitive strategy. No state university would agree to losing one part of its student body to a private university just because the latter wants higher market shares in order to survive. Setting up university ranking lists is a complex problem of competitiveness analysis in which the supply-side and the demand-side approach should be combined in a much more effective way than shown by the lists assessed. This combination needs to have an explicit strategic element to show how players coming from non-competitive environments want to compete.

The supply side should include human capital and financing. Financing was omitted from the two lists compared. Some other inputs such as technological levels or geographical locations could also complete the picture of the supply-side factors of university competitiveness.

The demand-side analysis of competitiveness usually includes elements such as output, market shares and possible substitution effects related to the concept of relevant market known from competition policy literature (Motta, 2004). The definition of these elements of competitiveness analysis for universities still needs clarification. The two ranking lists surveyed do not take the output side into account. University ranking lists could potentially use a simple output indicator established in MBA rankings, namely the percentage share of graduates reaching a threshold of annual income (usually USD 100 000) within one year from graduation. Such indicators are biased by the wage levels of labour markets taking these graduates. Furthermore, graduates in natural or social sciences usually continue their careers in jobs offering lower earnings than employment in business.

A reliable student related output indicator of higher education apt for international comparisons is not available yet. Many of the potential indicators are too strongly influenced by cross-country differences in labour market characteristics, unemployment levels or, simply, by the differences in geographical structures of output, for example, if 70 per cent of graduates from university A take jobs in Africa and the same percentage of B's graduates find work in North America.

Most ranking lists work with simplistic input and output indicators some of which are not education related at all. The most frequent indicators include elements of human capital (e.g. awards and data on teaching staff) on the input side and research performance on the output side (e.g. citations). Such choices of input and output indicators lack coherence because part of the teaching staff teaches in the first place instead of doing research as a priority.

## 10.8 OVERALL ASSESSMENT AND CONCLUSIONS

University ranking lists have gained increasing popularity in the press, but their value is questionable from the perspective of competitiveness analysis. The technology of university comparisons is quite rough as yet, and serious improvements are needed if really credible lists are to be produced.

Such lists are useful beyond doubt, namely from the perspective of understanding the process of globalisation of higher education, and the need for worldwide quality standards facilitating the choices of students. Such quality standards are being set up. They will be made general by increasing global competition between universities, but measuring quality and competitiveness is not the same thing at all. Neither of the ranking lists surveyed makes it clear whether it serves the measurement of quality or competitiveness. The proliferation of university ranking lists makes it likely that such comprehensive international comparisons are understood as some proxies of competitiveness lists.

The correctness of the ranking lists surveyed and their links to real university performance can be judged only in an intuitive way. This is due to the lack of any benchmark for comparing the lists. The tops of the lists surveyed may be correct because they reflect general beliefs on university strength to a large extent. These beliefs may be linked to real university strength or rather prestige, but prestige is created and potentially confirmed by ranking lists of the same kind. Harvard, Berkeley, Stanford, MIT, Princeton and Oxford (and a handful of other elite universities in the US and England) have top positions on both lists surveyed here, and also on a number of others. Therefore, the role of prestige in shaping university ranking lists is still an open question although the top sections of the lists can be intuitively accepted.

Ranking lists differ increasingly when moving down towards their respective midfields. A university with a relatively good research output and/or with a certain record of relationships with Nobel Prize winners or with good peer review results can almost surely get a place in the Top 100 of either list. It still remains open whether such a position reflects competitiveness, quality or both.

Competitiveness can be potentially improved by cutting prices (tuition fees in this market), but this may not help quality improvement. Furthermore, quality can be either measured or perceived. If it is measured then the reliability of the measures used is questionable, since most of these are based on either past individual scientific performance, or current research output, but this is much more a research- than a teaching-based picture. If, on the other hand, quality is perceived quality (as shown by peer reviews) then it remains open which factors have shaped these subjective judgments. Some universities obtained good peer review results based on the former composition of their research/teaching staff, and not at all on their current performance.

Research and teaching are two interrelated but distinct fields of competition between universities. A good performance in one does not automatically guarantee a similarly good showing in the other.[24] The lists surveyed tend to emphasise research performance which gives a strongly biased picture and helps such 'research universities' to good ranking positions which may not be the best choices for students wishing to do research only depending on their previous study experience.

In spite of all the serious shortcomings of the lists surveyed, they share an important merit. This is a strong warning signal to the European Union: the leading positions in the world of higher education belong almost exclusively to American universities. This speaks for an increased effort to improve both the teaching and the research performance of European universities. Although the midfields of the lists have to be treated with great caution, it also seems likely that a number of American universities well outside the Ivy League are much better financed and perform much better than their (still) prestigious counterparts in continental Europe. The overall quality improvement of higher education should be made a primary target for the Lisbon Agenda. The Agenda was reviewed in early 2005 with an increased emphasis on R&D but, in this context, it does not attach any special importance to the future of higher education in Europe.[25]

## NOTES

1.  The author thanks the National Fund for Research and Development, Gyöngyi Csuka, Adél Németh, Judit Ványai and Attila Varga for their support for carrying out this research, but all the remaining errors and omissions are his own responsibility.
2.  The author' affiliations: University of Pannonia, Veszprém, Hungary; Hungarian Academy of Sciences – University of Pannonia Networked Research Group on Regional Innovation and Development Studies.
3.  This market behaviour of universities is not widespread even in highly industrialised countries. Such an example is Sweden in comparison with the United States (Goldfarb and Henrekson, 2003).
4.  For a list of relevant publications see also Thursby (2000).
5.  There is increasing interaction between public and business-oriented research in many universities of the industrial world (see Schartinger et al., 2002).
6.  This corresponds to the ranking points of former World Champion Garry Kasparov who retired from active chess in 2005.
7.  The ranking list produced by Avery et al. (2004) puts only six universities in the group of the elite (above 2600), no more than 19 universities with Elo numbers (above 2200) can be considered 'competent amateurs' if the parallel with chess is pursued further, and merely 38 have Elo numbers above 2000 (Avery et al.,

2004, Table 3, 26). A wide range of internationally recognised universities would be, according to this methodology, very much outside the range of any kind of rational qualitative comparison with Harvard, Yale or Stanford. Therefore this ranking list is able to create a reasonable qualitative standard only for a few American universities. In case this were to be used for international comparisons a huge number of renowned European universities would get such Elo numbers which belong to beginners in chess.

8. The spread to Eastern Europe and the Third World happened around the year 2000 (see *Asiaweek*, 2000; Mihályi, 2002; Shanghai Jiao Tong, 2005).

9. Two searches made with different search engines on the Internet on 4 December, 2005 (for the same timespan of the past six months) produced approximately 858 000 and 156 000 hits for the *Times Higher Education Supplement* list, and 8 630 and 40 900 hits respectively for the list from Shanghai Jiao Tong University. A comparison with national lists and lists of MBA programmes showed that international ranking lists of universities are lagging much behind them regarding worldwide popularity. Two parallel searches conducted under the same conditions gave 4.35 million and 1.75 million hits for *The Financial Times*, 4.53 million and 3.0 million hits for *Business Week*, and 27.3 and 2.7 million hits for the US News and World Report lists, respectively. Even the two German lists produced significantly more hits than one of the two international rankings: 196 000 and 55 800 for the *Spiegel*, and 61 300 and 39 900 for the *DAAD-Die Zeit* list.

10. On the concept of research universities and the increasing anti-education bias of some of them see Feller et al. (2002).

11. On the problems of research strategies of PhD students with respect to their chances of finding jobs corresponding to their levels of education see Mangematin (2000).

12. Science and technology policies are being transformed in the transition countries with far-reaching implications for their higher education systems (see Radosevic, 2003; Török et al., 2005).

13. The Fields Medal is considered the equivalent of the Nobel Prize in mathematics. The Abel Prize and the Wolf Prize also enjoy a very high international reputation among mathematicians.

14. For example: the University of Szeged is ranked best in Hungary probably because it had the only Hungarian winner of the Nobel Prize in chemistry while still living in the country ... but back in 1937.

15. The strong overall relevance of this indicator cannot be questioned, but its accuracy depends on a number of non-scientific factors including personal networks for mutually generating citations and the chances of publishing widely read (and highly cited) periodicals as well. The main problem is not that such biases exist, but that they vary greatly across fields of science. The number of reputed periodicals is also high in some fields of science (as medicine or physics) and relatively low in others (e.g. in botanics or geology), not to mention the differences between chances of publishing in highly cited periodicals of natural and social sciences.

16. This picture is similar to the one presented by Avery et al., (2004) in that it shows the top American universities too strong to be compared with the rest of the world. Both ranking lists put a great weight on opinion surveys, i.e. the

subjective element of university comparisons. A Harvard graduate earns significantly more than a holder of the degree of a non-Ivy League American or a good European university.

17. The usefulness of citation indexes for measurement of institutional R&D performance is widely debated in literature. For recent analyses of inherent biases of publication and citation counts see Coupé (2003, 2004) and Simonovits (2005).

18. A case in point is the Lomonosov University Moscow (92nd on the list). It lost some valuable teaching staff due to the massive emigration of Russian scientists after 1990 (see Radosevic, 2003), but it still enjoys a good reputation owing to its performance during the Soviet period when it had a number of Nobel Prize-winning professors.

19. Latin America is not represented in the top 40 universities of the 'Rest of the World', and the Top 200 has no Latin American university either. On the problems of higher education in Latin America see Arocena and Sutz (2001).

20. For example, Helsinki University has the 129th position on the Top 200 list while it is the taillight (50th) of the Europe-50 list.

21. This list has a business school (ESCP-EAP) jointly operated by British, German, French, Spanish and Italian institutions in position 2, a Spanish business school (ESADE) placed 6th, and a Hungarian business university (Corvinus Budapest) ranked 23rd. These three countries are missing from the THES list of Europe-50 while their combined share amounts to about 10 per cent on the list of the best 25 business schools in Europe. To obtain this figure, it is supposed that the Spanish and Italian contributions to ESCP-EAP are proportionately accounted for. Furthermore, it is also interesting to see that French business schools are also strongly represented in the European Top 25. Their combined weight is more than 24 per cent in the Top 25 of European business schools while France has only a 10 per cent representation in the Top 50 of European universities.

22. Gross Expenditure on Research and Development.

23. On 'entrepreneurial universities' see Etzkowitz et al. (2000) and Etzkowitz (2003).

24. See on this Behrens and Gray (2001).

25. See European Commission (2005, 9) where only six lines are devoted to the strategic role of universities in the Lisbon Agenda in a document of ten pages.

# REFERENCES

Arocena, R. and J. Sutz (2001), 'Changing knowledge production and Latin American universities', *Research Policy*, 30, 1221-1234.

*Asiaweek* (2000), 'Asia's best univerities', www.asiaweek.com.

Avery, C., M. Glickman, C. Hoxby and A. Metrick (2004), 'A revealed preference ranking of U.S. colleges and universities', *NBER Working Paper* No. 10803, National Bureau of Economic Research, October, http://www.nber.org/papers/w10803.

Behrens, Teresa R. and Denis O. Gray (2001), 'Unintended consequences of cooperative research: impact of industry sponsorship on climate for academic freedom and other graduate student outcome', *Research Policy*, 30, 179-199.

Coupé, T. (2003), 'Revealed performances: worldwide rankings of economists and economics departments', *Journal of the European Economic Association*, 1, 1309-1345.

Coupé, T. (2004), 'What do we know about ourselves? On the economics of economics', *Kyklos*, 57 (2), 197-216.

Etzkowitz, H. (2003), 'Research groups and "quasi-firms": the invention for the entrepreneurial university', *Research Policy*, 32, 109-121.

Etzkowitz, Henry, Andrew Webster, Christiane Gebhardt and Branca Regina Terra (2000), 'The future of the university and the university of the future: evolution of ivory tower to entrepreneurial paradigm', *Research Policy*, 29, 313-330.

European Commission (2003a), 'Communication from the Commission. The role of the universities in the Europe of knowledge', COM 58.

European Commission (2003b), 'Education & Training 2010 – The success of the Lisbon Strategy hinges on urgent reforms', Draft joint interim report on the implementation of the detailed work programme on the follow-up of the objectives of education and training systems in Europe, COM 685 final.

European Commission (2005), 'Communication from the Commission to the Council and the European Parliament: common actions for growth and employment', The Community Lisbon Programme, COM 330 final.

Feller, Irwin, Catherine P. Ailes and J. David Roessner (2002), 'Impacts of research universities on technological innovation in industry: evidence from engineering research centers', *Research Policy*, 31, 457-474.

*Financial Times* (2005), 'Financial Times masters in management 2005. The Top 25 European masters in management programmes', Downloaded on November 22, 2005.

Godin, B. (2003), 'The emergence of S&T indicators: why did governments supplement statistics with indicators?', *Research Policy*, 32, 679-691.

Goldfarb, B. and M. Henrekson (2003), 'Bottom-up versus top-down policies towards the commercialization of university intellectual property', *Research Policy*, 32, 639-658.

Hargittai, István (2002), *The Road to Stockholm. Nobel Prizes, Science and Scientists*, Oxford University Press, Oxford, UK.

Krugman, P. (1994), 'Competitiveness: a dangerous obsession', *Foreign Affairs*, 2, 28-44.

Krugman, P. and M. Obstfeld (2000), *International Economics. Theory and Practice*, Fifth Edition, Addison-Wesley Publishing Company.

Laursen, Keld (2000), *Trade Specialization, Technology and Economic Growth. Theory and Evidence from Advanced Countries*, Edward Elgar, Cheltenham, UK and Northampton, MA, USA.

Mangematin, V. (2000), 'PhD job market: professional trajectories and incentives during the PhD', *Research Policy*, 29, 741-756.

Mihályi, P. (2002), 'Mit érnek a közgazdász diplomák?' (What are economics degrees worth?), *Figyelő*, 37, 46-54.

Moore, J.I. (2001), *Writers on Strategy and Strategic Management. Theory and Practice at Enterprise, Corporate Business and Functional Levels*, Second Edition, Penguin Business.

Morgan, A. (2000), 'Reform in Hungarian higher education', *International Higher Education*, Spring, http://www.bc.edu/bc_org/avp/soe/cihe/newsletter/News19/text15.html.

Motta, Massimo (2004), *Competition Policy: Theory and Practice*, Cambridge University Press, Cambridge, UK.

Radosevic, S. (2003), 'Patterns of preservation, restructuring and survival: science and technology policy in Russia in post-Soviet era', *Research Policy*, 32, 1105-1124.

Rodrigues, Maria J. (2003), *European Policies for a Knowledge Economy*, Edward Elgar, Cheltenham, UK and Northampton, MA, USA.

Rodrigues, Maria J. (2005), 'The debate over Europe and the Lisbon strategy for growth and jobs', Background paper for the Advisory Group, 'Social Sciences and Humanities in the European Research Area', Unpublished manuscript, Brussels.

Schartinger, Doris, Christian Rammer, Manfred M. Fischer and Josef Fröhlich (2002), 'Knowledge interactions between universities and industry in Austria: sectoral patterns and determinants', *Research Policy*, 31, 303-328.

Shanghai Jiao Tong University (2005), 'Academic ranking of world universities 2005', Institute of Higher Education, Shanghai Jiao Tong University.

Simonovits, A. (2005), 'Selection by publication in economics', *Acta Oeconomica*, 55 (3).

Thursby, Jerry G. (2000), 'What do we say about ourselves and what does it mean? Yet another look at economics department research', *Journal of Economic Literature*, XXXVIII (June), 383-404.

*Times, The* (2004), 'World university rankings', *The Times Higher Education Supplement*, November 5.

Török, Ádám, Balázs Borsi and András Telcs (2005), *Competitiveness in R&D. Comparisons and Performance*, Edward Elgar, Cheltenham, UK and Northampton, MA, USA.

UNCTAD (2002), 'World Investment Report 2002. Transnational corporations and export competitiveness', United Nations, New York and Geneva.

Vossensteyn, H. (2004), 'Fiscal stress: worldwide trends in higher education finance', *NASFAA Journal of Student Financial Aid*, 34 (1), 39-55.

# 11. Product differentiation or spatial monopoly? The market areas of Austrian universities in business education

## Gunther Maier

### 11.1 INTRODUCTION

The European Higher Education system is currently in a fundamental transition. In the Bologna declaration from 19 June 1999, 29 European countries agreed to reform their respective higher education systems to allow for better comparability of programmes and degrees, higher levels of mobility, and European standards. In the follow-up meetings in Berlin and Bergen this declaration has been substantiated by defining the process toward this goal and by setting a deadline. An important element of the process is the introduction of an undergraduate-graduate structure. By 2010, the European signatory countries intend to have established the 'European Space for Higher Education'.

At the European level, these activities are closely related to the creation of the 'European Research Area' (ERA) and to the Lisbon objectives, which state that the EU should become 'the most competitive and dynamic knowledge-based economy in the world' (EU, 2000). The university system is arguably the most important element in this strategy playing a crucial role in both the area of education and the area of research. The strategies implemented in this context clearly aim to make university output more comparable and to increase competition between the institutions.

The relationship between the European actions and the reform measures in the individual countries is not a simple one. For most countries, the European actions are both the result of and the framework for major restructuring efforts within their national higher education system. The Austrian university system, for example, has been experiencing a period of constant reform since at least the early 1990s.

The laws regulating organizational structure, the employment status of faculty, the structure and content of teaching all changed more than once in a period of just over ten years, most of the time exposing the sector to inconsistent regulations. In the early 1990s universities in Austria were part of public administration governed – at least formally – by the Ministry of Higher Education. Faculty members were civil servants appointed by that ministry. The basic structure of the teaching was defined by law.

Since the beginning of 2004 Austrian universities are independent legal entities, new faculty members are employees as in private companies, and teaching is largely decided within the universities, constrained only by a few general regulations. The relationship between government and universities has changed from command and control through line item budgeting to partners negotiating a contract that forms the basis of a guaranteed lump sum budget in exchange for a set of services.

All these measures are expected – and intended – to set off a process that American higher education underwent in the last half century. Hoxby (1997, p. 1) describes it as follows: 'Since 1940, American higher education has experienced a very significant change in market structure. Essentially, higher education has been transformed from a series of local autarkies to a nationally and regionally integrated market in which colleges face many potential competitors for inputs and consumers'. This is basically a spatial market argument, which she states in terms of trade between formerly isolated spatial units:

> If we open trade between many autarkies, each of which has colleges offering education of varying quality (producing vertically differentiated products), then theory predicts several reactions. Colleges' loss of market power over their local consumers causes a decrease in their rent and a corresponding increase in the average value (quality for cost) they offer students. Colleges' loss of local monopsony power generates an increase in the wages of college inputs. Since these inputs include students (students are simultaneously consumers of and inputs into education), high ability students are predicted to receive increased subsidies after geographic market integration. Moreover, average college quality should rise in the more integrated market. This is because any given investment in quality has higher returns in the market with open trade. Higher average quality is accompanied by a rise in tuition (though no decrease in value). (Hoxby, 1997, pp.1-2)

The US discussion is largely concerned with this rise in tuition (e.g., McMillen et al., 2005) and whether it is offset by an equivalent increase in the value of education (Leslie and Brinkman, 1988; McPherson and Shapiro, 1998; Brewer et al., 1999). In Europe, the issue is whether the

European university system has entered this process and, if so, how far it has progressed on this course. This is the question we will deal with in this chapter. We will frame the issue in terms of spatial competition between institutions and their corresponding market areas. However, our focus will only be on teaching, which is just one of the functions of universities (Goldstein et al., 1995). To keep the product reasonably comparable, we will analyse the segment of business education.

In section 11.2 of the chapter we will briefly sketch the relevant theory and apply it to the issue of a student's choice of university. Section 11.3 discusses the relevant institutional issues concerning the Austrian university system. Section 11.4 then presents the empirical analysis. The chapter ends with a summary and conclusions (section 11.5).

## 11.2 SPATIAL COMPETITION AMONG UNIVERSITIES

From the students' point of view, participation in higher education can be considered an investment in their own human capital. The money, time, and effort spent on higher education is expected to be repaid in a later career in the form of higher income and more attractive working conditions. So, the decisions made by students about whether or not to go to university, what subject to study and where, can be looked at in terms of a cost-benefit comparison.

While the costs are probably reasonably clear, the benefits are uncertain for a number of reasons:

1. They depend upon the market conditions at the time when the student enters the labour market, which are unforeseeable at the time the decisions have to be made;
2. They are not gained at once, but over the whole period the student will be active in the labour market. So, major parts of the expected benefits are not only years but decades in the future and therefore quite uncertain;
3. The human capital the student expects to accumulate during his or her education may not be the type of knowledge needed in the labour market of the future;
4. The student may not receive the type of education, training, and human capital he or she expects to get in the educational programme. These factors are difficult to observe and measure and in most cases only become apparent through consumption.

These factors make higher education choices particularly difficult and risky. For that reason, the reputation of universities plays a major role. On the one hand, students may decide between programmes based on reputation to overcome the problem of partly unobservable characteristics of the programmes. On the other hand, in their hiring decisions future employers may partly decide based on reputation to overcome the problem that the applicant's human capital, skills, and knowledge can only partially be observed.

In this investment decision, a high school graduate has to make at least three interdependent choices. He/she has to decide

1. whether to go to university or not;
2. which field of study to pursue; and
3. at which university to enrol.

Since we will only deal only with the third decision in the empirical analysis, the following discussion will also concentrate on this aspect.

All three decisions involve a spatial dimension. As far as the decision between universities is concerned, a larger distance between the student's home location and the location of the university will imply higher costs. But also the benefits may differ by the student's home location. If they intend to return to their home location after graduation, the different economic structures of regions will influence the estimated benefits. In Austria, for example, we find that students from rural regions are more likely to study agriculture, forestry, or veterinary medicine than students from urban areas. Since we only deal with the third decision this effect is not relevant in our case. However, there may be regional differences in the reputation of various universities, based, for example, on regional habits. We will come back to this point later.

So, given that the high school graduate has decided to go to university and to study business/economics, he or she has to decide between the various universities that offer this field of study. Based on the above arguments we can write the utility student $i$ from region $r$ will receive from studying at university $u$ as

$$U_{iu} = U_i(Rep_{ru}, Rep_{iu}, C_{ru}) \qquad (11.1)$$

We separate the reputation into a region-specific part ($Rep_{ru}$) and an individual-specific part ($Rep_{iu}$). In the end, the individual-specific part will become part of the random component and we will only deal systematically with the region-specific part.

From a choice-theoretic point of view, the prospective student compares these utility levels and selects the university that gives the highest utility. Because of this comparison, only the relative levels of reputation and costs play a role in the decision making at this level.

### 11.2.1 Spatial Competition – Homogeneous Goods

The theory of spatial competition looks at the competition between suppliers of a homogeneous good and takes into account the costs to transport the product from the supplier's location to the customer's location. Of course, we are well aware that university education is not a homogeneous good. Nevertheless, we will first base our discussion on this assumption and later analyse the consequences of relaxing this assumption.

When a supplier of the homogeneous good charges a price of $P$, the effective price $(P')$ for a customer located a distance $d$ away from the supplier is

$$P'(d) = P + td \qquad (11.2)$$

where $t$ is the transport costs per unit of distance (transport rate). Note that the effective price increases with distance. Consequently, with elastic demand the demand of the customer will decline with increasing distance from the supplier.

When there is more than one supplier, since the product is assumed to be homogeneous, the customer will always buy at that supplier that leads to the lowest effective price at his or her home location. This gives a well defined boundary between the market areas of suppliers at the location where their effective prices are identical. For two suppliers being a distance of $D$ apart and charging prices of $P_1$ and $P_2$, respectively, the market boundary will be at distance $d'$ given by

$$d' = \frac{tD + (P_2 - P_1)}{2t} \qquad (11.3)$$

When the two prices are equal, the market boundary is right in the middle between the two suppliers (see Figure 11.1).

The theory of spatial competition has derived a number of interesting results, not all of which are relevant in our context (see Greenhut and Ohta, 1975; Beckmann and Thisse, 1986; Greenhut et al., 1987; Beath and Katsoulacos, 1991). One relevant consequence is that in such a situa-

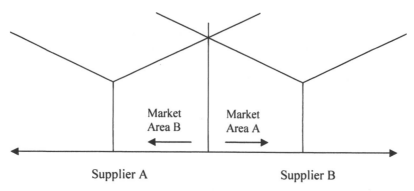

Supplier A                                    Supplier B

*Figure 11.1  Effective prices and market areas for homogenous goods*

tion the market areas are perfectly delimited. Up to the market boundary, all customers buy from the respective supplier, beyond the boundary none. Another relevant consequence is that competition in such a spatial market is limited to the area of the market boundary. When a supplier lowers the price, this will shift out the market boundary and a few customers will switch over to him from his competitor. The size of this effect depends upon the transport rate.

In the immediate vicinity, a supplier enjoys a captive market. Customers enjoy a low effective price because of the small distance to the supplier, but can switch to another supplier only at much higher costs. So, for the vicinity of his or her location the supplier enjoys a so-called spatial monopoly. In such a situation the supplier can set the price strategically like a monopolist, despite the fact that there is a homogeneous good and a number of competitors.

Since the product has been assumed to be homogeneous, the (effective) price is the only factor by which the supply of two suppliers may differ. Because of their identical characteristics, customers will always be indifferent between the products from one or the other supplier irrespective of their utility function. Consequently, effective price is the only differentiating factor.

## 11.2.2  Spatial Competition – Heterogeneous Goods

When the products are similar but heterogeneous, as is the case with higher education, customers may value them differently. When customers' preferences are heterogeneous as well, the market boundaries of the theory of spatial competition will get blurred. Some customers will prefer the characteristics of a distant supplier's product so much that they will buy it despite the higher effective price. Therefore, in the case of heterogeneous products, the market areas will not be perfectly delimited. We will expect the market share of a supplier to decline with distance from his or her location, but not to drop off sharply as in the homogeneous case.

Conceptually, we can characterize this situation by adding a random component ($R$) to the effective price equation:

$$P'(d) = P + td + R \qquad (11.4)$$

$R$ is a random variable which takes on a different value at every location and for every supplier. So, when the customer compares the effective prices of two suppliers (located a distance of $D$ apart), the result will depend upon the specific values of the random variables:

$$P_1'(d) - P_2'(D-d) = P_1 + td + R_1 - P_2 - t(D-d) - R_2$$
$$= P_1 - P_2 + t(2d - D) + R_1 - R_2 \qquad (11.5)$$

Analogous to equation (11.3) we can calculate the boundary of the market area in the case of heterogeneous products:

$$d' = \frac{tD + (P_2 - P_1) + (R_2 - R_1)}{2t} \qquad (11.6)$$

Since the random component enters this equation, the market boundary is not midway between the two suppliers, even when the prices are equal. Moreover, the random component is also assumed to vary with the customer location. Therefore, market areas are not clearly delineated any more. Because of the random component, a supplier may not attract a nearby customer, but still attract one who is located further away. This effect will be larger, the larger the random influence is as compared to the transport rate.

In reality the heterogeneity of products is not random, but typically targeted toward certain groups of potential customers. Suppliers differentiate their products strategically to attract customers with specific preferences. In the extreme, a number of suppliers may serve the same spatial market, but divide it up horizontally through product differentiation rather than vertically through spatial monopolies. In the empirical analysis of the chapter we will try to find out which situation better characterizes the higher education market for business/economics studies in Austria.

## 11.3 THE AUSTRIAN UNIVERSITY SYSTEM

Before we can turn to the empirical analysis, we will need to take a quick look at the Austrian university system (for details see Beerkens, 2003; Wadsack and Kasparovsky, 2004). In the introduction we have already sketched the transition of Austrian universities from public administration units under the direct command of the ministry to autonomous legal entities. This major step in the transformation of the Austrian higher education system, however, became effective in 2004, past the end of our observation period (1990-2002). The most fundamental changes in the Austrian universities happened after this regulation became effective. Therefore we do not expect any major direct implications for our analysis. What we do expect, however, are indirect effects due to anticipation of this transition.

Until July 2005, the Austrian universities were characterized by free access. Any Austrian student who passed the final high school exams (Matura) could sign up at any Austrian university and had to be accepted. Exceptions existed only for art schools and some areas had special requirements. For business/economics no such requirements existed. This is an important feature for our analysis. Because of free access the data about university choice of first year students really show the preferences and cost benefit calculations of the students and are not distorted by the acceptance policy of universities or some administrative unit.

Until autumn 2001 studying at an Austrian university was also free of charge. In the study year 2001-02 a moderate tuition fee was introduced which led to a temporal decline in student numbers. Since this tuition fee was introduced at all universities at the same time, it affected all of them simultaneously and therefore will not impact the students' comparison between universities.

An important factor of increased competition in Austrian higher education was the introduction of the so called 'Fachhochschule' system in 1993 with the first institutions going into operation in 1994. These institutions are subject to special regulations which some universities argue lead to unfair competition. Most importantly, 'Fachhochschulen' are not subject to the free access that universities have to provide. They screen applicants and on average accept less than half of them. On the other hand, they have less research function than the universities and are not allowed to award doctoral or PhD degrees.

## 11.4 EMPIRICAL ANALYSIS

In this section we will apply the conceptual arguments of section 11.2 to a segment of the Austrian university system. As mentioned in the introduction, we will concentrate on business education.

The two market structures, spatial monopoly and product differentiation, are characterized by distinctly different features. In the case of spatial monopoly, we expect to find spatially clearly delineated market areas with little differentiation between the various institutions. In the case of product differentiation, market areas would overlap strongly and institutions would be clearly differentiated, whereas in our context differentiation can be measured by differences in reputation. Based on observation and earlier investigation (Maier, 2003), we expect the Austrian university system to be closer to a spatial monopoly structure. Because of the series of reforms and the external pressures, however, we expect to see the system move away from the hypothesized spatial monopoly structure toward product differentiation.

We have information about students of Austrian nationality beginning their first degree programme at an Austrian university for the period between autumn 1990 and summer 2003. The data were provided by the Austrian Ministry of Higher Education and are based on information collected at the individual universities. The data refer to university students only and do not include students from Fachhochschulen (see section 11.3). Although we have information per semester, we aggregate them to study years. So, year 1990 consists of autumn term 1990 and spring term 1991; year 2002 consists of autumn term 2002 and spring term 2003.

When we aggregate the programmes offered at Austrian universities to broad categories, it turns out that over a fifth of the 300 000 beginning students of our observation period decided for business, economics or a similar subject (see Table 11.1). We refer to this category as 'business education'. At the programme level business is highly differentiated. The names of the programmes differ from one university to the other and some universities offer a set of business education programmes. This is quite different from areas like medicine or law, which are highly standardized and named identically at all Austrian universities offering such a programme. So, as far as naming the programmes is concerned,

*Table 11.1  First year students by study area (1990-2002)*

| Area of Study | No. of studies | Per cent |
|---|---|---|
| Business | 68 227 | 22.74 |
| Medicine | 43 454 | 14.48 |
| Technical Studies | 42 835 | 14.28 |
| Law | 34 853 | 11.62 |
| Philosophy | 28 206 | 9.40 |
| Languages | 24 364 | 8.12 |
| Computing | 15 187 | 5.06 |
| Land Studies | 7 544 | 2.51 |
| Religious Studies | 2 478 | 0.83 |
| Other | 32 852 | 10.95 |

*Source*: Austrian Ministry of Higher Education.

we see product differentiation between Austrian business education programmes.

Seven universities in Austria offered a business education programme during the whole time period of our analysis. Three of them are located in Vienna (University of Vienna, TU Vienna, WU Vienna), the remaining four can be found in Innsbruck, Klagenfurt, Linz and Graz (University). The TU Graz began offering a business education programme in 2002 and had only 0.22 per cent of all the students beginning business education. Since this option became available only in the last year of our study period, we excluded TU Graz from the analysis.

As can be seen from Table 11.2, WU Vienna (The Vienna University of Economics and Business Administration) is by far the largest business education institution in Austria. Aggregated over the 13 years of our observation period, it attracted over 40 per cent of all beginning business education students. Next in importance are the universities in Linz (16.7 per cent) and Graz (12.5 per cent).

Since we know the district of the home location of all the first year students − and all business education students among them − in our observation period, we can calculate the share of business students entering each of the seven universities for every Austrian district.[1]

As it turns out, for most districts there is one university that captures the absolute majority (over 50 per cent) of students. These universities clearly dominate the market for business education in these districts. The districts belong to the 'area of market domination' of these universities. When the dominating university captures over 50 per cent of the students we refer to it as 'strong market domination', when its share is below 50

*Table 11.2 First year business students by university (1990-2002)*

| University | No. of studies | Per cent |
|---|---|---|
| WU Vienna | 27 570 | 40.41 |
| Linz | 11 362 | 16.65 |
| Uni. Graz | 8 550 | 12.53 |
| Uni. Vienna | 8 168 | 11.97 |
| Innsbruck | 7 895 | 11.57 |
| Klagenfurt | 2 730 | 4.00 |
| TU Vienna | 1 799 | 2.64 |
| TU Graz | 153 | 0.22 |

*Source*: Austrian Ministry of Higher Education.

per cent, we call that 'weak market domination'. Two (University of Vienna and TU Vienna) of the seven universities in our set cannot dominate any one of the districts. They do not have an area of market domination.

When we map the areas of market domination calculated for the whole period of time, we get a striking picture (see Figure 11.2). The locations of the universities are marked by arrows.

Each one of the remaining five universities has its own area of strong market domination. It forms a contiguous region around the location of the respective university. For most of them, this area of strong market domination is also the most important source of students. As we can see from Table 11.3, four of the five universities recruit much more than three quarters of their Austrian first year students from this area surrounding their location. Only Klagenfurt has a much lower rate which results from the fact that this university has a small area of strong, but a large area of weak market domination.

For three of the universities (Linz, Klagenfurt, Innsbruck) also the area of weak market domination is contiguous around the location of the respective university. Only two districts, Salzburg Stadt and Hermagor, are not contiguous to the area of market domination to which they belong. From Salzburg Stadt 34.1 per cent of business education students begin their studies at WU in Vienna, as compared to 26.2 per cent in Innsbruck and 22.4 per cent in Linz. The business education students from Hermagor are almost evenly split between Graz (28.9 per cent), WU (27.2 per cent) and Klagenfurt (24.4 per cent).

The weak area of market domination is of overwhelming importance for all universities that can develop such an area. As shown in Table 11.4, the shares of first year students coming from these areas exceeds 80 per

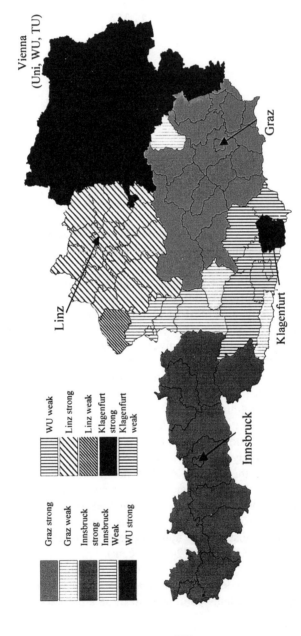

Figure 11.2 Areas of strong and weak market domination

*Table 11.3  Percentage of first year business students coming from the respective area of strong market domination by university (1990-2002)*

| University | Per cent |
|------------|----------|
| WU Vienna | 81.1 |
| Linz | 90.1 |
| Uni. Graz | 79.7 |
| Inssbruck | 78.8 |
| Klagenfurt | 41.6 |

*Note*: The percentages given in this table are relative to the number of first year business students entering the respective university. This differs from the threshold criterion for strong market domination, which is relative to the first year business students from a particular district. Universities are ordered as in Table 11.2.

*Source*: Austrian Ministry of Higher Education.

*Table 11.4  Percentage of first year business students coming from the respective area of weak market domination by university (1990-2002)*

| University | Per cent |
|------------|----------|
| WU Vienna | 82.2 |
| Linz | 91.8 |
| Uni. Graz | 82.4 |
| Innsbruck | 84.4 |
| Klagenfurt | 87.8 |

*Note*: Universities are ordered as in Table 11.2.

*Source*: Austrian Ministry of Higher Education.

cent for all universities. The highest percentage values, i.e. the most localized demand, exist for Linz and Klagenfurt.

From the analysis thus far we see clearly separated market areas around the locations of the universities and a significant share of student coming from these areas. This supports the hypothesis that the Austrian university system in business education is characterized by spatial monopolies. However, the analysis of the area of market domination concentrates only on the highest shares of students and ignores the lower shares. This discards part of the information that is available. In the following analysis we will take into account the full set of information.

From the analysis so far we know that for each university the highest shares of students are clustered around the location of the university. At a more general level, the hypothesis of spatial monopolies implies that districts with high student shares will be near other districts with high shares, and those with low student shares will be near ones with low shares. This relationship can be measured by Moran's I (see Anselin, 1988). So, if the hypothesis of spatial monopolies holds, we will see significantly positive values for Moran's I for all universities. In addition, the hypothesis implies that the districts with high shares will be clustered around the location of the respective university.

Table 11.5 shows the Moran's I statistics for all universities as well as the simulated means and standard deviations based on a random permutation approach (see Anselin, 2005). Clearly, all the statistics are positive and highly significant. This again supports the hypothesis at least insofar as spatial clustering is concerned.

*Table 11.5  Moran's I statistics for rates of first year business students by universities (1990-2002)*

| UNI | Moran's | Mean | Std dev. |
| --- | --- | --- | --- |
| Uni Vie | 0.77 | 0.080 | 0.047 |
| Uni Graz | 0.86 | 0.022 | 0.080 |
| Innsbruck | 0.93 | 0.004 | 0.083 |
| TU Vie | 0.63 | 0.013 | 0.065 |
| WU | 0.87 | 0.024 | 0.061 |
| Linz | 0.89 | −0.008 | 0.058 |
| Klagenfurt | 0.86 | −0.006 | 0.068 |

To verify that the markets are centered at the location of the respective university location, we compute local Moran statistics (Anselin, 1995, 2005) and map those values that are significant at the 5 per cent level (see Figures 11.3a, b, c). The areas marked as High-high show districts that have high numbers of first year business students at the respective university and also border districts with high values. The areas marked Low-low show just the reverse: districts with low numbers of first year students that border districts also with low numbers.

As we can see from these maps, for all universities the clusters of high local Moran statistics (meaning high shares of students in the district and in the neighboring districts) are indeed centered at the location of the respective university. Moreover, the clusters of low values (meaning low shares in the district and in the neighboring districts) are typically

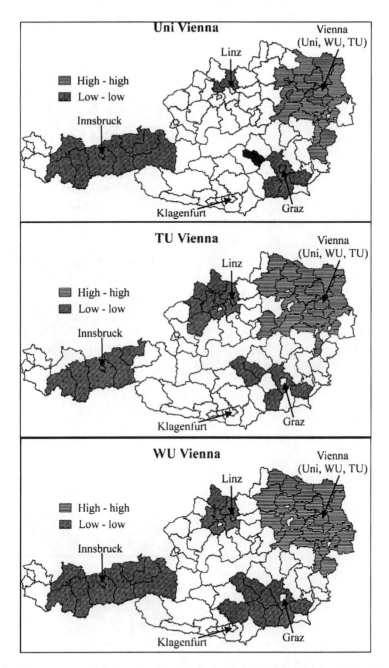

*Figure 11.3a Clusters in significantly high and low local Moran statistics*

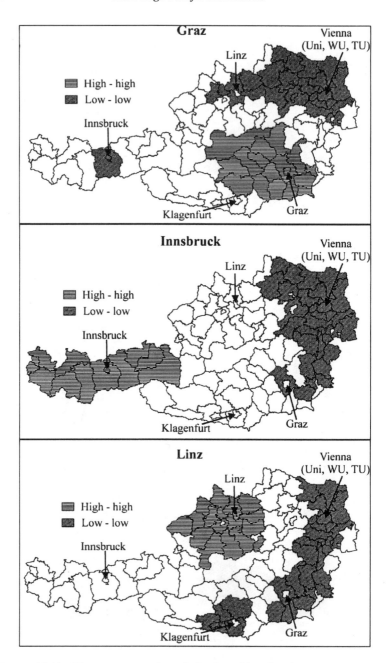

*Figure 11.3b  Clusters in significantly high and low local Moran statistics*

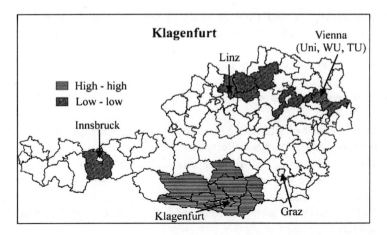

*Figure 11.3c Clusters of significantly high and low local Moran statistics*

centered at the locations of the competing universities. This shows again the strong spatial concentration of the market areas of Austrian universities. Only Klagenfurt is somewhat an exception as on the one hand it does not develop a cluster of low values in most cases, and on the other generates rather small clusters of low levels at its competitors' locations (and none at all in and around Graz). Klagenfurt, it seems, is a relatively weak competitor for business education students in Austria.

The analysis of the full range of student shares supports the results of the previous analysis of areas of market domination: in business education we see highly concentrated spatial markets for those Austrian universities that offer such programmes.

In the introduction we have mentioned the fundamental transitions in the European higher education system in general and the Austrian one in particular. This raises the question, whether due to these changes or their anticipation, the competitive position of Austrian universities has changed over the observation period. To check for this, we could repeat the analysis for each year of observation and check whether there are any significant changes over time. However, it is difficult to find significant changes in maps and Moran statistics. Therefore, we focus directly on the students' decision making for this step of the analysis.

The arguments put forward in section 11.2 provide the basis for this step of the analysis. There, we have argued that students decide between universities based on expected costs and benefits, where the latter are closely related to the reputation of the respective university. In the available dataset, we can identify the students' home location, their chosen university and field of study. Since we only look at business education, the latter category is irrelevant for the decision between the universities.

We model the decision of students between universities as a logit model with students as decision makers, universities as alternatives, and the following explanatory variables:

1. $D_u$ – a vector of alternative (i.e., university) specific dummy variables with WU as the baseline alternative;
2. $Dist_{ru}$ and $Dist2_{ru}$ – distance and distance squared between the student's home and university location;
3. $Dom_{ru}$ – a dummy variable which is 1 when the student's home location is in the area of strong market domination of the respective university (except WU).

The alternative specific dummy variables capture all university location-specific aspects of the decision. These are factors like the general reputation of the university, but also price level and quality of life of the respective city. The parameter values need to be interpreted relative to WU, our baseline alternative. With the distance variable we capture travel costs between home and university, but also factors like the need for accommodation at the university location and the associated costs that increase with distance, and a possible systematic decline of reputation of more distant universities. Because of this multitude of factors, we do not assume a linear relationship, but allow for a quadratic one by use of *Dist2*. The variable *Dom* is intended to capture any effect of the area of market domination in addition to the influence of distance. Since almost all districts belong to the area of strong market domination of one of the universities, we had to exclude one of the universities in order to guarantee convergence of the estimation procedure. If the hypothesis of a spatial monopoly structure holds, this variable will have to show a significantly positive parameter.

The alternative specific dummy variables also serve another purpose. From Table 11.2 we know that the largest share of business education students choose WU. This can be the result of a high reputation of this university as compared to others (reputation effect), but could also result from the fact that WU is located in the population centre of Austria so that many students choose this university because it is nearby (location effect). The dummy variables in the logit model only measure the reputation effect, and thus allow us to separate it from the location effect.

The results of the estimation for all 13 years are shown in Table 11.6.

All parameters are highly significant for all years. From the corrected rho-square value, which is calculated relative to the model with all parameter values equal to zero, we see that the model explains the decision of the students quite well.

Since we ran separate estimations for each year, the parameter estimates cannot be compared directly from one year to the other (Ben-Akiva and Lerman, 1985, Maier and Weiss, 1990). Moreover, the parameters of the alternative specific constants can only be compared relative to the baseline.

Nevertheless, our estimations give a number of interesting results:

1. Since all the parameters are highly significant, we can statistically confirm all three effects: a reputation effect, a distance effect, and a market domination effect. Likelihood ratio tests with constrained versions of the model show that the market domination effect contributes least to the full model. Distance effect and reputation effect are both extremely important for explaining the process. When we compare the two, it turns out that constraining the reputation effect reduces the explanatory power of the model more than constraining the distance effect.

2. Since all alternative specific constants are significantly negative for all years, we can confirm a reputation effect as compared to the location affect, and conclude that the reputation of WU, our baseline alternative, exceeds that of all its competitors over the whole observation period. Figure 11.4 shows the estimated reputation of the universities relative to that of WU. Although some universities (Uni-Vienna and Innsbruck; TU-Vienna and Linz) trade places for some years, the ranking is quite stable and identical in the first and in the last year of the observation period. While WU clearly has the highest level of reputation in all years, Klagenfurt clearly has the lowest.

3. Since none of the signs changes from one year to the other and none of the parameters becomes insignificant, we can conclude that qualitatively the results do not change over the observation period.

4. As far as the distance effect is concerned, we find a negative coefficient for *Dist* and a positive one for *Dist2* for all years. This implies a strong distance decay close to the university location that levels off at larger distances (see Figure 11.5). The minimum of this quadratic function is between 347 and 380 kilometres. The maximum distance possible between a district and a university location is 523.5 kilometres.

5. As far as the market domination effect is concerned, it applies only to

*Table 11.6 Estimation results of logit model*

| | 1990 | 1991 | 1992 | 1993 | 1994 | 1995 | 1996 | 1997 | 1998 | 1999 | 2000 | 2001 | 2002 |
|---|---|---|---|---|---|---|---|---|---|---|---|---|---|
| D(Uni-Vienna) | −2.07 | −1.39 | −1.46 | −1.02 | −0.98 | −0.99 | −0.76 | −0.95 | −1.16 | −1.18 | −1.01 | −1.11 | −2.17 |
| D(Klagenfurt) | −4.06 | −3.37 | −4.27 | −4.03 | −4.14 | −4.17 | −3.82 | −4.06 | −3.77 | −3.97 | −3.95 | −4.22 | −3.91 |
| D(TU-Vienna) | −2.71 | −2.76 | −2.84 | −2.88 | −2.82 | −3.07 | −2.84 | −2.84 | −2.79 | −2.44 | −2.16 | −2.54 | −3.37 |
| D(Graz) | −2.17 | −1.77 | −2.08 | −1.85 | −2.01 | −2.08 | −1.95 | −2.06 | −2.23 | −2.15 | −2.01 | −2.27 | −2.33 |
| D(Innsbruck) | −1.77 | −1.23 | −1.41 | −1.17 | −1.24 | −1.42 | −1.54 | −1.44 | −1.41 | −1.62 | −1.59 | −1.81 | −1.90 |
| D(Linz) | −2.51 | −2.29 | −2.37 | −2.04 | −2.40 | −2.53 | −2.24 | −2.39 | −2.75 | −2.75 | −2.64 | −2.98 | −2.96 |
| DIST(*100) | −3.03 | −2.72 | −3.25 | −3.32 | −3.28 | −3.26 | −3.22 | −3.19 | −3.20 | −3.37 | −3.15 | −3.30 | −2.96 |
| DIST2(*100000) | 4.22 | 3.80 | 4.60 | 4.75 | 4.73 | 4.54 | 4.24 | 4.34 | 4.52 | 4.70 | 4.33 | 4.34 | 3.98 |
| DOM | 1.07 | 1.28 | 1.12 | 0.77 | 0.99 | 0.98 | 0.89 | 1.02 | 1.08 | 1.01 | 0.90 | 0.68 | 0.83 |
| N | 5497 | 5916 | 5428 | 5182 | 4950 | 4873 | 4684 | 4624 | 5126 | 5432 | 6132 | 4948 | 5435 |
| Corr. rho-square | 0.61 | 0.57 | 0.6 | 0.57 | 0.56 | 0.56 | 0.56 | 0.56 | 0.56 | 0.57 | 0.52 | 0.52 | 0.61 |

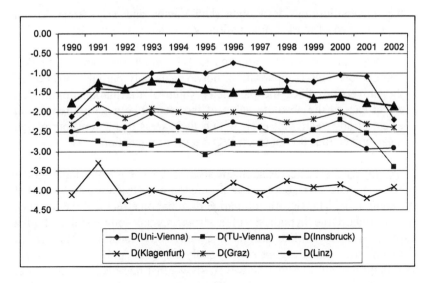

*Figure 11.4 Estimated reputation of universities*

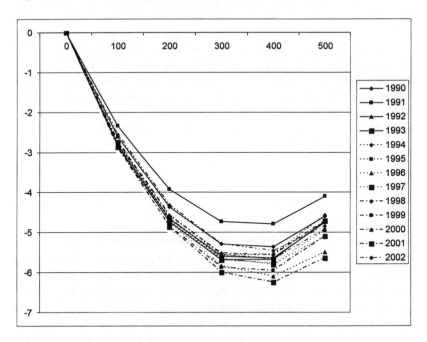

*Figure 11.5 Estimated distance decay of university choice*

the universities outside Vienna (Uni-Vienna and TU-Vienna do not have an area of market domination, WU-Vienna is excluded from this dummy variable). For them, the market domination effect compensates the reputation effect as compared to WU within their area of market domination. This compensation, however, is only partial. For none of the universities does the parameter value of the market domination effect exceed the absolute value of the respective negative reputation effect at any year.[2] So, even within their areas of market domination these universities do not reach WU's level of reputation.

As far as the basic research question is concerned, the logit model yields support for both spatial monopoly structure and market differentiation. The significant market domination effect and the distance effect support the hypothesis of a strong spatial component and of a captive student market around the university location. This result confirms the results of the earlier analyses. The significant and fairly important reputation effect, however, shows that there is also substantial differentiation between the universities competing for business students.

The analysis does not show any pronounced changes over time. From the analysis we do not see any indications for a transition from spatial monopoly to product differentiation or vice versa. The structure remains fairly stable over time and particularly when we compare the first and the last year of our observation period, we can see no significant differences. This is a little surprising in so far as some fundamental changes have occurred in the Austrian university system during this period or at least it became apparent that they will occur in the immediate future. Our results show that the market structure has not reacted to these changes or anticipated them.

## 11.5  SUMMARY AND CONCLUSIONS

In this chapter we have analysed the question whether universities in Austria act like spatial monopolists or like product differentiating suppliers in their competition for students. We analysed this question for the area of business education, because on the one hand it is the most important one in Austria and on the other is not structured by external regulations and therefore provides opportunities for product differentiation.

The analysis starts from the hypothesis that for historical institutional reasons competition between Austrian universities is limited, which lets us expect spatial monopolies, but that because of the substantial pressures resulting from changes in Austria and in the European Union, there should be a transition toward more direct competition and toward product differentiation, as has occurred in the USA over the last half century.

The empirical results provide evidence for spatial monopolies. Most universities offering business education have developed a sizeable area around their location, which they dominate. For these areas, the respective university is the most important provider of business education. At the same time, these areas are the most important source of students for most of the universities. This perception is also supported by an analysis of spatial clustering using spatial statistics methods.

While the results at the aggregate level only support the hypothesis of spatial monopolies, an analysis based on students' decisions gives more diverse results. On the one hand it does not contradict the hypothesis of spatial monopolies, but on the other hand provides strong evidence for substantial differences in reputation between the universities, indicating that they offer differentiated products. So, based on these results, we may say that while the Austrian university system is dominated by spatial monopolies, there is also substantial direct competition through product differentiation.

The one thing for which we could not find empirical evidence is the expected change over time. Over the 13-year period of our analysis, the structure remained fairly stable. In particular, there is no indication of any trend in one direction or the other. This could result from the fact that the most fundamental changes have occurred after the end of our observation period (although they could have been anticipated). Therefore, it will be interesting to repeat this analysis in a few years to see whether the expected changes have occurred once the structural changes were implemented.

## NOTES

1. There are 100 districts in Austria, the smallest one being Rust with 1 714, the largest one being Vienna with 1 550 123 inhabitants in 2001. The use of districts is determined by data availability.
2. Since the reputation dummies as well as the market domination dummies take on only the values 0 and 1, we can compare their parameter values for every year.

# REFERENCES

Anselin, L. (1988), *Spatial Econometrics*, Dordrecht: Kluwer.
Anselin, L. (1995), 'Local indicators of spatial association – LISA', *Geographical Analysis*, 27, 93-115.
Anselin, L. (2005), 'Exploring spatial data with GeoDa: a workbook', https://www.geoda.uiuc.edu/pdf/geodaworkbook.pdf (18 April, 2006).
Beath, J. and Y. Katsoulacos (1991), *The Economic Theory of Product Differentiation*, Cambridge: Cambridge University Press.
Beckmann, M.J. and J.-F. Thisse (1986), 'The location of production activities', in *Handbook of Regional and Urban Economics*, 1, Amsterdam: Elsevier.
Beerkens, E. (2003), 'Higher education in Austria, country report', Center for Higher Education Policy Studies, University of Twente, the Netherlands, http://www.utwente.nl/cheps/documenten/austria.pdf (18 April, 2006).
Ben-Akiva, M. and S. Lerman (1985), *Discrete Choice Analysis, Theory and Application to Travel Demand*, Cambridge: MIT Press.
Brewer, D.J., E.R. Eide and R.G. Ehrenberg (1999), 'Does it pay to attend an elite private college?', *Journal of Human Resources*, 34, 104-123.
EU (2000), 'Presidency Conclusions, Lisbon European Council, 23 and 24 March 2000', http://ue.eu.int/ueDocs/cms_Data/docs/pressData/en/ec/00100-r1.en0.htm (April 18, 2006).
Goldstein, H., G. Maier and M. Luger (1995), 'The university as an instrument for economic and business development: US and European comparisons', in D.D. Dill and B. Sporn (eds), *Emerging Patterns of Social Demand and University Reform: Through a Glass Darkly*, Oxford: Pergamon Press, pp. 105-133.
Greenhut, M.L. and H. Ohta (1975), *Theory of Spatial Pricing and Market Areas*, Durham NC: Duke University Press.
Greenhut, M.L., G. Norman and C.-S. Hung (1987), *The Economics of Imperfect Competition, A Spatial Approach*, Cambridge: Cambridge University Press.
Hoxby, C.M. (1997), 'How the changing market structure of U.S. higher education explains college tuition', *NBER Working Paper* w6323, National Bureau of Economic Research.
Leslie, L.L. and P.T. Brinkman (1988), *The Economic Value of Higher Education*, New York: Macmillan.
Maier, G. (2003), 'The market areas of Austrian universities', *SRE Discussion Paper* SRE-DISC.2003/02, Vienna University of Economics and Business Administration, http://www-sre.wu-wien.ac.at/sre-disc/sre-disc-2003_02.pdf (28 July, 2007).
Maier, G. and P. Weiss (1990), *Diskrete Entscheidungsmodelle, Theorie und Anwendung in den Sozial- und Wirtschaftswissenschaften*, Vienna, New York: Springer.
McMillen, D.P., L.D. Singell and G.R. Waddell (2005), 'Spatial competition and the price of college', http://ssrn.com/abstract=652083 (18 April, 2006).
McPherson, M.S. and M.O. Shapiro (1998), *The Student Aid Game*, Princeton: Princeton University Press.
Wadsack, I. and H. Kasparovsky (2004), 'Higher education in Austria', Bundesministerium für Bildung, Wissenschaft und Kultur, http://www.bmbwk.gv.at/medienpool/11790/hssystem_04e.pdf (18 April, 2006).

# 12. Higher education, graduate migration and regional dynamism in Great Britain

**Alessandra Faggian, Philip McCann and Stephen Sheppard**

## 12.1 INTRODUCTION

The relationship between higher education, aggregate economic growth and individual welfare is a complex topic which can be discussed from a variety of perspectives. In the case of the UK these discussions include an assessment of the real private returns to investments in human capital acquired in higher education (Blundell et al. 1997), a discussion of the cost-efficient structure of a publicly funded higher education system (Johnes 1997), the problem of the how higher education is to be funded (Dolton et al. 1997), and the contribution of higher education to national development (Chatterji 1998; Blundell et al. 1999). For regional economists, the role higher education plays in fostering specifically local economic development is of particular interest.

The aim of this chapter is to examine the relationship between tertiary educated human capital and regional performance. The way in which we do this, however, is rather different to traditional analyses, which tend to focus on regional multiplier impacts. Instead, we focus here on the extent to which the innovation dynamism of Britain's regions is interrelated with the interregional employment-migration behaviour of university graduates. In order to do this we analyse data from a large survey of British students with the aid of a Geographical Information System (GIS). This allows us to observe the spatial patterns of graduate migration behaviour from higher education and into first employment. We model these migration flows within a three-stage least squares simultaneous equation system by combining information on these graduate migration flows with information on the labour market and knowledge characteristics of UK regions. Our exploratory analysis suggests that the knowledge base of a region plays a role in attracting university graduates into a region, while these graduate inflows simultaneously play a role in promoting regional dynamism. As such, there is evidence of a cumulative causation mechanism mediated by labour migration. Once we control for these flows of highly mobile university graduates into employment, the evidence in favour of direct spillovers between university research and regional innovation is rather weaker than much of the literature suggests. In particular, while there is evidence of local spillovers for high technology industries, the evidence for manufacturing industry appears to be more of a London phenomenon.

The chapter is organised as follows: in section 12.2 we discuss the hypothesised links between higher education and regional performance; in section 12.3 we first outline the major features of Great Britain's interregional migration flows and then, within this context, we discuss the peculiarities of graduate migration flows; section 12.4 describes our data and methodology; section 12.5 presents our results, and section 12.6 provides a brief discussion and conclusions of our findings.

## 12.2  UNIVERSITIES AND REGIONAL PERFORMANCE

The traditionally dominant approach to understanding the relationship between higher education and regional development was to focus on the links between higher education institutions (HEIs) and their local regions. As such, this institution-region focus meant that the primary emphasis of this line of research tended to be on the local and regional multiplier effects associated with the HEIs, the analysis of which required estimating the local direct and indirect income-expenditure-employment effects of the higher education institutions (Florax 1992; Harris 1997). In the UK a wide variety of such studies have been undertaken in various countries, and Table 12.1 provides an overview of this research tradition.

The research outlined in Table 12.1 and the associated methodological approaches implied here accurately reflect the level of understanding of UK university-region links until relatively very recently. However, while it is true that universities are beneficial to local economies in terms of these local income and employment multiplier effects, regional multipliers are only part of a very much larger story. Universities are nowadays increasingly being characterised as regional knowledge assets which promote the growth of a region's local stock of human capital (Bradley and Taylor 1996; Bennett et al. 1995) and consequently a

*Table 12.1 Major contributions on the local impact of higher education institutions*

| Year | Authors | Case study | Methodology | Main findings |
|---|---|---|---|---|
| 1973 | Brownrigg M. | University of Stirling | Income multiplier | Value of multiplier on the Local Authority area (County of Clackmann, burghs of Stirling, Bridge of Allan, Doune and Dunblane) between 1.45 and 1.80. |
| 1988 | Lewis J.A. | Wolverhampton Polytechnic | Income and employment multipliers | Income and employment multipliers of respectively 1.70 and 1.57. |
| 1992 | Bleaney et al. | University of Nottingham | Keynesian and expenditure base multipliers | Expenditure base gross output multiplier of 1.021, expenditure base disposable income multiplier of 0.162, Keynesian gross output multiplier of 1.259 and Keynesian disposable income multiplier of 1.561. Final multiplier by combining the ones above: 1.059. Multipliers very sensitive to changes in the parameters. Local area: travel-to-work area of Nottingham. |
| 1993 | Armstrong H.W. | Lancaster University | Income multiplier | Value of multiplier on the local area (Lancaster and Morecambe council district) of 0.82. Low value, but the local area considered is very small (more leakages). |
| 1993 | McNicoll I.H. | Strathclyde University Review | Input-output regional multiplier | Output multiplier of 2.15, income multiplier of 1.66. |
| 1994 | CVCP | | | Report analysing the whole HE British system. Several case studies used. |
| 1997 | Harris R.I.D. | University of Portsmouth | Input-output multipliers | Output multiplier between 1.24 (Type I) and 1.73 (Type II). Local area defined by using postcodes (PO1-PO11). Survey data. |
| 1997 | Armstrong et al. | Lancaster University | 3 types of regional multipliers | Local gross output multiplier between 0.543 and 0.870, local disposable income multiplier between 0.314 and 0.439 and employment multiplier between 1.047 and 1.090. |

*Table 12.1 continued*

| Year | Authors | Case study | Methodology | Main findings |
|------|---------|------------|-------------|---------------|
| 1997 | Chatterton P. | University of Bristol | Total economic impact, regional Keynesian multipliers | Total income impact of £116.4 million on the Avon economy and £121.5 million in the South-West. Economic impact multiplier ranging from 1.23 to 1.30. |
| 2001 | Wardle P. (Canterbury City Council) | University of Kent (UKC), Canterbury Christ Church (CCCUC), Kent Institute of Art and Design (KIAD), Canterbury College (CC) | Gross output multipliers (*à la* Bleaney) | For every pound spent by each university, the percentage that finds its way into the local economy after allowing for multiplier effects is: 58% for UKC, 57% for CCCUC, 88% for KIAD and 49% for CC. |
| 2002 | HERDA-SW | 12 HEI located in the South-West | Total regional economic impact | Total direct and indirect injection to the regional economy of £1.3 billion and additional 40,000 jobs. |
| 2003 | EMUA | 9 HEI in East Midlands | Total regional economic impact | HEI in East Midlands add £1.15 billion to regional GDP (just over 2%) and 30,500 jobs (just under 2% of the regional employment). |

region's economic growth. In order to see why this is so we need to consider contemporary arguments regarding processes of innovation.

Over the last two decades there has been a significant growth in interest among regional economists and economic geographers regarding the relationship between innovation and regional economic performance. Following the work of key commentators (Ács and Audrestch 1990a, b; Porter 1990; Jaffe et al. 1993), innovation is regarded as the outcome of the interaction between human capital, knowledge spillovers and cumulative learning effects. From the perspective of economic growth in general, the basic interest in the innovation behaviour of firms comes from the fact that innovative firms and innovative industries are generally regarded as being not only at the forefront of the technological frontier, but also responsible for moving the technological frontier. The reason is that innovations in one industry tend to spread throughout the market environment into other industries. As such, developments within innovative firms and industries appear to have much wider impacts on the economy as a whole than just within the innovative firms themselves. One particular aspect of these innovation processes is that a disproportionately high share of innovations appear to come from small firms within these innovative industries (Ács and Audrestch 1990a, b), and these small firms are therefore the focus of much research. A second aspect of these innovation processes is that there is an apparent tendency of innovative firms to be clustered in particular locations (Castells and Hall 1994; Simmie 2001; Ács 2002; Van Oort 2004). Much of the interest in innovation on the part of regional economists and economic geographers focuses therefore on the location behaviour of small firm clusters. For reasons of data availability, the empirical literature on the geography of innovation tends to focus primarily on innovation as measured by either patent citations (Jaffe et al. 1993) or by innovation counts (Ács 2002), and also focuses on mainly manufacturing innovations, with much less evidence being available concerning the service industries (Gordon and McCann 2000).

The connection between the regional innovation literature and issues regarding higher education is that universities may play a crucial link in the promotion of regional innovation. The argument here is that in regions with research universities, the local research undertaken within the universities will tend to spill over to the local business community, and indeed, there is empirical evidence which indirectly supports this argument (Anselin et al. 1997; Simmie 1997, 2001; Caniels 2000). There are many commentators who suggest that the spatial behaviour of high technology industries, and particularly those dominated by networks of small firms, may find proximity to universities rather more advantageous than other more mature industries (Cooke 2007). If these spillover effects do operate, these in turn should increase the entrepreneurial base of the region, its innovation potential, or its attractiveness for inward investment. Yet, analysing these potential spillover effects is difficult, because any gains in human capital may also lead to changes in labour migration behaviour (Felsenstein 1995).

If local agglomerative forces act to encourage graduates to stay in a region, the growth in human capital fostered by the local higher education institutions may also engender further local growth in both public and private investments. On the other hand, if no such local agglomerative behaviour is evident, then many of the potential human capital gains will be lost to other regions, due to the subsequent migration behaviour of the graduates. These interregional flows of university graduate human capital therefore imply that the migration of human capital may be a very significant form of knowledge transfer both between regions as well as within regions. The net flow of graduates into or from a region will best indicate the extent to which a region is a net recipient of newly-acquired human capital. The rates of return to higher education will therefore differ according to the migration behaviour of the students both prior to, and subsequent to, higher education. The greater is the net inflow of newly-acquired human capital the greater will be the specifically local regional returns to higher education. On the other hand, where regions experience a net outflow of graduates, the lower will be the specifically local regional returns to higher education.

Very little research, however, has been undertaken to determine the role which labour migration plays in determining regional performance. Yet, the mobility of university graduates would appear to play an important role in the growth prospects of British regions. The reason is that there is a very high level of interregional mobility of university graduates entering into employment (Faggian and McCann 2006; 2009a, b, c; Faggian et al. 2006, 2007a). For example, in the buoyant London economy there are 40 per cent more university graduates employed in London than are actually educated in London, whereas in the economically weaker northern region of Yorkshire and Humberside, there are 40 per cent more graduates educated in Yorkshire and Humberside than are actually employed there (HMT-DTI 2001). This suggests that there are major flows of graduate human capital away from regions such as Yorkshire and Humberside into regions such as London and the South East, and these flows may themselves play a significant role in the persistent variations in UK regional growth performance.

Until now the interaction between regional innovation performance and the migration of graduate human capital is an issue on which there has been little or no serious empirical discussion. Therefore, what is not yet clear is whether innovation dynamism is a function of graduate migration, or alternatively whether graduate migration is a function of innovation dynamism, or whether each is dependent on the other.

## 12.3  UK INTERREGIONAL MIGRATION

Before we can explore the interrelationships between innovation and graduate migration econometrically, it is first necessary to review the key features of spatial labour market and migration behaviour in Great Britain for the population in general, rather than simply the particular sub-set of migration flows accounted for university graduates. This is so that we set up our migration model in such a way as to identify whether the interaction between the migration of graduate human capital and regional innovation performance plays any systematic role which is distinct from the more general features of British labour markets.

Most employment mobility within the UK is accounted for by intra-regional migration behaviour. Annually, less than 1 per cent of employment moves within the UK are accounted for by interregional (NUTS1) migration moves and the figure for unemployed people is only 2 per cent (HMT-DTI 2001). Mobility in the UK is therefore primarily a local phenomenon.

Where mobility is an interregional phenomenon, most UK evidence suggests that net migration flows for the population in general are generally towards areas of higher nominal wages (Gordon and Molho 1998; Molho 1995), although whether these flows are also to areas of higher real wages is much less clear (Shah and Walker 1983). However, UK migration flows are far too slow and weak to eliminate long-term interregional wage and unemployment differentials. Income per household differences of the order of some 40 per cent exist between the lowest and highest wage regions, and unemployment and productivity differences of over 100 per cent of the lowest level regions (ONS 1999). In order to explain these limited migration flows, it is often argued that UK interregional migration flows depend primarily on the number of job-matching opportunities available in each region (Gordon 1995), relative to the number of people seeking work. The implication here that interregional migration would appear to be primarily a consequence of, rather than a pre-condition for, a successful job-search (Jackman and

Savouri 1992a), except where travel-to-work areas straddle more than one region (Jackman and Savouri 1992b). On this argument, nominal UK interregional wage differences would therefore appear to reflect variations, constraints and segmentation in the spatial pattern of employment opportunities, which themselves may be related to the rank-order of the area within the national urban hierarchy, centred on London and its hinterland. In particular, interregional migration appears to exhibit the characteristics of an 'escalator' (Fielding 1992, 1993), whereby young highly skilled migrants from peripheral regions move to London and its hinterland region so as to enter high value graduate employment positions, only returning to more peripheral regions at a much later stage in their employment lives. In terms of the quality of employment opportunities these migration patterns can consequently be perceived of as reflecting a spatially segmented labour market. UK interregional migration behaviour therefore only appears consistent with the predictions of the human capital and search theory combination (Hughes and McCormick 1985) when these predictions are also set within a disequilibrium model of interregional migration, in which migration moves are also subject to the constraints imposed by spatial variations in regional employment opportunities (Gordon and Molho 1998).

For our purposes, this description of the UK labour market suggests that the likelihood of an individual graduate moving between areas will be positively related to the human-capital characteristics of the individual, as well as to interregional differences in both wages and employment opportunities. As such, interregional net graduate migration will tend be driven by the spatial distribution of expected regional wages. UK interregional graduate migration flows will therefore tend to be directed towards areas primarily with higher nominal wages, rather than necessarily areas of higher real wages, because areas with higher nominal wages will also tend to be areas in which the likelihood of employment in any given regional sector will also be higher.

In order to set graduate migration behaviour within the context of UK migration behaviour there is one additional issue that we need to consider here that relates to the initial costs of acquiring human capital. Simple models of human capital and migration are sensitive to variations in the costs of acquiring education. Therefore, observations of migration outcomes can only be interpreted within a human capital framework if we also have information regarding education costs. However, UK undergraduate students do not have to pay variable higher education tuition fees, as all course fees payable are set and subsidised by the national government. As such, for UK undergraduate students, the costs of UK higher education do not vary between the subject studied or institution attended. The only variation in education costs relates to the local costs of living, determined primarily by interregional differences in real estate prices. For our particular sample of students, therefore, the costs of higher education will vary only between location, and are neither subject- nor institution-specific. The result of this is that regional variations in average nominal wages will accurately reflect regional variations in the average real returns to university education (Naylor et al. 2001) associated with graduate migration. Therefore, this simplifies our interpretation of graduate migration behaviour relative to countries such as the USA where higher education fees vary according to institution or degree programme.

*Table 12.2  Regional retention of university graduates (NUTS1 level)*

| NUTS1 | Graduates finding employment in region of study (%) | | | |
|---|---|---|---|---|
| | 1996-97 | 1997-98 | 1998-99 | 1999-00 |
| North East | 54.22 | 58.01 | 63.15 | 56.39 |
| North West | 41.91 | 44.21 | 40.83 | 43.26 |
| Yorkshire and the Humber | 39.61 | 41.90 | 42.84 | 43.98 |
| East Midlands | 46.98 | 42.86 | 38.80 | 41.41 |
| West Midlands | 44.91 | 45.40 | 47.13 | 49.58 |
| Eastern | 46.40 | 43.30 | 51.48 | 53.79 |
| London | 69.19 | 69.27 | 70.50 | 72.58 |
| South East | 46.84 | 46.35 | 47.57 | 48.85 |
| South West | 40.04 | 44.17 | 40.89 | 44.83 |
| Wales | 60.09 | 58.87 | 56.78 | 55.45 |
| Scotland | 63.86 | 61.88 | 60.76 | 60.11 |
| No. of valid observations | 39,382 | 44,367 | 45,019 | 45,747 |

*Source*: Our elaborations on HESA data.

Table 12.2 presents the proportion of locally educated university graduates who remain in the same NUTS1 region as their university for employment after graduation. As we see, there is a clear 'London effect' in retaining graduates. The most surprising result is that while London has the highest graduate retention rate (on average around 70 per cent in the four-year period), the South East does not have a very high retention rate for its own graduates (only around 46-48 per cent). Conversely,

Scotland does retain a considerable percentage of its own graduates (the second highest graduate retention rate after London), despite not being a 'booming' region, although the reasons for this are probably related to the differences between the education system in Scotland and the rest of the UK (Faggian et al. 2007b).

A slightly more nuanced and subtle interpretation can also be gained by considering a more disaggregated spatial scale. Figure 12.1 depicts the

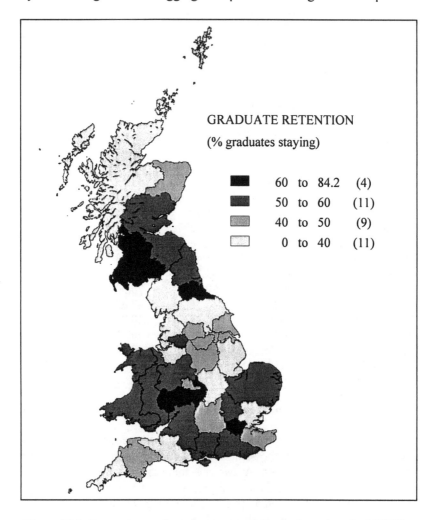

*Figure 12.1  Percentages of graduates remaining in the university NUTS2 region for employment*

proportion of locally-educated graduates who remain in the same NUTS2 region as their university for subsequent employment. As we see in Figure 12.1, apart from London, the NUTS2 regions which tend to retain a high proportion of locally-educated university graduates are the Glasgow region and Warwickshire, followed by regions in Wales and regions surrounding cities with high-ranked universities such as Manchester, Cambridge, Durham, Edinburgh and Southampton.

*Table 12.3 Regional net flows of university graduates (NUTS1 level)*

| NUTS1 | Net flow of graduates (absolute values) | | | |
|---|---|---|---|---|
| | 1996-97 | 1997-98 | 1998-99 | 1999-00 |
| North East | −384 | −64 | −51 | −268 |
| North West | −166 | −443 | −771 | −582 |
| Yorkshire and the Humber | −2081 | −1935 | −1433 | −1450 |
| East Midlands | 45 | 312 | 294 | −180 |
| West Midlands | −468 | −625 | −678 | −569 |
| Eastern | 656 | 515 | 560 | 529 |
| London | 1656 | 2293 | 2261 | 2465 |
| South East | 1377 | 981 | 994 | 1287 |
| South West | −189 | 36 | 86 | 123 |
| Wales | −324 | −382 | −621 | −713 |
| Scotland | −104 | −686 | −641 | −644 |

*Source*: Our elaborations on HESA data.

Table 12.3 presents the net inflows of graduates at NUTS1 level, defined as the total number of university graduates that are in-migrating into an area minus the total number of university graduates who are out-migrating from the area. When we look at regional net graduate flows rather than graduate retention rates, we see some noticeable differences. The most significant of these differences is with regard to the South East. While the South East does not have a very high retention rate of its own graduates (below 50 per cent), it is the second region of Great Britain, after London, for attracting graduates from elsewhere. Overall, it is clear that graduates tend to flow from northern and more geographically peripheral regions to southern and eastern regions, and in particular to the regions of London and the South East. This phenomenon has been quite persistent over time and appears to be a major contributor to current UK regional disparities (Faggian and McCann 2006; 2009a, b, c).

As above, we can also consider these issues at a finer spatial scale and Figure 12.2 depicts graduate net inflows at the NUTS2 level. As we see in Figure 12.2, it is the London region plus also particular NUTS2 sub-regions within the NUTS1 South East and NUTS1 East regions which exhibit the greatest net inflows of university graduates. The preliminary evidence we have so far from the above discussion and also from previous work (McCann and Sheppard 2001) suggests that graduate migration flows tend to be dominated by one of two types of movements. Firstly, a major aspect of graduate migration flows is from major urban

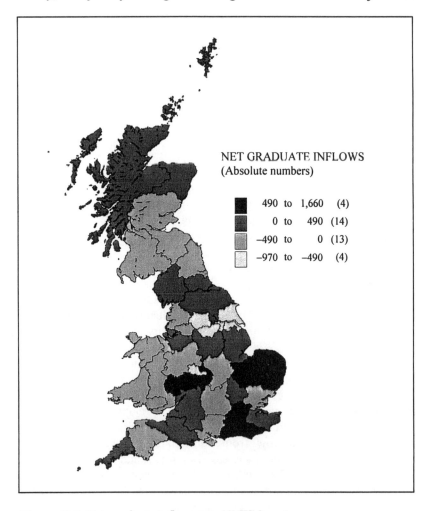

*Figure 12.2  Net graduate inflows into NUTS 2 regions*

centres to relatively local employment positions or from hinterland locations to the major metropolitan centres of the same region. These movements reflect the fact that major urban centres contain not only the majority of the UK's higher education institutions but also the majority of graduate job opportunities. Overlaying these local and hinterland migration patterns, however, are also enormous flows of graduates from all UK regions to London and the South East, and initial observations suggest that the nature and strength of these flows is directly related to distance from London (McCann and Sheppard 2001). There also appears to be a clear divide in terms of the quality of the graduate migrants. Movements to employment positions in London and the South East generally involve the migration of students from relatively higher quality universities than the more local movements into employment in other regions.

This dichotomous pattern of graduate migration, whereby many graduates remain in the same region as their higher education for employment, while many other graduates move to London and the South East, can be interpreted as being at least broadly consistent with the overall spatial patterns of UK interregional migration behaviour, in which London and the South East is regarded as being the centre of an escalator process. This is partly because university graduates make up a large proportion of UK interregional migrants, although the scale of the flows to London and the South East for university graduates are relatively much more important than the equivalent flows for the UK population as a whole. Therefore, in order to clearly identify the extent to which UK graduate migration flows contribute to regional innovation performance or vice versa we need to control for the aggregate spatial and structural features of UK interregional labour markets. This are the issues discussed in the following sections.

## 12.4  DATA AND METHODOLOGY

Our graduate labour market information comes from the HESA (Higher Education Statistics Agency) student leavers' questionnaire, and provides us with data on 187,474 university graduates for the year 2000. The HESA survey provides us with the postcode units details of the domicile, university and first full-time employment workplace locations of each student. This information allows students to be located with incredible precision[1] in each step of their migration process. The postcode units were transformed into spatial coordinates by associating each postcode unit with the latitude and longitude of its centroid, and the coordinates of the centroids were provided by the Geocode table. By setting this information within the Geographical Information System MAPINFO, we are able to model the interregional geographical flows of the graduates into employment after university. This allows us to identify the flows of human capital into a region.

In order to make use of the HESA data within a general migration discussion it is necessary for us to integrate the explicitly spatial data with local labour market wage and employment data. We adopt as our basic spatial unit of analysis the 54 local authority district-based counties of England and Wales defined by the 1974 Local Government Act, plus the nine regional councils of Scotland as defined in 1975.[2] Ideally, we would have liked to use either smaller areas of analysis, such as districts or wards, or alternatively the 1984 or 1998 travel-to-work areas, as the spatial unit of analysis for employment data. However, the NUTS2 areas are the lowest level of spatial disaggregation for which all of the relevant information required to to construct the maps are available.

So as to make full use of this HESA data within a general migration, human capital and innovation discussion it is also necessary for us to integrate the explicitly spatial data with local labour market data, regional industrial and geographical structural data, and regional innovation data. In order to do this, we adopt as our basic spatial unit of analysis the 36 NUTS2 areas of Great Britain. These are the smallest areas of spatial disaggregation for which all of the relevant data required are available.[3] We then use the GIS system to allocate each of the student domicile, education and employment locations to the respective NUTS2 regions.

In order to estimate the relationship between the flows of graduate human capital and the regional innovation performance of Britain's regions, we employ a simultaneous equation approach. The reason we employ such an econometric approach is that, from our previous discussion, there is no obvious dominant causality between labour migration and regional innovation, in that one may lead to the other. Therefore, it is necessary to estimate these relationships as simultaneous processes.

The simultaneous equation model we estimate captures the dynamic interrelationships between a region's inflows of graduate human capital, the region's stock of knowledge assets, and its innovation performance.

The structural equations of the model are:[4]

$$HK_i = \alpha_0 + \alpha_1 INN_i + \alpha_2 ILO_i + \alpha_3 WAGE_i + \alpha_4 CENT_i +$$
$$+ \alpha_5 JOBS_i + \alpha_6 QLIFE_i + \eta_i \qquad (12.1)$$

$$INN_i = \beta_0 + \beta_1 HK_i + \beta_2 RAEIND_i + \beta_3 SMAFIR_i +$$
$$+ \beta_4 NUNI_i + \beta_5 LQ_i + \varepsilon_i \qquad (12.2)$$

The model is based on the number of university graduates who enter into full-time employment after graduation in a particular destination region, which is a different region to both the domicile and the higher education region of the graduate. In the case of the UK, these graduates are found to be both the most mobile and the most highly qualified cohort of UK migrants (Faggian et al. 2006, 2007a). The number of such graduates in each destination region is then divided by the population of the destination region, in order to produce a standardized index *HK* which indicates the relative contribution to the local labour market of externally generated interregionally mobile human capital.

From our previous discussion regarding employment and migration, in equation (12.1) we can hypothesise that the ability of a destination region to attract mobile graduate human capital *HK* is a function of the number of the labour market features such as the number of job vacancies in a region *JOBS*, the level of regional unemployment *ILO*, the quality of life of the region *QLIFE*, the innovativeness of a region *INN*, the regional wage *WAGE*.

At the same time, we need to control for the dominant features of the UK spatial structure. On this point, the UK exhibits a very clear core-periphery structure in that UK regional rankings are highly correlated with proximity to London across a range of criteria, such as productivity, growth, activity rates, employment rates, credit availability (HMT-DTI 2001). As such, increasing distance from London *CENT* can be regarded as an indicator of both economic as well as geographical peripherality. Indeed, the centre-periphery hierarchical structure of the UK is so marked that distance from London *CENT* is the closest approximation of the GDP per head for a UK region (HMT-DTI 2001). As such our *CENT* variable controls for both the centre-periphery aspects of migration discussed above as well as the UK urban-regional hierarchy.

From our earlier arguments regarding the innovation dynamics of regions, the innovation performance *INN* of a region, as reflected in equation (12.2), is related to the level of human capital inflows *HK*, the number of small firms *SMAFIR* in the region, the number *NUNI* and quality *RAEIND* of the local universities, and the relative contribution *LQ* of manufacturing to the region, given that the regional innovation index we employ *INN* is the region's number of patents generated, all of which come under the broad SIC divisional classification of manufacturing.

The simultaneous equation model represented by equations (12.1) and (12.2) can be estimated with either a limited or full information method (Wooldridge 2002). The most popular technique for estimating simultaneous equations is the two-stage least square method (2SLS), which belongs to the limited information family. 2SLS is easy to implement, but provides inefficient estimates of the $\alpha$'s and $\beta$'s if the error terms $\varepsilon$ and $\eta$ are correlated. Since there is no theoretical reason to exclude a priori the existence of such correlation in our model, it may well be appropriate to estimate our equation using three-stage least squares (3SLS). 3SLS, developed by Zellner and Theil (1962), is a full information method and can be seen as a logical extension of 2SLS to which an additional step is attached. This extra step consists of the estimation of the variance-covariance matrix of cross-equation error terms, which is then used to correct the estimates of the parameters $\alpha$'s and $\beta$'s. 3SLS provides consistent and more efficient estimates than 2SLS since it incorporates the additional information on the structure of the error terms and is therefore also better for statistical inference. If there is no correlation or heteroskedasticity in the error terms, the 3SLS gives exactly the same results of 2SLS.

In order to examine whether the 3SLS method produces superior estimates to alternative techniques, we estimated the above model using both separate OLS regressions and also two-stage least squares (2SLS). A comparison of the results of these methods indicates significant differences between each of the approaches depending on their specification. This observation suggests that the full information 3SLS method is superior to the alternative techniques. The model system is therefore estimated using the 3SLS technique, and this allows for the interrelationships between the innovation performance of a region and human-capital migration behaviour to be analysed.

Table 12.4 gives a brief description of the variables used in the empirical analysis. The overall measure of regional innovation performance that we employ is *INN1*, which represents the number of EPO patent applications[5] per million inhabitants by NUTS2 region in 1999.[6] In addition, we also narrow the analysis down to the case of high technology industries, by employing a variable *INN2*, which represents the number of high technology EPO patent applications per million inhabitants by NUTS2 region in 1999. The regional knowledge asset

*Table 12.4 Variable descriptions*

| Name of variable | Definition |
|---|---|
| HK | Number of externally domiciled and externally educated university graduates entering into the destination region for employment after graduation, during the academic year 1999-00, as % of the total regional population |
| INN1 | Number of patent applications by NUTS2 region in 1999 (European Commission 2002)[7] per million inhabitants |
| INN2 | Number of high technology patent applications by NUTS2 region in 1999 (European Commission 2002) per million inhabitants |
| RAEIND | RAE quality index by NUTS2 area calculated as: $$\sum_{i=1}^{6} iX_i$$ where $X_i$ is the percentage of staff belonging to the $i$-th RAE category ($i$=6 equals RAE 5*) in 2000 (DFEE 2001) |
| SMAFIR | Percentage of regional firms with less than 50 employees in 1999 (ONS 2000) |
| NUNI | Total number of universities in the region |
| WAGE | Average gross weekly salary of managerial occupations in each region in 2000 (NES 2001)[8] |
| QLIFE | Inverse of average regional crime rate (crimes against the person) in 2000 (ONS 2001)[9] |
| JOBS | Number of vacancies per head of regional population in 1999 (ONS 2000) |
| ILO | ILO regional unemployment rate 1999 |
| LQ | Regional Location Quotient for Manufacturing Industry 1999 (NOMIS) |
| CENT | Distance between the centroid of each NUTS 2 region and the centroid of London |

indicators we employ are *RAEIND*, which represents the average RAE score of the local regional university departments and *NUNI*, which represents the number of universities in the region. The regional labour market indicators we employ are: *WAGE*, which represents the average gross weekly salary of workers in managerial occupations in each region in 2000; *JOBS*, which represents the total number of unfilled job vacancies in the region, standardised by dividing this by the total regional population in 1999. The regional industry structure indicators we employ are *SMAFIR*, which represents the percentage of regional firms which have less than 50 employees in 1999, and *LQ*, which is the 1999 regional location quotient for manufacturing, and the regional quality of life indicator we employ is *QLIFE*, which represents the inverse of the average crime rate (crimes against the person) in 2000. Finally, the regional geographical indicators we employ is *CENT*, which represents the distance between the centroid of each NUTS2 region and the centroid of London. The logic of including the variable *CENT* here is that, as we have seen, there is a large literature on the possible role played by geographical core-periphery structures in both knowledge spillovers and growth (Fingleton 2006) and in addition, *CENT* can also be regarded in part as an index of the position of the region within the UK urban hierarchy.

We estimate the simultaneous equation systems using firstly an indicator of overall regional innovation dynamism and also secondly, one which captures just high technology regional dynamism. The reason we do this is that as we have seen, there are many arguments which suggest that the spatial behaviour of high technology industries may be very different to that of other industries (Cooke 2007). In particular, high technology industries are regarded as being those industries which are particularly sensitive to the benefits from proximity to universities. Secondly, we also estimate the models with and without the London regions. This is in order to identify the extent to which any estimated results depend on the London 'escalator' phenomenon.

The data for the variables representing wages *WAGE*, quality of life *QLIFE*, and the regional proportion of small firms *SMAFIR*, all come from the UK Office for National Statistics. The data for the variables representing the number of job vacancies *JOBS* and the location quotient *LQ* of manufacturing come from NOMIS, the UK National On-line Manpower Information System, the university Research Assessment Exercise RAE data comes from the UK government Department for Education and Skills, and the data for the variables representing patent applications *INN* comes from the European Union Statistical Office.

## 12.5  RESULTS

Tables 12.5 and 12.6 present the 3SLS estimates for the interrelationship between the graduate human capital flows into a region *HK* and innovation, defined first as overall innovation dynamism *INN1* (Table 12.5) and then as high technology innovation *INN2* only (Table 12.6). Tables 12.7 and 12.8 present the same model as Tables 12.5 and 12.6 respectively, but after excluding London from the sample. As we see, the results are largely the same as with London included in the sample.

A common element in all four models is that the innovation dynamism of the region is positively related to the inflows of graduate human capital *HK* into the region. This result vindicates our original aim of including human capital inflows into the innovation analysis. The second common element in all four models is that human capital flows into a region are very highly and positively related to the availability of local employment opportunities, *JOBS*.

The first clear difference between the models (Tables 12.5 and 12.7) relating to all manufacturing industry and those restricted to high technology industries (Tables 12.6 and 12.8) is that while the human capital flows into a region *HK* for employment are positively related to the region's high technology innovation *INN2*, they do not seem to be

*Table 12.5  Three-stage least squares regression of innovation dynamism (including London)*

| Equation | | Obs | R-sq | chi2 | P |
|---|---|---|---|---|---|
| *HK* | | 34 | 0.7617 | 105.84 | 0.0000 |
| *INN1* | | 34 | 0.4843 | 31.41 | 0.0000 |
| | | Coef | Std. Err. | Z | P>\|z\| |
| *HK* | *INN1* | 2.844942 | 12.26614 | 0.23 | 0.817 |
| | *ILO* | −0.4549839 | 0.3144105 | −1.45 | 0.148 |
| | *WAGE* | −0.0020383 | 0.0049594 | −0.41 | 0.681 |
| | *CENT* | −0.0086345 | 0.0054977 | −1.57 | 0.116 |
| | *JOBS* | 0.0572907 | 0.0114714 | 4.99 | 0.000 |
| | *QLIFE* | 0.0080662 | 0.0499929 | 0.16 | 0.872 |
| | *_cons* | 6.918832 | 3.872262 | 1.79 | 0.074 |
| *INN1* | *HK* | 0.0163914 | 0.0036391 | 4.50 | 0.000 |
| | *RAEIND* | −0.0648912 | 0.0297636 | −2.18 | 0.029 |
| | *SMAFIR* | 1.773615 | 0.6263394 | −2.83 | 0.005 |
| | *NUNI* | −0.0028294 | 0.0035756 | −0.79 | 0.429 |
| | *LQ* | −0.1504215 | 0.0570324 | −2.64 | 0.008 |
| | *_cons* | 2.051105 | 0.6483906 | 3.16 | 0.002 |

*Table 12.6  Three-stage least squares regression of high technology innovation dynamism (including London)*

| Equation | | Obs | R-sq | chi2 | P |
|---|---|---|---|---|---|
| HK | | 34 | 0.7219 | 96.63 | 0.0000 |
| INN2 | | 34 | 0.4644 | 26.08 | 0.0001 |
| | | Coef. | Std. Err. | Z | P>\|z\| |
| HK | INN2 | 0.0705395 | 0.034203 | 2.06 | 0.039 |
| | ILO | −0.3699423 | 0.176488 | −2.10 | 0.036 |
| | WAGE | −0.0070936 | 0.006561 | −1.08 | 0.280 |
| | CENT | −0.0028189 | 0.004567 | −0.62 | 0.537 |
| | JOBS | 0.0461399 | 0.008792 | 5.25 | 0.000 |
| | QLIFE | 0.0311383 | 0.045684 | 0.68 | 0.495 |
| | _cons | 7.8842020 | 3.559384 | 2.22 | 0.027 |
| INN2 | HK | 2.54277 | 1.142107 | 2.23 | 0.026 |
| | RAEIND | 18.87822 | 9.070771 | 2.08 | 0.037 |
| | SMAFIR | 551.79170 | 189.888700 | 2.91 | 0.004 |
| | NUNI | 0.16209 | 1.122745 | 0.14 | 0.885 |
| | LQ | 0.61306 | 17.113030 | 0.04 | 0.971 |
| | _cons | −563.06020 | 196.162000 | −2.87 | 0.004 |

*Table 12.7  Three-stage least squares regression of innovation dynamism (excluding London)*

| Equation | | Obs | R-sq | chi2 | P |
|---|---|---|---|---|---|
| HK | | 33 | 0.7603 | 106.61 | 0.0000 |
| INN1 | | 33 | 0.4893 | 35.43 | 0.0000 |
| | | Coef. | Std. Err. | Z | P>\|z\| |
| HK | INN1 | 4.523572 | 11.4964300 | 0.39 | 0.694 |
| | ILO | −0.4803254 | 0.2915383 | −1.65 | 0.099 |
| | WAGE | −0.0008599 | 0.0048925 | −0.18 | 0.860 |
| | CENT | −0.0074463 | 0.0051297 | −1.45 | 0.147 |
| | JOBS | 0.0573471 | 0.0108977 | 5.26 | 0.000 |
| | QLIFE | 0.0982274 | 0.1114624 | 0.88 | 0.378 |
| | _cons | 4.954681 | 4.1036870 | 1.21 | 0.227 |
| INN1 | HK | 18.33969 | 3.6903940 | 4.97 | 0.000 |
| | RAEIND | −36.41934 | 30.1383500 | −1.21 | 0.227 |
| | SMAFIR | −1988.06100 | 619.7437000 | −3.21 | 0.001 |
| | NUNI | −17.01612 | 7.8128800 | −2.18 | 0.029 |
| | LQ | −135.61320 | 55.7443600 | −2.43 | 0.015 |
| | _cons | 2147.92400 | 633.6025000 | 3.39 | 0.001 |

*Table 12.8   Three-stage least squares regression of high technology innovation dynamism (excluding London)*

| Equation | | Obs | R-sq | chi2 | P |
|---|---|---|---|---|---|
| HK | | 33 | 0.7236 | 98.07 | 0.0000 |
| INN2 | | 33 | 0.4297 | 21.89 | 0.0005 |
| | | Coef. | Std. Err. | Z | P>|z| |
| HK | INN2 | 0.0733212 | 0.0329651 | 2.22 | 0.026 |
| | ILO | −0.4033686 | 0.1740553 | −2.32 | 0.020 |
| | WAGE | −0.0050869 | 0.0063361 | −0.80 | 0.422 |
| | CENT | −0.0015395 | 0.0048690 | −0.32 | 0.752 |
| | JOBS | 0.0478110 | 0.0084633 | 5.65 | 0.000 |
| | QLIFE | −0.1275639 | 0.1181428 | −1.08 | 0.280 |
| | _cons | 4.9992820 | 3.6075250 | 1.39 | 0.166 |
| INN2 | HK | 2.8869070 | 1.1796750 | 2.45 | 0.014 |
| | RAEIND | 21.1385900 | 9.7863940 | 2.16 | 0.031 |
| | SMAFIR | 477.6758000 | 192.0284000 | 2.49 | 0.013 |
| | NUNI | −2.3406100 | 2.4348040 | −0.96 | 0.336 |
| | LQ | 0.3747747 | 17.0208400 | 0.02 | 0.982 |
| | _cons | −499.8263000 | 196.6822000 | −2.54 | 0.011 |

affected by the overall innovation dynamism measured by *INN1*. This implies that a process of cumulative causation operates specifically between high technology innovation and graduate inflows, rather than between the overall innovation dynamism of a region and graduate inflows.

A second difference concerns the role of small firms *SMAFIR*. This is positively related to innovation for high technology industries but negatively related to overall manufacturing innovation.

In terms of the role of universities, while research quality *RAEIND* is always positively related to high technology innovation, it is either negative or insignificant with respect to overall innovation. Moreover, the number of universities is also always either negative or insignificant with respect to innovation.

The fact the results are largely the same, irrespective of whether London is included or excluded from the sample, confirms that an endogenous process of feedback between inflows of graduate human capital and high technology innovation dynamism operates in all UK regions, while no such process operates for all manufacturing. Relatively larger net inflows of human capital help foster high technology regional innovation, and high technology regional innovation further encourages such human capital inflows. A process of circular and cumulative causation appears to be operating in the case of high technology industries, but this is not case for all manufacturing industries. For all manufacturing industries, inflows of graduate human capital are still essential for innovation.

## 12.6 DISCUSSION AND CONCLUSIONS

There are various important observations to emerge from this exercise. Firstly, and most importantly, models of regional innovation which do not control for interregional human capital flows are likely to be mis-specified. The migration of human capital plays an essential role, over and above any other possible knowledge transfers mechanisms, such as local knowledge spillovers between firms or between universities and firms.

Secondly, our 3SLS simultaneous equation approach confirms that in terms of UK regional knowledge flows, a process of circular and cumulative causation operates at the NUTS2 level between high technology innovation and human capital inflows, but not with respect to all manufacturing industries. Moreover, this migration mechanism operates for models which either include or exclude London. As such, this cumulative causation phenomenon appears to be generally applicable to all high technology sectors in all regions.

Thirdly, the small firm sector and university research quality play a crucial role in the regional innovation performance of high technology industries, but this is not the case for all manufacturing industries, where their effect is either negative or insignificant. As such, the evidence in favour of direct local spillovers between university research and regional innovation appears to be significantly weaker than much of the literature would suggest.

By way of conclusions we note that although the migration of embodied human capital appears to be a major regional knowledge transfer mechanism, it has largely been overlooked in most of the knowledge spillover literature. While discussions of the role played by small firm clusters and universities are very much at the forefront of current thinking on regional dynamism (Cooke 2007), the role played by labour migration is largely ignored. In the exploratory modelling exercise undertaken here where we simultaneously control for each of these various issues, our UK results suggest that this human capital interregional migration mechanism is critical for all industries and for all spatial structures. Moreover, the consistent importance of this *inter*regional knowledge transfer mechanism is in contrast to the importance of the *intra*regional role played by small firms and universities, the importance of which appears to be related only to high technology industries. The fact that graduates are produced by universities therefore suggests that as far as innovation is concerned, UK universities play more of an interregional role than an intraregional role. Obviously, confirmation of this will require more detailed data analysis than has been undertaken here (Faggian and McCann 2009a, b), although these are the clear implications of this exploratory analysis. If subsequent work indeed confirms our initial findings, it would imply rather different policy approaches to UK university funding than those which are currently popular, which focus almost exclusively on the promotion of intraregional spillovers.

## NOTES

1. In the UK there are more than 2,000,000 postcode units, which means we are able to locate a student with a precision of hundreds of metres.
2. In England, seven of these counties are the metropolitan county councils covering the largest urban agglomerations of over one million people. In Scotland, the three separate island councils are combined into a single council for the purposes of our analysis. The average employment size of the areas is 330,825.
3. The 36 NUTS2 regions have an average population of 1.66 million, and range in population from 440,000 to 3.5 million. The populations of 26 of the 36 NUTS2 areas range between 1 million and 2.6 million. Our spatial areas are therefore comparable in both scale and spatial definition to the Functional Urban Regions (FURS) employed by Cheshire et al. (1988) and Cheshire and Carbonaro (1995), and are also broadly comparable with Metropolitan Statistical Areas in the USA. We exclude the Highlands and Islands NUTS2 region in this analysis, because their ILO unemployment data are not available for this particular region for this particular year. This should not affect the overall results because the flow of university graduates into this area is so tiny.
4. Where the subscript $i$ =1,..., 36 identifies the NUTS2 regions in Great Britain.
5. European Patent Office Data is largely comparable both in terms of its technology characteristics (Caniels 2000) and also its spatial distribution to the US patent office data for Europe (Cantwell and Iammarino 2003).
6. Using patent application data obviously means that many service-related activities are ruled out. However, the strengths and weaknesses of using patent data as a measure of innovation have been extensively discussed elsewhere (Caniels 2000), and using these data is entirely appropriate for high technology industries.
7. As registered with the International Patent Office.
8. Data comes from NES (2000) the New Earnings Survey, Office for National Statistics, London.
9. Office for National Statistics, London.

# REFERENCES

Ács, Z.J. (2002), *Innovation and the Growth of Cities*, Edward Elgar: Cheltenham, UK and Northampton, MA, USA.

Ács, Z.J. and D.B. Audrestch (1990a), *Innovation and Small Firms*, MIT Press, Cambridge, MA.

Ács, Z.J. and D.B. Audrestch (1990b), *Innovation and Technological Change: An International Comparison*, University of Michigan Press, Ann Arbor.

Anselin, L., A. Varga and Z.J. Ács (1997), 'Local geographical spillovers between university research and high technology innovations', *Journal of Urban Economics*, 42, 422-448.

Armstrong, H.W. (1993), 'The local income and employment impact of Lancaster University', *Urban Studies*, 30 (10), 1653-1668.

Armstrong, H.W., J. Darrall and R. Grove-White (1997), 'The local economic impact of construction projects in a small and relatively self-contained economy – the case of Lancaster University', *Local Economy*, August, 146-159.

Bennett, R., H. Glennerster and D. Nevison (1995), 'Regional rates of return to education and training in Britain', *Regional Studies*, 29 (3), 279-295.

Bleaney, M.F., M.R. Binks, D. Greenaway, G.V. Reed and D.K. Whynes (1992), 'What does a university add to its local economy?', *Applied Economics*, 24 (3), 305-311.

Blundell, R., L. Dearden, A. Goodman and H. Reed (1997), *Higher Education, Employment and Earnings in Britain*, The Institute for Fiscal Studies, London.

Blundell, R., L. Dearden, C. Meghir and B. Sianesi (1999), 'Human capital investment: the returns from education and training to the individual, the "Firm, and the Economy" ', *Fiscal Studies*, 20 (1), 1-23.

Bradley, S. and J. Taylor (1996), 'Human capital formation and local economic performance', *Regional Studies*, 30 (1), 1-14.

Brownrigg, M. (1973), 'The economic impact of a New University', *Scottish Journal of Political Economy*, 20 (2), 123-139.

Caniels, M.C.J. (2000), *Knowledge Spillovers and Economic Growth*, Edward Elgar: Cheltenham, UK and Northampton, MA, USA.

Cantwell, J.A. and S. Iammarino (2003), *Multinational Corporations and European Regional Systems of Innovation*, Routledge, London.

Castells, M. and P. Hall (1994), *Technopoles of the World: The Making of 21st Century Industrial Complexes*, Routledge, London.

Chatterji, M. (1998), 'Tertiary education and economic growth', *Regional Studies*, 32 (4), 349-354.

Chatterton, P. (1997), 'The economic impact of the University of Bristol on its region', www.bris.ac.uk/Publications/Chatter/impact.htm.

Cheshire, P.C. and G. Carbonaro (1995), 'Urban economic growth in Europe: testing theory and policy prescriptions', *Urban Studies*, 33 (7), 1111-1128.

Cheshire, P.C., D. Hay, G. Carbonaro and N. Bevan (1988), 'Urban problems and regional policy in the European Community', Commission of the European Communities, Brussels.

Committee of Vice-Chancellors and Principals (CVCP) (1994), *Universities and Communities*, CVCP, London.

Cooke, P. (2007), *Growth Cultures: The Global Bioeconomy and its Regions*, Routledge, London.

DFEE (2001), *Research Assessment Exercise: The Outcome*, Department for Education and Employment, London.

Dolton, P.J., D. Greenaway and A. Vignoles (1997), '"Whither higher education?"', an econometric perspective for the Dearing Committee of Inquiry', *Economic Journal*, 107, 710-725.

EMUA (2003), 'The impact of the East Midlands higher education institutions on the regional economy', East Midlands Universities Association and East Midlands Development Agency, Loughborough, Leicestershire.

European Commission (2002), *Regions: Statistical Yearbook 2002*, Brussels.

Faggian, A. and P. McCann (2006), 'Human capital flows and regional knowledge assets: a simultaneous equation approach', *Oxford Economic Papers*, 52, 475-500.

Faggian, A. and P. McCann (2009a), 'Human capital, graduate migration and innovation in British regions', *Cambridge Journal of Economics*, 33 (2), 317-333.

Faggian, A. and P. McCann (2009b), 'Universities, agglomeration and graduate human capital mobility', *TESG Journal of Economic and Social Geography*, forthcoming.

Faggian, A. and P. McCann (2009c), 'Human capital, and regional development', in R. Capello and P. Nijkamp (eds), *Regional Dynamics and Growth: Advances in Regional Economics*, Edward Elgar, Cheltenham, UK and Northampton, MA, USA.

Faggian, A., P. McCann and S.C. Sheppard (2006), 'An analysis of ethnic differences in UK graduate migration', *Annals of Regional Science*, 40 (2), 461-471.

Faggian, A., P. McCann and S.C. Sheppard (2007a), 'Some evidence that women are more mobile than men: gender differences in UK graduate migration behaviour', *Journal of Regional Science*, 47 (3), 517-539.

Faggian, A., P. McCann and S.C. Sheppard (2007b), 'Human capital, higher education and graduate migration: an analysis of Scottish and Welsh students', *Urban Studies*, 44 (13), 2511-2528.

Felsenstein, D. (1995), 'Dealing with "induced migration" in university impact studies', *Research in Higher Education*, 36 (4), 457-472.

Fielding, A.J. (1992), 'Migration and social mobility: South East England as an escalator region', *Regional Studies*, 26 (1), 1-15.

Fielding, A.J. (1993), 'Migration and the metropolis: recent research on the causes and consequences of migration to the Southeast of England', *Progress in Human Geography*, 17 (2), 195-212.

Fingleton, B. (2006), 'The new economic geography versus urban economics: an evaluation using local wage rates in Great Britain', *Oxford Economic Papers*, 58 (3), 501-530.

Florax, R. (1992), *University: A Regional Booster*, Avebury, Aldershot.

Gordon, I.R. (1995), 'Migration in segmented labour markets', *Transactions of the Institute of British Geographers*, NS, 20, 139-155.

Gordon, I.R. and P. McCann (2000), 'Industrial clusters: complexes, agglomeration, and/or social networks', *Urban Studies*, 37 (4).

Gordon, I.R. and I. Molho (1998), 'A multi-stream analysis of the changing pattern of inter-regional migration in Great Britain', *Regional Studies*, 32 (4), 309-323.

Harris, R.I.D. (1997), 'The impact of the University of Portsmouth on the local economy', *Urban Studies*, 34 (4), 605-626.

HERDA-SW (2002), 'The economic impact of higher education in the South West region: full report', Exeter, Devon.

HMT-DTI (2001), 'Productivity in the UK: 3 The Regional Dimension', H.M. Treasury and Department for Trade and Industry, London.

Hughes, G. and B. McCormick (1985), 'Migration intentions in the UK. Which households want to migrate and which succeed?', *Economic Journal*, 95, Supplement, 113-123.

Jackman, R. and S. Savouri (1992a), 'Regional migration in Britain: an analysis of gross flows using NHS central register data', *Economic Journal*, 102, 1433-1450.

Jackman, R. and S. Savouri (1992b), 'Regional migration versus regional commuting: the identification of housing and employment flows', *Scottish Journal of Political Economy*, 39 (4), 272-287.

Jaffe, A.B., M. Trajtenberg and R. Henderson (1993), 'Geographic localization of knowledge spillovers as evidenced by patent citations', *Quarterly Journal of Economics*, 108, 577-598.

Johnes, G. (1997), 'Costs and industrial structure in contemporary British higher education', *Economic Journal*, 107, 727-737.

Lewis, J.A. (1988), 'Assessing the effect of the polytechnic, Wolverhampton, on the local community', *Urban Studies*, 25, 53-61.

McCann, P. and S.C. Sheppard (2001), 'Higher education and regional development', in D. Felsenstein, R. McQuaid, P. McCann and D. Shefer (eds), *Public Investment and Regional Development*, Edward Elgar, Cheltenham, UK and Northampton, MA, USA.

McNicoll, I.H. (1993) 'The impact of Strathclyde University on the economy of Scotland', Department of Economics, University of Strathclyde.

Molho, I. (1995), 'Spatial autocorrelation in British unemployment', *Journal of Regional Science*, 35 (4), 641-658.

Naylor, R., J. Smith and A. McKnight (2001), 'Determinants of occupational earnings: evidence for the 1993 UK university graduate population from the USR', *Oxford Bulletin of Economics and Statistics*, 63 (1), 29-60.

ONS (1999), 'Regional trends', 34, Office for National Statistics, London.

ONS (2000), 'Regional trends', 35, Office for National Statistics, London.

ONS (2000), 'New Earnings Survey', Office for National Statistics, London.

ONS (2001), 'Regional trends', 36, Office for National Statistics, London.

Porter, M. (1990), *The Competitive Advantage of Nations*, Free Press, New York.

Shah, A. and M. Walker (1983), 'The distribution of regional earnings in the UK', *Applied Economics*, 15, 507-519.

Simmie, J. (1997), *Innovation, Networks and Learning Regions*, Atheneum Press, London.

Simmie, J.M. (2001), *Innovative Cities*, Spon, London.

Van Oort, F. (2004), *Urban Growth and Innovation*, Ashgate, Aldershot.

Wardle, P. (2001), 'The economic impact of four large education institutions on the Canterbury District Economy', Canterbury City Council.

Wooldridge, J.M. (2002), *Econometric Analysis of Cross Section and Panel Data*, MIT Press, Cambridge MA.

Zellner, A. and H. Theil (1962), 'Three-stage least squares: simultaneous estimation of simultaneous equations', *Econometrica*, 30 (1), 54-78.

PART FOUR


University-based regional development: the
experience of lagging areas in Asia, Europe and
North America

# 13. Barriers against the transfer of knowledge between universities and industry in newly-industrialised countries: an analysis of university-industry linkages in Thailand

## Daniel Schiller, Björn Mildahn and Javier Revilla Diez

## 13.1 INTRODUCTION

Like many of the other so-called tiger states of Southeast Asia, Thailand succeeded in keeping its gross domestic product growing at a rate of 6 to 10 per cent over several decades, thus rising to the status of a newly-industrialised country. Even after the Asian crisis, the economy swiftly returned to the path of growth. The reasons for the success of this process of economic recovery and industrialisation included, among others, the existence of a stable, liberal macro-economic framework and the exploitation of comparative advantages, particularly the relatively low cost of the production factor labour (Kraas, 1996; Schätzl, 2000, 234f). Confronted by rising wages and increasingly stiff competition for foreign direct investment after the liberalisation of China, Thailand has not as yet succeeded to an adequate extent in advancing structural reforms aiming at more sophisticated knowledge-based technologies and enhancing the country's endogenous innovation potential. In technological terms, therefore, its growth remained a superficial phenomenon (Arnold et al., 2000).

In the future, acquiring or imitating foreign technologies alone will not be enough to enable enterprises to build up their own technological capabilities and produce innovations, so that they may hold their own in competition with enterprises from other, more advanced emerging countries. In the long run, success can only be assured by manufacturing companies, company-oriented service providers, and scientific as well as governmental institutions forming an efficient innovation system based on suitable institutional arrangements that permit joint interactive learning.

Within such an innovation system, research institutions and, more importantly, state universities represent essential sources of knowledge. By tradition, they produce qualified human capital as well as scientific knowledge. Moreover, the academic services they offer on the market either by themselves or in cooperation with business enterprises enable them to interact directly with other players within the innovation system, thus ensuring the transfer of knowledge between universities and enterprises.

Against this background, this chapter investigates the barriers that impede the transfer of knowledge between universities and enterprises in newly-industrialised countries, using the regional innovation system of Bangkok as an example. The following hypotheses will be discussed theoretically and analysed empirically:

1. Unlike the fully-formed innovation systems of the industrialised countries, fragmented innovation systems in emerging countries impede the transfer of knowledge between universities and enterprises.
2. Although the industrial sector in Thailand is growing dynamically in quantitative terms, no corresponding improvement in technological capability has resulted so far.
3. Structural changes in the scientific sector failed to keep pace with the dynamism of the economy; university research as a whole is too remote from technology.
4. Because of the different specialities of the two sectors, the potential for cooperative relations between universities and enterprises is low at the moment.
5. Cooperative relations between universities and enterprises are weak even within the regional innovation system of Bangkok.


## 13.2 TRANSFER OF KNOWLEDGE BETWEEN UNIVERSITIES AND ENTERPRISES IN NEWLY-INDUSTRIALISED COUNTRIES

Today, regional-innovation researchers largely agree that new products, production processes, and forms of organisation are launched on the market as the result of an interactive process (Kline and Rosenberg, 1986). A process of innovation implies a close mutual exchange between different departments within an enterprise (R&D, production, marketing, etc.), other enterprises (suppliers, customers, competitors), company-oriented service providers, public research institutions and universities (Nelson, 1993; Lundvall and Johnson, 1994). Interactions between these players are guided by both formal and informal rules within a so-called innovation system (Edquist, 1997; Freeman, 2001). Because of the different geographical and sectoral reaches of institutional framework conditions and innovation fabrics, innovation systems tend to interpenetrate at various levels – global, national, regional, local and sectoral (Bunnell and Coe, 2001). The nature and intensity of the interactions that go on between players crucially influence the innovative performance of the enterprises that belong to a given innovation system (Nelson, 1993).

Cooperative relations between universities and enterprises intensified in recent years, a fact that can be explained by two mutually interactive processes. Universities and other public research institutions were constrained to develop new external sources of funds as public funding for the science system dwindled. Industrial contract research constitutes one optional source of funds (Schmoch, 1999). For this purpose, innovation systems have to be configured in an interactive way. Universities have to be transformed in a way that allows them to work interactively with external partners at all stages of the innovation process instead of trying to produce innovations *for* industry. Knowledge is no longer transferred by a one-way road leading from public research institutions to business enterprises but through a mutual exchange which also operates in the other direction (Cabo, 1999).

At the same time, enterprises find that the importance of university knowledge is growing because their parent economy is increasingly knowledge-based (OECD, 1996). Rapid changes in technology and market conditions call for higher innovation rates and shorter lead times for the development of products and processes. Strategies pursued by enterprises to accelerate innovation processes include outsourcing research and development activities and forming strategic cooperation. Universities and other public institutions may potentially act as partners in this respect.

### 13.2.1 New Forms of Knowledge Production

In recent years, new forms of knowledge production have evolved that are designated as Mode 2 by Gibbons et al. (1994). Mode 2 is characterised by growing diversity in the localisation of research activities and the enhanced importance of interdisciplinary research. The interplay between universities, enterprises and the state, all playing different roles in each innovation strategy, is called a 'triple helix' by Etzkowitz, a structure within which different spheres overlap and 'hybrid' organisations form increasingly (Etzkowitz et al., 2005). Examples include the establishment of technology-transfer departments at universities, the creation of incubators for technology-based enterprises, and the establishment of science parks or venture-capital enterprises. As the universities expand their entrepreneurial activities, their capability to transfer technology to enterprises increases as well, which leads to partial superposition of the functions of universities and enterprises in a process of innovation.

### 13.2.2 Innovation Systems in Developing and Newly-industrialised Countries

Innovation systems in countries like Thailand, where industrialisation was delayed, and in western industrialised nations differ in their R&D regime. The impact of these differences on the manner in which technology is transferred is significant. While R&D is now a crucial competitive factor in the industrialised nations, the proportion of enterprises operating their own R&D activities is markedly lower in the developing countries. To operate their own R&D, enterprises in Thailand must enhance their own technological capabilities, for which they need well-trained human capital and a functional innovation system.

### 13.2.3 Absorption Capacities and Technological Capabilities in Enterprises and Universities

Much of the technological knowledge needed in developing countries is available in industrialised states. This is why innovation systems in developing countries crucially depend on obtaining access to transnationally available knowledge and capital. At the same time, enterprises and research institutions in these countries need to improve their absorption capacities to utilise fresh technological knowledge (Asheim and Vang, 2004; Cohen and Levinthal, 1990). Such absorption capacities enable enterprises to identify relevant university knowledge and translate it into company-specific competitive advantages (Rothaermel and Thursby, 2005).

In the past, many countries pursued the strategy of acquiring fresh technological knowledge through the import of capital goods, reverse engineering, or licensing agreements. Together with other states in Southeast Asia, Thailand pursued the strategy of attracting foreign direct investment by granting taxation privileges. In Thailand, however, the key objective was to create jobs rather than to improve the technological capabilities of local enterprises, as it was in Singapore and elsewhere (Wong, 1999). Yet technological capability-building is the key to product and process improvement. Building such capabilities often proves a difficult undertaking, and many companies failed in an attempt to catch up with technologically advanced competitors (Hobday, 1995; Bell and Pavitt, 1995).

At the same time, the absorptive capacities of universities in developing countries have to be strengthened. A sufficient level of absorptive capacities at the university level is a prerequisite for their interactions with the international scientific community and the transformation of university research into industry-relevant applications. A close interaction with industrial partners is necessary at each stage during the process of academic capacity building. These issues and the related higher education reform is discussed in more detail in Schiller (2006) and Schiller and Liefner (2007).

### 13.2.4 University-Industry Cooperation

There are many forms of university-industry cooperation, from *ad hoc* consultation to the formation of research consortiums. Refusing to classify the isolated exchange of knowledge between individuals as a cooperative relationship, Inzelt (2004) argues that such interaction should be institutionalised to qualify. It is the lack of efficient institutions in developing countries, where legal systems are often weak, which makes it difficult to conduct effective transactions and conclude contracts on research cooperation. Because of this, most of the relations entered into are informal and based on mutual trust (Knack and Keefer, 1997). Nevertheless, even the importance of such isolated informal interactions should not be underestimated because this kind of relationship may form the starting point for the development of more sophisticated cooperative relations.

The low level of R&D activities in Thai enterprises affects their cooperation potential. When enterprises do not conduct any R&D of their own and use universities as vicarious research institutions, the development of an effective innovation system with its own 'technological culture' becomes highly improbable (Lall, 2002). A number of case studies have demonstrated that corporate technological endeavours may be supported but not substituted by governmental and academic activities (Nelson and Rosenberg, 1993).

Whether or not an enterprise is inclined to conduct R&D jointly with a university depends on certain specific characteristics, including the knowledge base of the enterprise in question, the technology intensity of its parent industry, and the corporate culture of innovation (Faulkner and Senker, 1995). Successful R&D cooperation calls for a balance of interests between individuals and organisations, each with its own specific capabilities, and each guided by different motivations (Johnson and Johnston, 2004).

Cooperation among players may take place at various levels – between individuals, groups, or institutions – and may differ in its geographical reach – local, regional, national or international (Inzelt, 2004). Cooperative relationships may be effected through different transfer channels (consultation, licensing, contract research, joint research, spin-offs). The nature and intensity of cooperative relationships within a system of innovation crucially influences the processes that govern the generation, distribution, and application of knowledge (ibid.).

## 13.3  PROJECT BACKGROUND AND METHODOLOGY

The Thai innovation system was first examined by German economic geographers in 2000 and 2002 within the framework of the Thailand R&D/Innovation Survey (TIS). Conducted in cooperation with local partners, this innovation survey followed the methodology used in similar projects implemented in eleven European regions within the framework of the DFG (German Research Foundation) programme 'Technological Change and Regional Development in Europe' (Schätzl and Revilla Diez, 2000) as well as in another DFG project addressing regional innovation potential and innovative networks in Singapore and Penang, Malaysia (Kiese, 2004; Stracke, 2003). One of the essential conclusions of the TIS was that both in Thailand as a whole as well as in the metropolitan region of Bangkok, the level of corporate R&D activity is not only much lower than in the European regions investigated but also falls short of that prevailing in the reference regions of Southeast Asia (Kiese, 2003; Schiller, 2003). In Thailand, even multinational enterprises contribute much less towards the national innovation potential than in other regions of Southeast Asia (Berger, 2007).

Lastly, the DFG project 'Public Research Institutions in Thailand' investigated the extent to which Thailand's universities might contribute towards strengthening the country's innovation potential by cooperating with business enterprises. In the period from July to October 2004, the authors interviewed 72 department-level institutions at five Thai universities as well as 34 enterprises from the manufacturing sector, using partially standardised questionnaires. The interviews took about one hour. All enterprises and three of the universities interviewed were located in the region of Bangkok, while the remaining two universities were located in the north and the northeast of the country.

University data is based on interviews with professors and administrators at five universities in Thailand. Three of them are located in Bangkok: Chulalongkorn University (CU), Kasetsart University (KU) and King Mongkut's University of Technology Thonburi (KMUTT). Two of them are regional universities: Chiang Mai University (CMU) and Khon Kaen University (KKU). The universities selected for the case studies are regarded as nationally outstanding in science and technology research and teaching. They are therefore expected to possess the highest potential impact at the regional level. The selection of interviewees was based on their experience and involvement with industry. The survey did not aim at measuring the impact of universities by a representative sample of interviews, but at learning about the process of regional involvement of universities in a developing country. However, this method has its limitations, as it might underestimate less successful attempts to work with industry. The large number of interviews conducted with professors who cooperate with private companies (n=72) and of identified cooperation projects (n=136) from a wide field of disciplines allows descriptive methods of analysis to be applied.

Secondary information on the Thai higher education system is based on fiscal data from the Bureau of the Budget and the higher education statistics from the Commission on Higher Education, both of which are published annually. Their data have been extended by information from the five universities surveyed. However, it has to be taken into account that official data collection in Thailand is not as sophisticated as in Western countries. In order to allow more precise interpretation, figures have been checked for reliability during interviews with university presidents and vice presidents.

As much of the manufacturing industry of Thailand is located in the extended region of Bangkok, an analysis of the regional innovation system of Bangkok reflects the situation prevailing in Thailand as a whole in many respects. In this context, the boundaries of the extended Bangkok region (EBR+) follow the demarcation suggested by Schiller (2003) that was adopted by Kiese (2003). The EBR+ comprises Bangkok itself, the five provinces surrounding the Bangkok metropolitan region, the three provinces of the eastern seaboard region, and the neighbouring province of Ayutthaya towards the north. In 2003, 75 per cent of the added value created by the manufacturing industry was generated in this region (NESDB, 2004). Moreover, all major political control centres as well as a large proportion of the country's training and research institutions are located in the region as well.

## 13.4 THE COMPATIBILITY OF THAILAND'S ECONOMIC AND SCIENTIFIC SYSTEM

### 13.4.1 Economic Development

Before the Asian crisis, Thailand's economy expanded from 1990 to 1996, with the GDP growing at annual rates ranging between 5.9 and 11.2 per cent (NSO, 2003). Having declined swiftly immediately after the Asian crisis in 1997, Thailand's economy returned to the path of growth in 1999, expanding again at rates ranging from 1.8 to 6.1 per cent (World Bank, 2005). This economic catching-up process went hand in hand with a profound structural change. Between 1981 and 2001, the share of agriculture in the GDP was halved, dropping from 21.4 to 10.2 per cent. Conversely, the industry's share increased from 30.1 to 40 per cent during the same period. This increase in the importance of the industrial sector is based on the consistently swift expansion of the manufacturing industry, which grew at an average rate of 5.8 per cent from 1991 to 2001. Even within the manufacturing industry, structural changes are evident. Between 1975 and 1998, the share of the food, wood, paper, and other resource-based industries fell from 50 per cent to 25 per cent. During the same period, knowledge-based sectors such as medicine, computers, and computer accessories quadrupled their share from 3 to 13 per cent (UNIDO, 2002).

Foreign direct investments (FDI) provided a major impetus for Thailand's economic development, growing markedly from the mid-1980s onwards. Next to domestic technological activities, FDI play an important role in promoting the spread of technological capabilities, given adequate absorption capacities among local enterprises (Dhanani and Scholtès, 2002; Woo, 2004). In the beginning, foreign capital was mostly invested in the primary, steel, and petrochemical industries as well as in infrastructural projects (Brooker Group, 2002). In the 1990s, investments began to focus on more knowledge-based sectors such as hard-disc production. Even now, however, these industries largely depend on imports of purchased materials and services. Thailand's speciality is assembling rather than manufacturing these products.

## 13.4.2 R&D in the Economy

The development of technological capabilities failed to keep pace with the swiftness of industrial development and structural change. Throughout the 1990s, the share of R&D expenditures in the GDP stagnated at 0.10 to 0.15 per cent. An increase to 0.26 per cent was logged only very recently (2002) (NRCT, 2004; IMD, 2004). Western industrialised nations as well as emerging countries in Asia, such as Korea, Taiwan, and Singapore, spend as much as 3 per cent of their GDP on R&D. Although the share of privately-financed R&D grew from 11 per cent in 1997 to around 40 per cent in 2002, more than half of Thailand's R&D expenditures are still financed by public funds (ibid.).

Comparative R&D and innovation surveys in Europe and Asia have shown that absorptive and technological capacities in Thailand are rather low (e.g. Revilla Diez and Kiese, 2006; Hennemann and Liefner, 2006; Revilla Diez and Berger, 2005; Kiese, 2003). This can be shown by several indicators, e.g. number of companies performing R&D or innovation activities. The innovation activities of Thai enterprises focus on the acquisition of machines and equipment, on design, licence purchases and training, rather than formal R&D activities. Similarly, the output of patents and innovations in Thailand is considerably lower than in other countries of Southeast Asia. One important reason for the low level of corporate R&D activity is a prevailing lack of scientifically and technically qualified human capital which results in low absorptive capacities (World Bank, 2005).

## 13.4.3 Research and Development in the Science System

Thailand's science system is comprised of 24 state universities conducting graduate as well as postgraduate research and teaching (MUA, 2002). Next to these, there are 54 private universities as well as a multitude of universities of applied sciences, most of which specialise in Bachelor studies (Krongkaew, 2004, 2). Extramural governmental research is concentrated in four research institutes belonging to the National Science and Technology Development Agency (NSTDA). In addition, some ministerial departments operate their own minor research Facilities.

As in many other developing and emerging countries, a major proportion of the R&D work conducted by state institutions and universities is in applied research (NRCT, 2000, 2004). More recently, universities have been expanding their R&D activities, particularly in the field of basic research. The recent increase in corporate R&D expenditures is mainly due to enhanced experimental development activities. The marked application orientation of research in the public sector and the growing development activities in the corporate sector might increase the knowledge transfer potential. However, as R&D volumes in Thailand remain low in absolute figures, any funds available need to be focused on and employed in no more than a few areas.

A breakdown of R&D expenditures by scientific fields shows that the state, the universities, and business enterprises specialise in noticeably different fields (cf. Table 13.1). While universities traditionally cover a wide range of research, government R&D focuses on agricultural sciences. The increase in engineering R&D on the part of the enterprises is duplicated to some extent in the public research sector. University R&D, on the other hand, shows a markedly slower structural change towards more application-oriented disciplines. Universities only spend somewhat more than one fifth of their R&D budget, usually low in the first place, on engineering sciences. In terms of human resources, public R&D presents a similar picture, the consequence being that, considering the input in the public research sector, technology-related fields of science that might be of interest to business enterprises are underrepresented (NRCT, 2004).

The output of the science system is commonly analysed on the basis of bibliometric data. In this case, the analysis was performed using the multi-disciplinary database Science Citation Index (SCI) (for methodological constraints applying to SCI use, see, *inter alia*, Legler et al., 2000, 48f). For the period from 2002 to 2004, publications in Thailand were subdivided into five scientific fields.

In the last few years, Thailand's publication activity as registered in the SCI increased markedly. Between 1977 and 1990, the number of publications per year doubled for the first time, increasing from about 250 to more than 500. In 1998, more than 1,000 articles involving Thai authors were published in SCI journals, and the number of articles published rose beyond the 2,000 mark for the first time in 2003. Even this powerful growth, however, falls short of the performance of other newly-industrialised countries in Southeast Asia (NIW et al., 2002). Today, most Thai publications by far deal with medical subjects. On the other hand, the rate of expansion is greatest in engineering and natural

Table 13.1 *Percentage of R&D expenditures by scientific field 2001 (1997)*

| Scientific fields | Total | | Government | | Universities | | Enterprises | |
|---|---|---|---|---|---|---|---|---|
| Total (in million Baht) | 11,065 | (4,811) | 5,020 | (2,667) | 1,950 | (1,897) | 4,095 | (247) |
| Natural science | 6.8 | (19.2) | 7.4 | (24.1) | 13.2 | (13.4) | 3.1 | (11.8) |
| ICT | 3.7 | (2.3) | 5.3 | (1.3) | 3.6 | (3.5) | 1.7 | (2.9) |
| Engineering | 41.9 | (12.7) | 16.8 | (6.9) | 21.0 | (18.6) | 82.6 | (28.8) |
| Health science | 7.9 | (15.3) | 10.5 | (19.1) | 14.1 | (11.4) | 1.8 | (3.5) |
| Agricultural science | 28.8 | (30.3) | 50.4 | (42.3) | 19.9 | (14.9) | 6.7 | (19.5) |
| Social science | 9.4 | (18.8) | 8.8 | (5.7) | 22.9 | (35.7) | 3.7 | (30.4) |
| Humanities | 1.5 | (1.4) | 0.9 | (0.6) | 5.4 | (2.4) | 0.4 | (3.1) |

*Notes:* 1997: 1 USD = 25 Baht, 2001: 1 USD = 44 Baht.

*Source:* NRCT 2000, 2004.

sciences, while agricultural sciences are stagnating.

An index of specialisation for SCI publications is calculated to highlight the relevance of certain fields of science in a particular country. Therefore, the percentage of national publications within a given segment is related to the corresponding global percentage figure. Then, figures are transformed so that resultant values range between −100 and +100. Positive values indicate that the degree of specialisation in this field of science is above the world average. The calculation of the index is based on equation (12.1) as suggested by NIW (2002, 2007).

$$ SI = \tanh\left\{ \log\left[ \left( x_{vk} / \sum_{j=1}^{J} x_{jk} \right) \Big/ \left( \sum_{i=1}^{I} x_{vi} / \sum_{i=1}^{I}\sum_{j=1}^{J} x_{ij} \right) \right] \right\} \quad (13.1) $$

for country $v$ and field of science $k$.

Like some other second-generation tiger states, Thailand's degree of specialisation is highest in the agricultural sciences. There is, however, no discernible focus on any of the industry-related engineering sciences that might be compared to those of other emerging countries. Yet it is remarkable that Thailand alone among the Asian emerging countries speci alises in medicine and life sciences more than the average (Table

*Table 13.2  Specialisation of newly-industrialised countries in Asia by scientific fields[1]*

|  | Agricultural sciences | Medicine | Engineering | Life sciences | Natural sciences |
|---|---|---|---|---|---|
| Thailand | +47 | +22 | +11 | +26 | −51 |
| 1st Generation NICs[2] | −38 | −34 | +71 | −26 | +41 |
| 2nd Generation NICs[3] | +81 | −36 | −14 | −37 | −2 |
| China | −64 | −88 | +47 | −72 | +71 |
| India | +45 | −80 | + 8 | −63 | +40 |

*Notes:* [1] Thailand: 2002-04; other countries 1996-2000.
     [2] Korea, Taiwan, Singapore, Hong Kong.
     [3] Malaysia, Philippines.

*Sources:* own calculation based on SCI EXPANDED; NIW et al. (2002).

13.2). Further applications of the specialisation index can be found in NIW et al. (2002, 2007).

However, publications alone provide no indication of the technological benefits of scientific insights. It is patents that indicate whether scientific knowledge is being successfully translated into technology-related applications (NIW et al., 2002, 28). For Thailand, patent indicators may be dispensed with as the number of international patent applications still is too low. This shows that the growing number of scientific publications did not result in new technologies and patents as much as it did in the first-generation tiger states and elsewhere. Whether and to what extent this scientific output nevertheless provided impulses for technological development will be investigated in the following section.

## 13.5 UNIVERSITY-INDUSTRY COOPERATION IN THAILAND

The impact of the imperfect compatibility of science and the economy within the regional innovation system of Bangkok will be examined in detail on the basis of a characterisation of the actual scope of cooperation between universities and enterprises.

Thailand's university landscape is undergoing a profound change at the moment. Because of its elitist roots and the predominance of teaching, the university system was embedded in the innovation system only through its training function right up to the 1990s. Academic services were mainly provided free of charge for governmental authorities. In recent years, government subsidies stagnated or declined, while universities had to meet more stringent requirements owing to increasing enrolment figures and other factors. There are long-term plans to convert all universities into autonomous institutions. This is why options of commercialising and opening up the transfer of knowledge through organisational innovations are being debated at many universities. At the same time, many professors were induced in the past to carry out informal projects of their own to enhance their personal income, as the wages paid within the university system are low (Kirtikara, 2001, 2004). Furthermore, the acceptance of change is affected by cultural factors. Thus, university teachers enjoy social privileges as public education is rooted in the Buddhist monastery system, and all university graduates are honoured by the King in person.

### 13.5.1 Characteristics of Cooperating Enterprises

In the corporate sector, a rethinking process began after the Asian crisis of 1997. It is true that most companies still procure technologies (e.g. machines and software) as well as production licences from providers abroad. On the other hand, relatively small local enterprises in particular are now unable to pay foreign consultancy firms because of the massive devaluation of the Thai currency, the Baht. Their interest in utilising new services offered by the universities has grown correspondingly.

As discussed in the theoretical literature, not all enterprises are equally receptive towards cooperating with universities and research institutions (Bell and Pavitt, 1995). As Figure 13.1 shows, these differen-

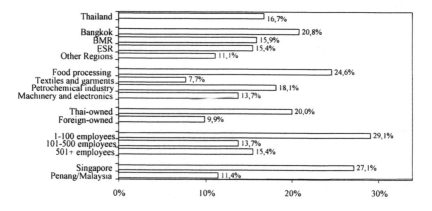

*Sources:* Thailand R&D/Innovation Survey 2002; Singapore and Penang State Innovation Survey 2000.

*Figure 13.1 Share of innovating companies with intensive linkages to public research organisations in Thailand, Singapore and Penang/Malaysia*

ces are distinguishable in Thailand as well. Most of the universities' cooperation partners come from the manufacturing industry. Within this heterogeneous sector, it is mainly the companies that manufacture food and kindred products that cooperate with universities. Cooperation with enterprises from the textile and primary industries is very limited, despite the great importance of these sectors in Thailand. However, many of these companies lack the requisite capacity to absorb external knowledge. The chemical industry as well as the mechanical and electrical engineering sectors are dominated by relatively large enterprises whose inclination to cooperate is generally low. However, the first and last-named industries are dominated by local enterprises and foreign companies, respectively, which explains why the proportion of cooperating enterprises is different in each. Enterprises in which foreign owners hold majority shares tend to use sources of knowledge in their respective home countries. Small and medium-sized enterprises (SMEs) cooperate considerably more frequently with research institutions than bigger companies.

### 13.5.2 Corporate Requirements Regarding University Cooperation

The requirements applied by companies to cooperation with universities are often the same they would habitually apply to cooperation with other enterprises. They include, for instance, compliance with project completion deadlines and exclusive rights to utilise the results of cooperation. This calls for fostering a relationship of mutual trust between the partners. In addition, differences between various enterprise types can be distinguished. Multinational enterprises frequently use cooperation to gain access to university graduates and influence the content of their training. SMEs, on the other hand, frequently expect cooperative projects to support them in the development of products and processes.

### 13.5.3 Forms of Cooperation

The theoretical debate emphasises that, while there is a multitude of ways in which universities and enterprises may interact, only a few cases go beyond mere individual contacts and involve genuine research work (Inzelt, 2004). Most cooperation projects in Thailand merely involve informal and personal contacts as well as services that do not include detailed research. There are some cases in which turnkey research results were either sold outright or licensed to business enterprises. Cooperation resulting from teaching plays a certain role. Contract or joint research is confined to a few projects. Table 13.3 depicts the differences that exist between several fields of science.

A comparison of the forms of interaction prevailing in the different target sectors of cooperation shows that more recent industries, such as mechanical and electrical engineering as well as the chemical and pharmaceutical industry, tend to prefer research-oriented forms of coope ration, whereas traditional branches, such as the food, textile, and

*Table 13.3 Modes of interaction between universities and industry by scientific field (in %)*

| Mode of interaction | Total | Engineering | Natural sciences | Agricultural sciences | Health sciences | Others |
|---|---|---|---|---|---|---|
| Advisory services | 49.3 | 52.2 | 34.5 | 64.3 | 35.3 | 56.3 |
| Technical services | 34.6 | 45.7 | 20.7 | 50.0 | 23.5 | 12.5 |
| Informal meetings | 19.9 | 13.0 | 24.1 | 21.4 | 23.5 | 25.0 |
| Licensing | 16.9 | 6.5 | 10.3 | 14.3 | 64.7 | 12.5 |
| Contract research | 15.4 | 23.9 | 24.1 | 3.6 | 5.9 | 6.3 |
| Selling of products | 8.1 | 2.2 | 13.8 | 10.7 | 17.6 | – |
| Joint conferences | 8.1 | 6.5 | 3.4 | 14.3 | – | 18.8 |
| Continuous education programmes | 8.1 | 6.5 | 3.4 | 10.7 | 5.9 | 18.8 |
| Internship programmes | 7.4 | 15.2 | 3.4 | 7.1 | – | – |
| Joint research | 6.6 | 13.0 | 6.9 | – | 5.9 | – |

*Notes*: Multiple answers possible, table includes modes used by more than 5 % of all respondents.

*Source*: Own data.

primary industries, predominantly use consultation services that relate to human resources. The mechanical and electrical engineering industries are the only ones in which 30 per cent of all projects involve joint and/or contract research. One tenth of these projects result in joint patents, and researchers are beginning to plan the establishment of spin-offs. In the chemical and pharmaceutical industries, straight licences on university research form the most important pathway of knowledge transfer, being used in more than two thirds of all projects. Moreover, these are the only industries in which long-term transfers of personnel between enterprises and universities are of any significance, being implemented in more than one fifth of all projects. Enterprises that do not belong to the manufacturing sector predominantly use consultation and further education services offered by universities. The high proportion of linear transfers (mainly through licensing) suggests that scientific capabilities in the health science sector are not reflected in the corporate sector. Obviously, the absorption capacity required for interactive cooperation is lacking.

### 13.5.4 The Regional Reach of Cooperation

One point of particular geographical importance is the integration of cooperation within the regional innovation system of Bangkok. Cooperation patterns differ markedly, depending on the location of the university in question (cf. Figure 13.2). In regional terms, the cooperation of all Bangkok universities are concentrated in the EBR+. This trend is less marked in the light industry as well as in sectors outside the manufacturing industry. Likewise, universities located away from Bangkok find most of their partners in the extended Bangkok region, particularly in the mechanical and electrical engineering sectors. As regional partners are especially hard to find in the chemical and pharmaceutical industry, they tend to look for cooperation partners abroad.

### 13.5.5 Typical Cooperation Obstacles

Interviews with players about the cooperation obstacles perceived by them led to relatively profound conclusions about cooperation barriers. It is obvious that most of the obstacles named at universities relate to the corporate side (cf. Figure 13.3). In interviews with enterprises, most of the defects addressed were on the university side. This game of claim and blame shows clearly that in many cases, cooperation partners do not

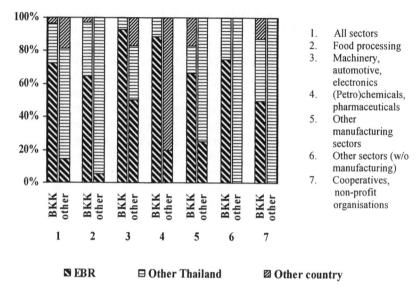

*Source*: Own data.

*Figure 13.2  Regional distribution of university-industry linkages by
         location of university and sector of industrial partner (share
         of linkages by region in %)*

understand each other's specifics.

Professors blame enterprises for not being prepared to cooperate because they distrust the universities and are unwilling to pay in advance for uncertain research results. Moreover, there are many areas in which suitable partners are hard to find because many Thai enterprises still have no R&D activities of their own. The survey of the corporate sector showed that enterprises with advanced technological capabilities find it similarly hard to identify suitable partners at Thai universities because of the prevailing lack of material and human resources. This confirms that the two sectors specialise in different things. On the one hand, there are many enterprises whose ability to absorb the results of university research is insufficient, while on the other, the universities are not yet equipped to offer adequately qualified results to large or multinational enterprises.

Within the universities themselves, professors as well as entrepreneurs would like to see bureaucratic processes simplified and rules for coope- ration projects formulated clearly. This might help to resolve time- and

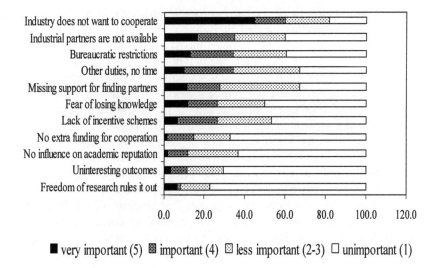

■ very important (5)  ▓ important (4)  ▨ less important (2-3)  □ unimportant (1)

*Source*: Own data.

*Figure 13.3  Importance of limitations for university-industry linkages at Thai university departments*

interest-related conflicts between the academic core competences of research and teaching on the one hand and the provision of external services on the other. At the moment, these fundamental problems relating to the support of cooperation are more significant at the universities than the creation of specific incentive systems.

## 13.6  SUMMARY

Within the regional innovation system, the scope of R&D and innovation activities by all players is markedly smaller than in the advanced emerging countries of Asia. This fragmentation reduces the potential for interaction among innovative players. More recently, however, R&D expenditures have been increasing, especially in the corporate sector. On the other hand, less than one fifth of the technologically-advanced, innovative enterprises describe themselves as cooperating intensely with universities. The intensity of cooperation with universities varies among sectors and seems to depend on corporate characteristics, e.g. size or ownership. Companies located in Bangkok are most receptive towards cooperating with universities, whose importance in the corporate process of innovation begins to decline in the neighbouring provinces.

Thai research institutions mainly specialise in applied research. The more technology-related engineering sciences, being underrepresented in comparison with other emerging countries in Asia, were unable to keep pace with the growth of modern industries. Because of their traditional orientation towards teaching, R&D at universities is growing more slowly overall than the commitment of the private economy in this field. Very likely, only a few top-flight facilities at universities and research institutions will reach a level of research high enough to render them interesting to large or foreign enterprises. Cooperation with SMEs, which is less research-intensive, is hampered mainly by defective absorption capabilities and financial capacities. As research expands in the sector of science, the potential supply of scientific services is improving slowly. New incentive systems have induced some universities to begin trying harder to commercialise their research. Further investigation will be required to identify the impact of these changes on the universities' commercialisation strategies as well as on the transfer of knowledge within the innovation system.

At the moment, most cooperation between science and the economy that is to be found within the regional innovation system of Bangkok is based on personal contacts and operates without an elaborate institutional framework. Genuine research cooperation is lacking. The differences that exist in the cooperation potential of individual industries and fields of science are reflected in the intensity and quality of cooperation. There is a lack of confidence-building communication among players. For a final evaluation of the cooperation scene, the data now available will have to be analysed further.

All in all, the cooperative relations maintained by universities are strongly embedded in the region. This equally applies to all industries within the regional innovation system of Bangkok. Universities located at the periphery mainly seek to cooperate with local SMEs. As far as the more modern sectors of the economy are concerned, however, the lack of regional partners forces them to look for suitable enterprises either in Bangkok or, given adequate excellence, among companies abroad. Within Thailand as a whole, the regional innovation system of Bangkok offers the best opportunities for cooperation between universities and enterprises.

### 13.6.1 Action Recommendations

Action recommendations for more efficient UIL in Thailand comprise the three actors: industry, universities and government. The government is an important facilitator and funding body to initialise more intensive cooperation and to increase absorptive capacities at this early stage of technological development.

*Industry*: enterprises in Thailand should step up their endeavours to build their own technological capabilities so as to achieve the competence to cooperate. At the same time, the content of university teaching and research should be harmonised with corporate requirements more than hitherto. At the moment, absorption-capability building is greatly hampered by a lack of adequately qualified university graduates. Thus, companies need to become more eager to invest in long-term innovation activities and human capital building.

*Universities*: the facilitation of university-industry linkages should be included in the design of concepts for higher education reform. (1) Universities have to possess academic capabilities that are relevant to industrial needs. (2) Bureaucratic obstacles should be gradually abolished and incentive structures built up at the universities to make the establishment of cooperative relationships easier. Both measures have to be included in the ongoing process of transforming Thai universities in autonomous organisation which has been hampered repeatedly.

The evaluation of professors should no longer be based exclusively on their academic excellence but also on other indicators such as, for instance, the success of their cooperative relationships. Cooperation is greatly impeded by a lack of mutual trust among the players. Termed exchanges of employees between enterprises and public research institutions might be a way of building mutual trust and learning more about the others' research needs.

*Government*: neither universities nor private companies in Thailand are yet equipped with sufficient resources to build up absorptive capacities or research excellence by themselves. The successful technological catch-up process of Korea, Taiwan and Singapore suggests that national governments have to provide initial funding and incentives for linkages between both sectors to a certain degree.

Public funding for universities is necessary for the development of more advanced academic capabilities, e.g. the latest equipment for research and teaching, staff development, etc. Scaling down public funding for universities might be an efficient way to foster own income generation by universities in developed higher education systems, but these activities might not be related to academic excellence or industrial needs in developing countries. Budget cuts in Thailand have mainly resulted in additional teaching programmes and higher tuition fees. Instead, additional public funding has to be made available to increase the responsiveness of existing academic units and to set up new structures that are directly related to industrial needs. Emphasis should be placed on S&T fields that are the most obvious partners for industrial R&D.

A promising approach for university-related support of UILs could be the establishment of university-industry cooperative research centres. Such centres might be a way to deepen already existing UIL activities of research units at Thai universities. A quite successful example is the Thai Higher Education Development Project that has been funded by the Asian Development Bank. During the first phase, the academic capabilities of seven newly founded centres have been increased significantly. However, a second phase should ensure the sustainability of these centres by providing more explicit support and incentives for UIL activities.

The analysis of public support schemes for corporate R&D has shown that there are quite a lot of support programmes ranging from (co-) funding of R&D projects or technical services to general tax deductions for R&D expenditures. However, these schemes are either quite small or too bureaucratic and thus only used by a few companies. Thus, such programmes have to be equipped with additional budgets and have to be implemented and managed more efficiently.

A promising approach of industry-related support for UILs could be the promotion of R&D and innovation in sectors of the Thai industry that face thresholds in their technological upgrading process. The most successful example in Thailand is the hard disk drive industry. Within the so-called HDD cluster the Thai government provides sector-specific funding via the National Electronics and Computer Technology Center (NECTEC) which is matched with funding provided by several foreign and local HDD producers and suppliers located in Thailand. The money is used to jointly solve technical problems of HDD cluster members and to develop skills for the HDD industry in cooperation with Thai universities and NECTEC itself.

Finally, enterprises located at the periphery find it even more difficult to build up cooperative relationships because the innovation system concentrates on Bangkok. To enhance Thailand's competitiveness as a whole, steps to improve cooperative relationships should not be confined to Bangkok alone.

## ACKNOWLEDGEMENT

The authors highly appreciate the funding by the German Research Foundation.

## REFERENCES

Arnold, E., J. Bessant, M. Bell and P. Brimble (2000), *Enhancing Policy and Institutional Support for Industrial Technology Development in Thailand*, Brighton, SPRU, Centrim, Technopolis.

Asheim, B. and J. Vang (2004), 'What can regional systems of innovation and learning regions offer developing regions?', paper presented at the second Globelics conference in Beijing, October.

Bell, M. and K. Pavitt (1995), 'The development of technological capabilities', in I. ul Haque (ed.), *Trade, Technology and International Competitiveness*, World Bank, EDI Development studies, Washington, pp. 69-101.

Berger, M. (2007), *Upgrading the System of Innovation in Late-Industrialising Countries. The Role of Transnational Corporations in Thailand's Manufacturing Sector*, Münster, Wien, Lit-Verlag.

Brooker Group (2002), *Foreign Direct Investment: Performance and Attraction. The Case of Thailand*, Bangkok.

Bunnell, T.G. and N.M. Coe (2001), 'Spaces and scales of innovation', *Progress in Human Geography*, 4 (25), 569-589.

Cabo, P.G. (1999), 'Industrial participation and knowledge transfer in joint R&D projects', *International Journal of Technology Management*, 3-4 (18), 188-206.

Cohen, W.M. and D.A. Levinthal (1990), 'Absorptive capacity: a new perspective on learning and innovation', *Administrative Science Quarterly*, (35), 128-152.

Dhanani, S. and P. Scholtès (2002), 'Thailand's manufacturing competitiveness: promoting technology, productivity and linkages', *UNIDO, SME Technical Working Paper Series*, Wien.

Edquist, C. (1997), 'Systems of innovation: technologies, institutions, and organizations', *Science, Technology and the International Political Economy Series*, London, Pinter.

Etzkowitz, H., J.M.C. de Mello and M. Almeida (2005), 'Towards "meta-innovation" in Brazil: the evolution of the incubator and the emergence of a triple helix', *Research Policy*, 4 (34), 411-424.

Faulkner, W. and J. Senker (1995), *Knowledge Frontiers. Public Sector Research and Industrial Innovation in Biotechnology, Engineering Ceramics and Parallel Computing*, Oxford, Clarendon Press.

Freeman, C. (2001), 'The learning economy and international inequality', in D. Archibugi and B.A. Lundvall (eds), *The Globalising Learning Economy*, Oxford, Oxford University Press, pp. 147-162.

Gibbons, M., C. Limoges, H. Nowotny, S. Schwartzmann and P. Scott (1994), *The New Production of Knowledge: The Dynamics of Science and Research in Contemporary Societies*, London, Sage.

Hennemann, S. and I. Liefner (2006), 'Hochtechnologieunternehmen im chinesischen Innovationssystem – Empirische Ergebnisse aus Beijing und Shanghai' (High-tech companies in the Chinese innovation system – empirical finding from Beijing and Shanghai), *Zeitschrift für Wirtschaftsgeographie*, 1 (50).

Hobday, M. (1995), *Innovation in East Asia: the Challenge to Japan*, Aldershot and Edward Elgar, Cheltenham, UK and Northampton, MA, USA.

Institute for Management Development (IMD) (2004), *World Competitiveness Yearbook*, Lausanne.

Inzelt, A. (2004), 'The evolution of university-industry government relationships during transition', *Research Policy*, 6-7 (33), 975-995.

Johnson, W.H.A. and D.A. Johnston (2004), 'Organisational knowledge creating processes and the performance of university-industry collaborative R&D projects', *International Journal of Technology Management*, 1 (27), 93-114.

Kiese, M. (2003), 'Regionale Innovationspotenziale in Südostasien aus der Sicht einer "neuen" Wirtschaftsgeographie' (Regional innovation potentials in South-East Asia from the perspective of a new economic geography), *Zeitschrift für Wirtschaftsgeographie*, 3/4 (47), 196-214.

Kiese, M. (2004), 'Regionale Innovationspotenziale und innovative Netzwerke in Südostasien: Innovations- und Kooperationsverhalten von Industrieunternehmen in Singapur' (Regional innovation potentials and innovative networks in South East Asia: innovation and cooperation patterns of manufacturing companies in Singpore), *Hannoversche Geographische Arbeiten*, Band 56, Münster, Hamburg.

Kirtikara, K. (2001), 'Higher education in Thailand and the National Reform Roadmap. Bangkok', paper presented at the Thai-US Education Roundtable.

Kirtikara, K. (2004), 'Transition from a university under the bureaucratic system to an autonomous university', Office of the National Education Council, Bangkok.

Kline, S.J. and N. Rosenberg (1986), 'An overview of innovation', in R. Landau and N. Rosenberg (eds), *The Positive Sum Strategy: Harnessing Technology for Economic Growth*, Washington, National Academy Press, pp. 275-305.

Knack, S. and P. Keefer (1997), 'Does social capital have an economic payoff? A cross-country investigation', *Quarterly Journal of Economics*, 4 (112), 1251-1288.

Kraas, F. (1996), 'Thailand – ein Newly Industrialized Country? – Die industrielle Entwicklung seit Ende der achtziger Jahre' (Thailand – a newly industrialized country? – industrial development since end of the 1980s), *Zeitschrift für Wirtschaftsgeographie*, 4 (40), 241-257.

Krongkaew, M. (2004), 'The promises of the new university financing system in Thailand: the income contingent loan (ICL) scheme', paper presented at the Monthly Workshop by the Monetary Policy Division, Bank of Thailand, Bangkok, January.

Lall, S. (2002), 'Science and technology in Southeast Asia', *EU Strata Meeting*, Brussels.

Legler, H., M. Beise, B. Gehrke, U. Schmoch and D. Schumacher (2000), *Innovationsstandort Deutschland. Chancen und Herausforderungen im internationalen Wettbewerb* (Germany as a location for innovation activities: opportunities and threats in the global competition), Landsberg/Lech, Verlag Moderne Industrie.

Lundvall, B.A. and B. Johnson (1994), 'The learning economy', *Journal of Industry Studies*, 2 (1), 23-42.

Ministry of University Affairs (MUA) (2002), *Thai Higher Education in Brief*, Bangkok.

National Economic and Social Development Board (NESDB) (2004), *Gross Provincial Product 2003*, Bangkok.

National Research Council of Thailand (NRCT) (2000), *National Survey on R&D Expenditure and Personnel of Thailand 1987-1996*, Bangkok, National Research Council of Thailand.

National Research Council of Thailand (NRCT) (2004), *Thailand's 2001 Research and Development Indicators*, Bangkok.

National Statistical Office (NSO) (2003), *Thailand Development Indicators 2003*, Bangkok.

Nelson, R.R. (1993), *National Innovation Systems: A Comparative Analysis*, New York.

Nelson, R.R. and N. Rosenberg (1993), 'Technical innovation and national systems', in R.R. Nelson (ed.), *National Innovation Systems: A Comparative Analysis*, New York, Oxford University Press, pp. 3-21.

Niedersächsisches Institut für Wirtschaftsforschung (NIW), Fraunhofer-Institut für System- und Innovationsforschung and Deutsches Institut für Wirtschaftsforschung (2002), 'Aufhol-Länder im weltweiten Technologiewettbewerb' (Catch-up countries in the global competition for technology), *Spezialstudie zum Indikatorenbericht zur Berichterstattung zur technologischen Leistungsfähigkeit Deutschlands 2001*, Hannover.

Niedersächsisches Institut für Wirtschaftsforschung (NIW), Fraunhofer-Institut für System- und Innovationsforschung and Deutsches Institut für Wirtschaftsforschung (2007), 'Die Bedeutung von Aufhol-Ländern im globalen Technologiewettbewerb' (The role of catch-up countries in the global competition for technology), *Studien zum deutschen Innovationssystem, 21*, Hannover.

OECD (1996), *The Knowledge-based Economy*, Paris.

Revilla Diez, J. and M. Berger (2005), 'The role of multinational corporations in metropolitan innovation systems – empirical evidence from Europe and South East Asia', *Environment and Planning A 37*, 10, 1813-1835.

Revilla Diez, J. and M. Kiese (2006), 'Scaling innovation in Southeast Asia: empirical evidence from Singapore, Penang (Malaysia) and Bangkok', *Regional Studies*, 40 (9), 1005-1023.

Rothaermel, F.T. and M. Thursby (2005), 'University-incubator firm knowledge flows: assessing their impact on incubator firm performance', *Research Policy*, 3 (34), 305-320.

Schätzl, L. (2000), *Wirtschafsgeographie 2 – Empirie* (Economic Geography 2 – Empirics), Paderborn, UTB.

Schätzl, L. and J. Revilla Diez (2000), 'DFG Schwerpunktprogramm "Technologischer Wandel und Regionalentwicklung in Europa". Kurzfassungen der Forschungsvorhaben' (DFG priority program 'Technological Change and Regional Economic Development in Europe', project summaries), *Hannoversche Geographische Arbeitsmaterialen*, Band 24, Hannover, Leibniz Universität Hannover.

Schiller, D. (2003), *Technologische Leistungsfähigkeit Thailands* (Technological capabilities of Thailand), unpublished diploma thesis, Dept. of Economic Geography, Univ. Hannover.

Schiller, D. (2006), 'Nascent innovation systems in developing countries: university responses to regional needs in Thailand', *Industry and Innovation*, 13 (4), 481-504.

Schiller, D. and I. Liefner (2007), 'Higher education funding reform and university-industry links in developing countries: the case of Thailand', *Higher Education*, 54 (4), 543-556.

Schmoch, U. (1999), 'Interaction of universities and industrial enterprises in Germany and the United States – a comparison', *Industry and Innovation*, 1 (6), 51-68.

Stracke, S. (2003), *Technologische Leistungsfähigkeit im Innovationssystem Penang, Malaysia* (Technological capabilities in the innovation system of Penang, Malaysia), Doctoral thesis, Hannover, Leibniz Universität Hannover.

UNIDO (2002), 'Thailand's manufacturing competitiveness: promoting technology, productivity and linkages', *SME Technical Working Papers Series*, 8-94, Wien.

Wong, P.K. (1999), 'National innovation systems for rapid technological catch-up: an analytical framework and a comparative analysis of Korea, Taiwan and Singapore', *DRUID Summer Conference on National Innovation Systems, Industrial Dynamics and Innovation Policy*.

Woo, W.T. (2004), 'The economic impact of China's emergence as a major trading nation', paper presented at the second WTO, China, and Asian Economics conference in Beijing, June.

World Bank (2005), *Thailand Economic Monitor*, Bangkok.

# 14. Knowledge-based local economic development for enhancing competitiveness in lagging areas of Europe: the case of the University of Szeged

## Imre Lengyel

### 14.1 INTRODUCTION

One of today's most exciting questions about lagging regions is related to the role universities can play in the improvement of regional competitiveness. This question is especially important for the new member states of the EU, since between 2007 and 2013 they receive significant subsidies from the European Union's regional development funds to improve the competitiveness of their regions. The analysis of this issue calls for clarifying various questions for the less developed regions. What do we mean by regional competitiveness and how can it be described and measured? Do the economic, social and institutional background of a region and its cultural characteristics influence the strategies of universities? Which development strategy of universities can most significantly improve regional competitiveness in the lagging region?

This chapter introduces the most important conclusions of the analysis and discusses our proposal for the bottom-up local economic development strategy of the University of Szeged that has been built on the development strategy of the region. The University of Szeged has an outstanding scientific capacity. Several research institutes operate in Szeged including the Biological Research Centre of the Hungarian Academy of Sciences. This excellent scientific capacity is located in an underdeveloped region. The GDP per capita (PPS) of the Southern Great Plain Region reached only 41 per cent of the EU-25 average in 2003. Do the economic development strategies of universities depend on the development level of the region and is it possible to use the same strategy in underdeveloped regions as in developed ones?

For the interpretation of regional competitiveness we introduce a model that offers a complex frame for the measurement and enhancing of competitiveness. The so-called 'pyramid model' of regional competitiveness is suitable for the systematization of both regional planning and bottom-up local economic development ideas, so consequently it was also applied to the Southern Great Plain region and Szeged subregion. A special version of the pyramid model can be designed for this subregion, the elements of which are built upon the real opportunities of the so-called 'knowledge-transfer' region type.

After outlining our analytical framework, the pyramid model of regional competitiveness, this chapter first assesses the level of competitiveness of the Southern Great Plain (NUTS2) region where the University of Szeged is located, then it examines the most important attributes of Szeged and its (LAU1) subregion. A complex methodology was used throughout our economic analysis to underlie the local strategy for improving regional competitiveness. On the one hand it included cluster mapping based on the location quotient method. On the other hand, with the help of a questionnaire, it surveyed local companies to characterize university-industry relations and the needs of enterprises. Furthermore, it mapped the scientific capacities of both the university and local research institutes. Interviews were also carried out with experts of local innovative companies to collect their suggestions concerning the development of the university. Consequently, this chapter provides a review of the major results of our survey dealing with the local economic development role of the university and discusses the fundamental aspects of its development strategy.

## 14.2 THE PYRAMID MODEL OF REGIONAL COMPETITIVENESS AND THE TYPES OF REGIONS

Successfulness in competition, or in other words competitiveness, has been one of the key concepts over the past two or three decades partly due to the increasing global competition (Camagni 2002). It is a fashionable term, the use of which seems nowadays to be nearly obligatory. In Iain Begg's (1999, p. 795) apt formulation: 'improved competitiveness, as we all know, is the path to economic nirvana'. Competitiveness as a collective notion has been in use for a long time; still it is difficult to define. It basically means the inclination and skill to compete, the skill to win position and permanently stay in competition, which is marked primarily by successfulness, the size of market share and the increase of business success (Kitson et al. 2004).

Two major issues emerged in the debates aiming at the interpretation of regional competitiveness: on the one hand, how to define competitiveness and what are the indicators to measure it? On the other hand, how can competitiveness be improved, and which economic development programmes may be regarded as successful? These two questions usually lie in the background of other professional debates too; while representatives of academic economics concentrate on the first one, experts of economic policy tend to focus on the second one. Questions of interpretation, measurement and economic development programmes related to the concept of competitiveness receive much attention in countries and regions as well.

There were a number of attempts to define the new notion of competitiveness according to new global competition conditions in the mid 1990s. Based on theoretical considerations both Krugman (1994) and Porter (1990) claim that at the macroeconomic level the growth rate and the level of productivity are suitable to describe the economic category that is otherwise called competitiveness. As Porter puts it: 'the only meaningful concept of competitiveness at the national level is national productivity' (1990, p. 6). So in their approach it is the effectiveness of producing internationally marketable products and services that generally defines competitiveness. Porter has argued that export-oriented (traded) regional clusters are capable of improving competitiveness and therefore proposed a cluster-based approach to regional economic development (Porter 2003b). The most important findings of the abundant literature dealing with the competitiveness of countries may also be applied to interpret the competitiveness of regions (Budd and Hirmis 2004; Camagni 2002; Gardiner et al. 2004; Malecki 2002, 2004).

There is an abundant literature on competitiveness with certain well-known approaches,[1] out of which the concept of so-called standard competitiveness especially common in the European Union, seems adequate in the case of the regions not only for scientific analyses but also for economic development applications. The standard (extended) notion of regional competitiveness in EU regional policy is as follows (EC 1999, p. 75): 'The ability of companies, industries, regions, nations and supra-national regions to generate, while being exposed to international competition, relatively high income and employment levels'. Otherwise it is described as 'high and rising standards of living of a nation with the lowest possible level of involuntary unemployment, on a sustainable basis' (EC 2003).

The concept of standard competitiveness is partly linked to the thought of economic growth; therefore, it also builds on theoretical economics, although it also has strong economic development aspects that bring it close to the question of business sciences as well. The aforementioned standard definition refers to 'relatively high income' that can be measured by per capita GDP and the GDP growth rate. A high employment level is in turn indicated by the rate of employment. These two indicators can be measured independently from one another, but as it is well known the per capita GDP can also be expressed as follows (EC 1999, p. 75; Lengyel 2004):

$$\frac{GDP}{total\ population} = \frac{GDP}{employment} \times \frac{employment}{working-age\ pop.} \times \frac{working-age\ pop.}{total\ population}$$

The first fraction on the right-hand side of the formula is approximately equal to labour productivity and the second to the rate of employment. The third fraction, the age distribution of the population, only changes slowly. It can nevertheless play an important role in some regions with smaller populations.

Hence the *substance of regional competitiveness*: the economic growth in the region, which is generated by both a high level of labour productivity and a high level of employment (EC 2003). In other words, competitiveness means economic growth driven by high productivity and a high employment rate. Thus given the standard definition of competitiveness, no unique indicator of regional competitiveness can be found. It is interpreted rather as a combination of closely related, easily-measurable and unambiguous traditional economic categories:

1. per capita GDP of the region (otherwise regional growth),
2. labour productivity of the region,
3. employment rate of the region,
4. economic openness (international competition) of the region (exports and tourism).

These indicators of revealed competitiveness[2] enable us to measure competitiveness fairly precisely (Lengyel 2003; Lukovics 2006a). The notion of competitiveness obtained in this way cannot be used, however, to identify factors responsible for regional competitiveness or areas which are to be strengthened or developed by regional development policies and programmes for improved competitiveness.

For the interpretation of regional competitiveness we established the pyramid model that offers a complex frame for the measurement and improvement of regional competitiveness (Lengyel 2000, 2004). The pyramid model of regional competitiveness attempts to provide a systematic account of these means and to describe the economic development programmes of improved competitiveness (Figure 14.1). 'This model is useful to inform the development of the determinants of economic viability and self-containment for geographical economies' (Pike et al. 2006, p. 26).

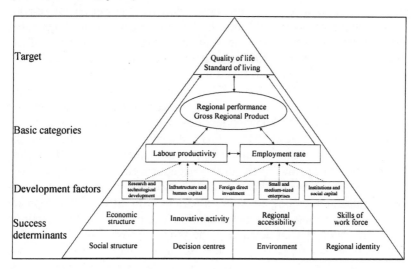

*Source*: Lengyel (2000, 2004).

*Figure 14.1 The pyramid model of regional competitiveness*

As depicted in Figure 14.1 the components of long-term success ('success determinants') can be found in the base of the pyramid,[3] the middle layer consists of the development factors (programmes) or 'key drivers' such as research and technological development (RTD), small and medium-sized enterprises (SME), foreign direct investment (FDI), infrastructure and human capital, institutions and social capital. Improving these individual development (programming) factors forms the object of regional policies. The basic categories (indicators of revealed competitiveness) included in the standard definition of competitiveness are located one level higher, while the standard of living and welfare of the region's population, the ultimate objective, forms the peak of the pyramid (Parkinson et al. 2006, p. 67).

The pyramid model can be widely used and is applicable to all basic territorial units, for instance to NUTS2 regions, and functional sub-regions. It is in essence a means to assess economic growth, while it also constitutes the objectives of economic development strategy under global and changing circumstances. It is necessary that the economic development strategy programmes improve competitiveness depending on the types of the regions.

Tödtling and Trippl (2005, p. 1209) describe three types of regions by problem areas and regional innovation deficiencies: peripheral regions (organisational thinness), old industrial regions (lock-in) and fragmented metropolitan regions. During the research to create the foundation of EU regional policy between 2007 and 2013, four 'theoretical' region types were distinguished based on two dimensions, the population density and the growth rate of GDP (Martin 2003, pp. 6-23): non-productive regions, regions as production sites, regions as sources of increasing returns and regions as hubs of knowledge. Based on the characteristics of competitive advantages, Porter (2003a) distinguishes three phases in the development of countries built upon one another such as the factor-driven, the investment-driven (or efficiency driven) and the innovation-driven phases. Today, the knowledge-based economy strongly transforms the specialization among a country's regions with different development levels also changing the former characteristics of territorial competition (Malecki 2004). Consequently, the three phases of competitive development should be specified based on the processes of the knowledge-based economy by using the specialization of the postfordist economy (Cooke 2001; Lengyel, B. 2005; Lengyel 2003).

The three phases of competitive development designed for countries can also be applied to the case of subnational regions (Lengyel 2003). In the case of competitive regional development the innovation-driven phase's competitive advantages are derived from knowledge creation, while competitive advantages of the investment- and factor-driven phases originate from mere adaptation of knowledge (Porter 2003a). Less developed, lagging regions are in an exposed situation, certain features of the knowledge-based economy are present, but neofordist characteristics are decisive (Asheim 2001). In harmony with the phases of competitive development three types of postfordist regions can be distinguished (Lengyel 2006b):

1. *neofordist regions* associated with the factor-driven phase (regions with low income and input cost),
2. *knowledge transfer regions* associated with the investment-driven phase (regions with medium income and efficiency) and
3. *knowledge creation regions* associated with the innovation-driven phase (regions with high income and unique value).

Neofordist and knowledge transfer regions differ from knowledge creation regions not only in terms of the sources of competitive advantages, but also because they are economically exposed and fragile (Lengyel 2006b). Decision centres of global companies hardly occur in less developed regions, so they demand knowledge less; the executive type activities of global companies are more likely to be present here. Besides assembly plants, units of global companies selling products and performing service activities on the local market, local branches of international banks and insurance companies, and sometimes subsidiaries engaging in minor research activities also operate here. Naturally, most regions are 'mixed', but while neofordist and knowledge transfer activities and companies also exist in knowledge creation regions, the number of firms based on knowledge creation is close to zero in neofordist regions.

The *neofordist region* is underdeveloped, it corresponds to a semi-periphery, the generated income (GDP/habitant) is low, and the economy is typically in the factor-driven phase. The development of infrastructure is insufficient, the education level of the labour force is low, the members of company management are not competitive internationally and part of the qualified labour force and talented young people leave the region. Improving hard infrastructure (e.g., transportation network) and attracting branches of global companies with low local taxes and wages are at the centre of local development polices.

*Knowledge transfer regions* are usually medium-level developed. The most important goal of economic development lies in continuing the structural change by keeping existing companies and creating work places with higher value added. These regions are in the investment-driven phase, they have traded large companies with local headquarters, which already have a network of local SMEs as their contractors. Transportation infrastructure is developed, therefore the improvement of the local business environment is in focus. The education level of the labour force and the training structure already correspond to the needs of the local economy while retraining programmes and courses to improve managerial skills are frequent.

In *knowledge creation regions* economic output is high, these regions are in the innovation-driven phase and the regional centres of significant global companies are situated there. Administration is decentralized and a cluster-based economic development is set partly to improve the business environment necessary to strengthen the competitive advantages of global companies with local headquarters. Developing the background of innovation capacities is in focus, scientific parks, universities, knowledge spillovers, incubator programmes, venture capital and other schemes have an important role (Varga 2000, 2003).

Concerning the three region types reviewed above, different development strategies must be applied for each of them which means that the improvement of competitiveness demands different measures based on the different types of regions. These steps correspond to the phases of competitive regional development and at the same time indicate that competitiveness can be improved only with the help of complex programmes. The pyramid model systematizes those economic development priorities adjusted to real social-economic situations and achievable aims of different region types. Improvement of regional competitiveness depends on the consistent realization of these development strategies. Similarly opportunities of universities in knowledge-based local economic development depend on the types of regions (Boucher et al. 2003).

## 14.3 COMPETITIVENESS OF THE SOUTHERN GREAT PLAIN REGION AND THE SZEGED SUBREGION

The change in Hungary's political and economic systems took place in 1989 and 1990. The transition first brought about significant economic fallback. In 1993 the GDP per capita fell back to 82 per cent of the 1989 value and exceeded it only in 2000. In 1996 real income was only 85 per cent of the figure in 1989, exceeding the value of the 1980s only in 2002 (HCSO 2004).

Changes in the Hungarian regional policy followed the transition with delay (Enyedi 1996, 2004; Horváth 2005; Nemes Nagy 2001; Rechnitzer et al. 2005). The Act on Local Governments passed in 1990 granted significant freedom to towns and villages, certain elements of the former state property (schools, hospitals, municipal flats, etc.) were also transferred to them even though they received hardly any budgetary sources for their maintenance (Lengyel 1993). On the other hand, the power of counties was remarkably reduced, which practically made the regional level disappear and left Hungary consisting of 'the alliance of 3200 municipalities', or more precisely, 'the central power and its 3200 subjects'. The act on regional development designed on the basis of the EU's regional policy and institutions was passed in 1996. Each of the 19 (NUTS3) counties established a 'County Territorial Development Council', whose members also include the representatives of the larger cities. In 1999 seven NUTS2 planning regions were established. In these regions 'Regional Development Councils' coordinate regional development concepts and development programmes of the counties in compliance with the expectations of the EU's regional policy (Horváth 2004; Rechnitzer 2000).

The Southern Great Plain region is situated in the south-eastern border area of the country, and includes three counties: Bács-Kiskun, Békés and Csongrád (Figure 14.2). The region's area is 18,000 square kilometres and it has a population of 1.4 million. The economy and settlement structure of this region is defined by the fact that it is situated in the Great Plain divided by the river Tisza. Szeged, Hungary's fourth most densely populated city, which is situated approximately 180 kilometers south from Budapest, is the centre of both the Southern Great Plain region and Csongrád county. Szeged has a population of 160,000 people and together with its subregion constitutes the labour market of approximately 250,000 people. Trans-European transport corridors pass

*Figure 14.2   Regions of Hungary and counties of Southern Great Plain
region*

the city, and the motorway towards Romania and Serbia reached the city in December 2005.

The situation report[4] of the Southern Great Plain region and the systematization of the proposals concerning the competitiveness of the region were carried out on the basis of the pyramid model. The revealed competitiveness figures, the development factors and the success determinants of the region were examined,[5] and it was pointed out, that this region is considered to be a neofordist type (Lengyel 2006b; Lukovics 2006b). These figures provide only an approximate picture of regional competitiveness and do not explain its actual level. However, they are suitable to outline the possibilities and limits of the university's role in economic development. Since 69 per cent of the students of the University of Szeged come from the Southern Great Plain region and 23 per cent are from the subregion of Szeged, data is reviewed both on the three counties and the subregion.

Table 14.1 provides data on the GDP, the employment and the level of education of Hungarian regions. GDP per capita varies significantly

*Table 14.1  Main data on GDP, employment and education level in Hungarian regions*

| Regions, Counties | GDP per capita (EU25=100) (%) 2003 | Average GDP growth (%) 1995-2001 | Employ-ment rate (%) 2004 | Unemploy-ment rate (%) 2004 | Higher educational attainment (%) 2002 |
|---|---|---|---|---|---|
| Hungary | 59.9 | 4.0 | 56.8 | 6.1 | 14.3 |
| Central Hungary | 96.5 | 5.2 | 62.9 | 4.6 | 21.5 |
| Central Transdanubia | 55.4 | 4.6 | 60.3 | 5.6 | 12.0 |
| Western Transdanubia | 64.4 | 4.3 | 61.4 | 4.6 | 12.2 |
| Southern Transdanubia | 42.9 | 2.6 | 52.3 | 7.3 | 10.8 |
| Northern Hungary | 38.3 | 2.3 | 51.6 | 9.7 | 11.3 |
| Northern Great Plain | 39.1 | 3.0 | 50.4 | 7.2 | 11.5 |
| Southern Great Plain | 40.7 | 1.6 | 53.6 | 6.3 | 10.5 |
| Bács-Kiskun | 39.8 | 1.7 | 56.0 | 7.1 | 9.0 |
| Békés | 36.4 | 1.5 | 49.9 | 6.5 | 8.1 |
| Csongrád | 46.0 | 1.6 | 53.9 | 4.9 | 12.5 |

*Source*: HCSO (1999, 2004) and EC (2004).

spatially. While Central Hungary (including Budapest) is near the average of the EU-25, most of the regions hardly reach even the half of the EU average. Dynamic growth is witnessed only in three regions: Central Hungary, Central Transdanubia and Western Transdanubia. These three regions with dynamically growing economies constitute one block geographically situated in the northwest of Hungary between Budapest and the Austrian border. Data clearly show that in Hungary there are great and constantly growing territorial differences among the regions.

The growth rate of the Southern Great Plain region is the lowest in Hungary, showing a considerable lag behind that of the EU-25. This is the only Hungarian region where the process of catching up has not yet started. In the rank based on the GDP per capita of the EU's 254 NUTS2 level regions, the Southern Great Plain region has the 242nd place, hence it is considered to be a less developed one in European terms too. All of the three counties in the Southern Great Plain region are underdeveloped; the economic output of Csongrád county reaches 46 per cent of the EU-25 average and only slightly exceeds that of the other two counties.

In Hungary the employment situation has improved parallel with the economic growth beginning in 1995. However, in 2004 the employment rate (56.8 per cent) still shows considerable lag behind the EU-15 rate (64.2 per cent) and the EU-25 rate (62.8 per cent). The territorial differences within the country were similar in the case of employment and in terms of economic output. In the three developed regions employment reached 60 per cent in 2004, while the same figure was 50 per cent in the four less developed ones. Also the rate of unemployment in the developed regions was approximately 5 per cent in contrast to the less developed regions with values ranging between 6 and 10 per cent. After 1995 labour productivity improved in all of the regions, almost parallel with the growth rate of the GDP per capita. Between 1995 and 2002 Central Hungary experienced a growth of 50 per cent, while in the Southern Great Plain this was 24 per cent and 29-42 per cent in the rest of the regions. So the growth was almost twice as fast in the developed regions as in the less developed ones. Consequently, the territorial differences apparent in labour productivity are rapidly increasing in Hungary. In Csongrád county labour productivity has changed similarly to the average of the Southern Great Plain.

Performance in global competition and openness can be described well with the help of data on exports and on international tourism. In 2003 the three developed regions produced 70 per cent of the Hungarian industrial exports, while the contribution of the Southern Great Plain region was only 6 per cent. Despite this low share, 37 per cent of the sales revenue of the industrial companies situated in the Southern Great Plain region derives from exports, while the same figure in the case of the sales revenue of companies in Csongrád county is 30 per cent. Tourists visiting Hungary spent 70 per cent of the tourism nights in the three developed regions and only 2 per cent in the Southern Great Plain region.

The basic figures of regional competitiveness show that the growth and competitiveness of Hungary's economy depends on three regions (precisely four counties) and mainly on the economy of the capital. The growth of the other four regions is slow; their employment and labour productivity is equally low. The Southern Great Plain region is less developed, the growth and employment rate of its economy is very low, showing hardly any improvement. Production of the companies situated here mainly target the national market, export is not significant and foreign tourists spend few nights here.

Although the weak competitiveness of the Southern Great Plain region has many reasons, only the two that influence the development strategy of the university are underlined here. The first one is that structural change has not occured in the region yet. The second one is the low educational level of the workforce. In the Southern Great Plain region 14.2 per cent of the employed population works in agriculture with low effectiveness. In the other Hungarian regions agriculture has already lost its importance and is replaced by the industry or service sectors with much greater labour productivity. The other important factor is the educational level of employees (Table 14.1). In the Southern Great Plain region the rate of people in the age group of 15-64 with higher education degrees is only 10.5 per cent while the EU-25 average is 20.8 per cent (EC 2004). Therefore, one critical point of the development is caused by the lack of qualified workforce, which is obviously connected with university education and postgraduate programmes.

Concerning the number of people employed in research, the R&D expenditures and the number of domestic patents, the Southern Great Plain has an outstanding position following the capital (Central Hungary) among the Hungarian regions (Table 14.2). In the Southern Great Plain region the majority of research and development is performed in Csongrád county, where a relatively large number of people work in research centres. Besides, the rate of people with scientific qualifications (PhD) is higher in Csongrád county than in the capital. The proportion of research and development expenditures compared to the GDP is 1.5 per cent in the county, which positions this county first in the country. This means that in terms of R&D Csongrád county excels in Hungary.

*Table 14.2 R&D and patenting in Hungarian regions*

| Region, county | Calculated staff number of R&D units (per 10,000 employment) 2003 | Persons with scientific degree (per 10,000 employment) 2003 | R&D expenditures per GDP (%) 2004 | Number of domestic patents 2000-2004 |
|---|---|---|---|---|
| Central Hungary | 148 | 52 | 1.4 | 1071 |
| Central Transdanubia | 37 | 10 | 0.6 | 136 |
| Western Transdanubia | 28 | 8 | 0.3 | 66 |
| Southern Transdanubia | 52 | 18 | 0.4 | 82 |
| Northern Hungary | 36 | 14 | 0.3 | 96 |
| Northern Great Plain | 54 | 22 | 0.7 | 121 |
| Southern Great Plain | 59 | 22 | 0.8 | 191 |
| Bács-Kiskun | 26 | 4 | 0.4 | 53 |
| Békés | 22 | 6 | 0.4 | 20 |
| Csongrád | 137 | 58 | 1.5 | 118 |

*Source*: HCSO (2004).

The characteristics of the Szeged subregion[6] are different from those of the other subregions in Csongrád county (Table 14.3). Services are dominant as two-thirds of the workforce are employed here, and the density of population is high in the Szeged subregion. The entrepreneurial spirit is vibrant; the number of active enterprises considerably exceeds the national average (Lengyel 2003). The rate of employees with higher education degrees is high, 24.3 per cent in the Szeged subregion. The employment rate is higher while that of unemployment is lower than in the other subregions of the county. Personal income in Szeged significantly exceeds the national average. Approximately 95 per cent of the researchers living in Csongrád county together with research expenditures are located in this subregion.

Szeged and its subregion is much more developed than Csongrád county and the Southern Great Plain; what is more, strong scientific

*Table 14.3  Figures of the subregions in Csongrád county*

| Subregions | Employm. rate (population aged 15-59, %), 2001 | Active economic corporations per 1000 inhabitants 2003 | Highly educated persons (% of total employm.) 2001 | Employm. by sector (% of total), 2001 | | |
|---|---|---|---|---|---|---|
| | | | | Agri-culture | Indus-try | Servi-ces |
| Csongrádi | 50.0 | 69 | 15.9 | 16.8 | 35.3 | 47.9 |
| Hódmezőv.i | 56.4 | 77 | 14.8 | 10.8 | 38.0 | 51.1 |
| Kisteleki | 53.5 | 57 | 8.0 | 33.3 | 26.3 | 40.5 |
| Makói | 51.2 | 56 | 10.8 | 15.0 | 33.1 | 51.9 |
| Mórahalmi | 56.5 | 49 | 7.2 | 33.8 | 21.2 | 45.0 |
| Szegedi | 61.3 | 103 | 24.3 | 5.8 | 27.0 | 67.1 |
| Szentesi | 58.6 | 69 | 12.6 | 21.3 | 25.7 | 53.0 |

*Source*: HCSO (2004).

institutes are also present here. It appears that the city has the human and scientific potential to develop a knowledge-based economy with the help of the EU's regional financing sources available in the period of 2007-2013. Therefore, Szeged is considered to be a *knowledge transfer subregion*[7] (Lengyel 2006a, Lengyel and Lukovics 2006), the development opportunities of which can be provided also based on the pyramid model (Figure 14.3). A special version of the pyramid model can be designed, the elements of which are built upon the real opportunities of the given region type and may contribute to improving the competitiveness of the region (Lengyel 2003). In each region type the basic categories and success factors of competitiveness are the same, however, the elements of the basic factors are different in each sub-type.

After reviewing data on the competitiveness of the Southern Great Plain region and the most important data of the Szeged subregion, the following must be underlined as the foundational criteria of the university's local economic development strategy:

1. The university operates in a less developed *neofordist region* with slow economic growth where the workforce is less educated and the employment rate is low.
2. Szeged and its subregion constitute an 'isle' in this region: the number of enterprises, education level, employment rate, scientific entrepreneurial spirit and incomes are high.
3. Szeged subregion is a *knowledge transfer region:* human resources are

qualified, the rate of services is outstanding, the number of people with scientific degrees, R&D units and expenditures, and the number of patents are high.

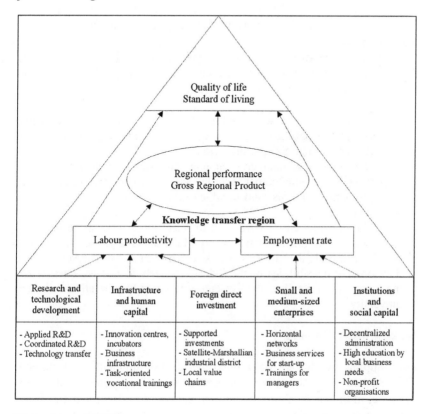

*Source*: Lengyel (2003).

*Figure 14.3 Enhancing competitiveness of a knowledge transfer region by the pyramid model*

## 14.4 UNIVERSITY-INDUSTRY RELATIONS IN THE CSONGRÁD COUNTY AND SZEGED SUBREGION

The development strategy of the university is greatly determined by the agglomeration, development level, economic structure and workforce demand of the region (Varga 2002). Before designing the university's local economic development strategy we surveyed the economy, the

traded business sectors of the region and the university-industry relations
(Inzelt 2004). The data in the previous chapter clearly showed that the
Southern Great Plain is a less developed agrarian region, however Szeged
and its area is considered to be a knowledge transfer subregion.
Therefore, our empirical analysis focused on Csongrád county and
Szeged subregion.

In 2005 various *empirical surveys*[8] were conducted in order to identify
the specialization of the county's and Szeged subregion's economy to
reveal the potential strategic business sectors linked to the university:

1. A cluster mapping was performed to find out which clusters (or potential
   clusters) and dynamically developing industry branches are present in
   Csongrád county and Szeged subregion.
2. We completed questionnaires among companies with at least 50
   employees to see which other local industry branches they are connected
   to (input-output analysis) and in which other areas they could cooperate
   with the university.
3. From the potential clusters we tried to select those that can be linked to
   the scientific and institutional capacity of the university. For this purpose
   we interviewed the leaders of key companies and analysed the
   development possibilities of the industry branches. And we interviewed
   the managers of small innovative enterprises that have relations with the
   university.

### 14.4.1  Cluster Mapping

For mapping the clusters we applied the location quotient (LQ) method.
This method has frequently been used to identify regional clusters, for
instance, in the USA, United Kingdom and Sweden (Bergman and Feser
1998; Feser 2003; Porter 2003b). In our case its use focuses on three
aspects (Patik 2005; Patik and Deák 2005):

1. According to the employment LQ method: based on the number of
   employees in companies with local headquarters to find out whether
   the given industry branch represents a higher rate of employment in
   the Csongrád county (and Szeged subregion) compared to the entire
   country. Furthermore, we considered the weight and significance of
   the industry branch as well as the number of employees working in the
   given sector.
2. LQ analysis was also performed based on the number of enterprises to
   decide whether there is an industry branch where the number of

operating enterprises is relatively higher in this county than in the country and if the number of enterprises is increasing in this sector.
3. Export was also examined to find out whether the share of the given industry branch in the export of the county (subregion) exceeds the national rate.

Mapping the clusters brought up serious problems concerning data registry. At present the Hungarian Central Statistical Office ranks industry branches in a 4-level system, therefore, the special branches indicate highly complex activities. Export data are collected for only industrial companies with at least 50 employees while not for services, for instance. In the county the number of company sites with headquarters in other counties and of employees working there is high, but these divisions were not included in the survey (Patik and Deák 2005). Due to these problems cluster mapping was only used as a preliminary filter.

The analysis of industry branches in Csongrád county according to the employment LQ analysis had the following results based on LQ>1.55: animal husbandry, metallurgy and mechanical engineering, water supply, construction industry, vegetable and fruit trade, real estate sector, R&D in the field of technical and natural sciences. In the case of LQ>1.25 no significant differences were seen in terms of the industry branches. Due to their export capacity the following emerged: meat industry, textile and furniture industry, rubber and plastic industry, instrument production. In Szeged and its subregion according to LQ>1.55: construction industry, real estate sector and education.

Mapping the potential clusters confirmed that the county's economy enters the competition only on the national and local market. No cluster formation could be revealed from the local relations of the industry branches either. In the industry branches outstanding according to the LQ index (LQ>1.55) the number of enterprises is low, the number of their employees is not significant either and does not reach the critical mass. Unfortunately, Hungarian statistical data is of much lower quality than the US data, severely limiting its use for cluster analysis with LQ index. For example, it does not include biotechnology or heath care services as activities, therefore, the LQ index is not enough to reveal potential innovative clusters at the level of counties or subregions.

### 14.4.2 Questionnaires

We asked companies with at least 50 employees and headquarters in

Csongrád county to complete questionnaires; 170 appraisable ones (25 per cent) were returned out of the 670 that had been sent out, which was a representative sampling (Bajmócy 2006). Our form was based on the Porter's diamond model but we added local features (Porter 2003a). The main areas of the questionnaire included the attributes of company strategy, the most important elements of the local (county-level) business environment, local relations within the industry branches, leaders' local informal relations, relations with the university and research institutes and the needs for university enterprise development services.

The evaluation of the questionnaire revealed the following most important conclusions applicable for designing the local economic development strategy of the university:

1. Only one-third of the local enterprises use a strategy based on innovation and these are mostly smaller enterprises.
2. The university relations of the companies are weak, only one-third of them cooperate occasionally with some university unit (market assessment, training, preparing grant proposals, etc.) and approximately half of them only have informal relations with the university.
3. Industrial companies were examined separately, but no industry branch and larger company was found that had ties with the university or local research institutes.

### 14.4.3 Interviews with Corporate Leaders and Innovative Small Enterprises

Analysing the position of industry branches within the county based on the LQ index as well as the changes in their market segment and ties with university training, we did not find any traded branch that could make the economic growth of the area more dynamic. The interviews completed with the leaders of these industry branches also confirmed that no cooperation is planned with the university in the field of research and development (Bajmócy 2007). As one of the strategic industry sectors closely connected to university institutes, R&D in the field of technical and natural sciences is obviously different from other branches.

In Szeged and its subregion 30 innovative small enterprises were interviewed that have regular relations with the university or research institutes (Bajmócy 2007). Most of these enterprises were founded two or three years earlier and are based on an innovative idea or modern business service and most of them cooperate with university researchers as well. These small enterprises develop dynamically; they rely on the university's scientific capacity and regularly use laboratories and instrument parks. Two-thirds of them are linked to a certain field of biotechnology and health industry.

The cluster mapping, the questionnaires and the interviews allowed us to formulate the following three conclusions:

1. In Csongrád county the production of large companies mainly targets the national market – these companies compete with cost advantages, and do not devote money to research or management trainings. These companies have informal relations with the university; they hardly have any business or research and development assignments.
2. In Csongrád county there are no clusters, significant specialization or agglomeration and companies have weak business relations within the county.
3. A new type of enterprise has also emerged that has strong ties with the university and local research institutes, engages in innovative activities and has a growing number of collaborations with international partners. The workforce employed by these small enterprises is highly qualified although the number of employees is still small.

Our empirical surveys confirmed that in a less developed region large companies mainly compete on the local and national market. The industrial companies operating in Csongrád county mainly manufacture mass products, for which they buy technologies instead of developing them. The specialization necessary for the more dynamic economic growth of the region and the county, the range of companies reaching the critical mass in a specific activity and agglomeration are missing, so no sign of cluster formation is apparent. Innovative small enterprises in the field of biotechnology and the health industry connected to the university have only emerged in the few last years, however their number is small and they have few employees.

Consequently, in the Csongrád county and Szeged subregion there is no industry branch and companies with which the university could cooperate in forming innovative clusters and new profitable workplaces. A fundamental question lies in whether the biological and life sciences capacity of the university and local research institutes, laboratories and instrument parks are suitable for the formation of future knowledge-intensive clusters growing out of the spin-off from biological and health industry companies linked to the university.

## 14.5  BIOPOLIS: THE KNOWLEDGE-BASED LOCAL DEVELOPMENT STRATEGY OF SZEGED

In the Southern Great Plain region there is only one university, which is located in Szeged. The University of Szeged is one of the most important higher education institutes of Hungary, and one of the most acknowledged universities in Eastern-Central Europe according to the Jiao Tong list (University of Shanghai). It has 30,000 university students, 11 faculties (medicine, science, business, education, etc.) and runs 68 university and 66 college departments. Six to seven thousand students graduate every year. It employs 7000 people, of which 1700 are lecturers-researchers and 770 have a scientific qualification (PhD). Seventeen doctorate (PhD) schools work at the university with 800 doctorate students. The university scientific work-groups have significant international connections and remarkable research results. Among other things, they submitted 11 successful proposals in the EU research framework programmes from 2002 to 2005.

The research institute network of Szeged is the second largest in Hungary, after Budapest. Beside the university research centres, the Biological Centre of the Hungarian Academy of Sciences is also located in Szeged, which gained the title of Centre of Excellence of the EU in 2000. There are 220 scientific researchers working at the Centre studying several fields of biotechnology and life sciences. Szeged also accommodates the Institute for Biotechnology of Zoltán Bay Foundation of Applied Research and the Cereal Research Non-Profit Company, both of which are closely connected to biotechnological researches. The university-industry relationships are increasingly flourishing, for instance, the SOLVO Biotechnology Corporation (located in Szeged) was the winner of the Hungarian Innovation Grand Prize in 2005, and promising spin-off companies are being established one after the other.

After Budapest, Szeged obtained the highest number of patents both in Hungary and abroad. Each year, 1500-1600 scientific publications are written at Szeged and published in well-known international journals. Almost two-thirds of research is being done in the field of biotechnology and life sciences.

Szeged subregion is a knowledge-transfer region on an economic, social and institutional basis. Based on the survey and the analysis of the scientific potential of the university and other research institutes of Szeged it is possible that successful small enterprises will emerge in Szeged connected to different fields of biotechnology and health industry to create an innovative cluster. However, due to the various areas of education involving many students and the concentrated research topics of the university, not only research and development, but other associated knowledge-intensive business services like patent-related services, venture capital, R&D marketing, business consulting and controlling may develop.

The university's development ideas appear in the knowledge transfer development concept[9] of Szeged (Lengyel 2006a). This concept was developed together by the university and the city, and is built on biotechnology and the health industry as dynamic directions of research. This is why it is called BIOPOLIS. It basically involves the conscious development of an innovative biotechnology cluster, for which financial sources are available from the EU between 2007 and 2013 (Figure 14.4).

The knowledge transfer local economic development programme of Szeged, using the existing capacities, is focused around biotechnology. There are five 'legs', five priorities of this strategy:

1. Knowledge-based economy: creating biotechnology (with health industry) cluster and entrepreneurship,
2. Economic development: improving business environment,
3. Human resources: education and training,
4. Infrastructure: research and industrial parks,
5. Environment: sustainable development.

Scientific research in active cooperation with companies is carried out in three subfields: the health industry, environmental industry and agrarian bio-innovation industry. Material sciences and the software industry comprise partially independent developmental directions which are necessary for the application of the results of biotechnological researches. In the course of the implementation of development, it must be taken into consideration that the present knowledge-based economy background of the institutes of Szeged have remarkable capacity, but currently their scientific results are not exploited sufficiently, therefore the technology and knowledge transfer institutions must be strengthened. So the key issues are enforcing enterprise development and encouraging investments.

The BIOPOLIS programme may be sorted by the basic (programming) factors of the pyramid model of the knowledge transfer region (see Figures 14.3, 14.4 and Table 14.4). So, in our opinion the BIOPOLIS strategy is able to enhance the competitiveness of the Szeged subregion and the Southern Great Plain region.

At present, the BIOPOLIS strategy is being developed and is in the

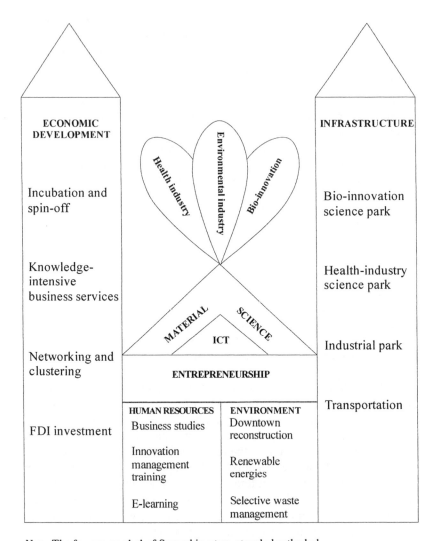

*Note*: The famous symbol of Szeged is a two-steepled cathedral.

*Figure 14.4  The structure of the Szeged BIOPOLIS strategy*

stage of negotiations and discussion with regional and national organizations. In February 2006 both the draft of the National Development Plan and the development strategy of the region included the BIOPOLIS programme. Hungary is also in the process of designing its 'development pole' (or 'competitiveness pole') programme that

*Table 14.4  Szeged BIOPOLIS: the strategic programmes and projects by pyramid model*

| Development factors (strategic programmes) of knowledge transfer region | Projects |
|---|---|
| Research and technological development | Southern Great Plain Neurobiological Knowledge Centre (established in 2004) |
| | Southern Great Plain Life and Materials Science Cooperative Research Centre (SLMRC) (established in 2004) |
| | Environmental and Nanotechnological Regional University Knowledge Centre (established in 2005) |
| | Geothermic Coordination and Consulting Centre |
| | Software Innovation and Research Centre |
| Infrastructure and human capital | Health Industrial Scientific Park and Healthcare Incubation Programme |
| | Environment Technology Experimental Development Field |
| | Bio-innovation Park, inside Bio-innovation Incubator |
| | Enterprise Development Training Centre |
| | Transport accessibility (airport) and logistic centre (established in 2005) |
| Foreign direct investments | Health Industrial Scientific Park |
| | Bio-innovation Park and Incubator (established in 2005) |
| | Industrial park for manufacturing companies |
| | Investment encouraging programme |
| | Complex marketing strategy |
| Small and medium-sized enterprises | Healthcare Incubation Programme |
| | Bioindustrial Preincubation Programme (established in 2005) |
| | Enterprise Development Training Centre |
| | Spin-off encouraging programme (established in 2005) |
| | Regional risk capital fund |
| | Development of professional and management trainings |
| Institutions and social capital | Healthcare Centre of Szeged (region) |
| | Entrepreneurial University |
| | Complex marketing strategy |
| | Network cooperation programme inside the region |
| | 'Intelligent university' development of digital device and content |

focuses on supporting the knowledge-based economy, research and development and improving university-industry relations in the five largest university towns and also in Budapest. Szeged's competitiveness strategy includes the BIOPOLIS programme.

## 14.6 SUMMARY

There is abundant literature on competitiveness with certain well-known approaches, out of which especially the concept of standard competitiveness common in the European Union seems adequate in the case of the regions not only for scientific analyses but also for economic political applications. The concept of standard competitiveness is partly linked to the thought of economic growth, although it also has strong regional economic development aspects that bring it close to the questions of business sciences as well. For the interpretation of regional competitiveness a pyramid model was established that offers a complex frame for the measurement and improvement of competitiveness. It not only makes a proposal concerning the indicators applicable for measuring competitiveness, but also systematizes economic development programmes depending on the types of regions.

The Southern Great Plain region falls into the less developed category of the European regions since its GDP per capita is 41 per cent of the EU-25 average. Therefore this area is expected to receive special assistance from the EU's Structural Funds. With the help of these financial sources it is possible to improve the presently insufficient infrastructure (road system, sewage disposal, etc.). A significant amount of money will also be allocated to improve the region's competitiveness: to create profitable workplaces, and to increase the labour productivity and the rate of employment. This means that working out well-designed programmes and implementing them based on local cooperation can accelerate the region's economic growth and improve its standard of living.

The University of Szeged operates in this peripheral, less developed agrarian region with a stagnating economy. Two-thirds of its students come from this region. However, the city hosting the university greatly differs from this rural environment, since the preparedness of the workforce and the quality of research makes the development of a knowledge-based economy possible. This also means that in the development concept of the region the innovation-oriented development of the university must be harmonized with the different type of development of the other, more rural areas of the region. Due to the weak competitiveness of the region the university is expected to establish partnerships with few companies in the area.

The university and the city's other scientific institutes are acknowledged internationally. Szeged hosts significant R&D and knowledge-intensive SMEs have appeared here. One of Eastern Europe's best universities with a great number of lecturers and students can be found here. Prepared and internationally competitive human resources including scientific lecturers, students and talented young professionals are concentrated in the city. Innovative small enterprises in the field of biotechnology and the health industry have emerged connected to the university in recent years, which can be created into a potential 'innovative' cluster.

In Szeged the BIOPOLIS programme is part of the city development concept and it was also included in regional and national plans in 2006. BIOPOLIS covers cluster-based local economic development based on biotechnology and the health industry, in which the university provides a key role. This strategy greatly depends *inter alia* on the development level of the region, the local infrastructure, the scientific capacity of the university, the economic structure of the city and the local social dialogue. Parallel with training programmes, enterprise development and technology transfer agencies, this programme will develop the missing infrastructure (scientific parks, incubation centres) over the course of the following years. This strategy was based on the 'knowledge-transfer' type of the pyramid model of regional competitiveness.

## NOTES

1. Just to give an example I refer to the World Competitiveness Yearbook (IMD 2003, p. 702) where two types of definitions are currently used. According to the academic definition 'competitiveness of nations is a field economic knowledge, which analyses the facts and policies that shape the ability of a nation to create and maintain an environment that sustains more value creation for its enterprises and more prosperity for its people'. On the other hand following the business definition 'competitiveness of nations looks at how nations create and maintain an environment which sustains the competitiveness of its enterprises'.
2. 'Together, productivity and employment rate are measures of what might be called "revealed competitiveness", and both are central components of a region's economic performance and its prosperity (as measured, say, by GDP per capita), thought obviously they say little about the underlying regional attributes (sources of competitiveness) on which they depend' (Gardiner et al. 2004, p. 1049).
3. This model was constructed to describe the economic development strategy for improving the competitiveness of the Southern Great Plain (Lengyel 2000). The five development factors (priorities of the regional development strategy) of the pyramid model underlying competitiveness were included in the Sixth Periodic Regional Report of the EU (EC 1999). Success determinants with an indirect, often spontaneously, long-term impact on regional competitiveness cover a wide range of variables. Enyedi (1996, pp. 62-64) lists ten important determinants underlying regional success, Begg (1999, pp. 802-804) highlights four

determinants and the Sixth Periodic Regional Report of the EU (EC 1999, p. 80) mentions four determinants as well.

4. In 2005 the situation report of the economy in the Southern Great Plain region was conducted by the Institute of Economics and Economic Development, University of Szeged under the supervision of Imre Lengyel, with help from Miklós Lukovics and Réka Patik.
5. In the situation report we described each basic category, development factors and success determinants with at least three or four variables.
6. Csongrád county includes 7 (LAU1) subregions that broadly constitute the commuting areas: Csongrádi, Kisteleki, Hódmezővásárhelyi, Makói, Mórahalomi, Szegedi and Szentesi.
7. We used 63 indicators by the pyramid model (Lengyel and Lukovics 2006).
8. Empirical surveys were conducted under Imre Lengyel's supervision, and made by Zoltán Bajmóczy, Norbert Buzás, Szabolcs Deák, Szabolcs Imreh, Balázs Lengyel, Miklós Lukovics and Réka Patik.
9. A preliminary version of this concept was conducted by Imre Lengyel and Márton Vilmányi, the strategic manager of the university.

# REFERENCES

Asheim, B.T. (2001), 'Localised learning, innovation and regional clusters', in A. Mariussen (ed.), *Cluster Policies – Cluster Development?*, Nordregio Report, 2, Stockholm, 39-58.

Bajmócy, Z. (2006), 'Egyetemi üzleti inkubáció lehetőségei elmaradott térségekben' (Opportunities for the university business incubation in the less favoured areas), *Tér és Társadalom*, 3, 31-47.

Bajmócy, Z. (2007), *A technológiai inkubáció elmélete és alkalmazási lehetőségei hazánk elmaradott térségeiben* (The theory of technology business incubation and opportunities of application in the less favoured regions of Hungary), PhD Dissertation, University of Szeged, 187.

Begg, I. (1999), 'Cities and competitiveness', *Urban Studies*, 5-6, 795-809.

Bergman, E.M. and E.J. Feser (1999), 'Industrial and regional clusters: concepts and comparative applications', Regional Research Institute, West Wirginia University, The Web Book of Regional Science.

Boucher, G., C. Conway and E. Van der Meer (2003), 'Tiers of engagement by universities in their region's development', *Regional Studies*, 9, 887-897.

Budd, L. and A.K. Hirmis (2004), 'Conceptual framework for regional competitiveness', *Regional Studies*, 9, 1015-1028.

Camagni, R. (2002), 'On the concept of territorial competitiveness: sound or misleading?', *Urban Studies*, 13, 2395-2411.

Cooke, P. (2001), *Knowledge Economies. Clusters, Learning and Cooperative Advantage*, Routledge, London.

EC (1999), 'Sixth periodic report on the social and economic situation and development of regions in the European Union', European Commission, Luxembourg.

EC (2003), 'European competitiveness report 2003', European Commission, Brussels.

EC (2004), 'A new partnership for cohesion: convergence – competitiveness – cohesion. Third report on economic and social cohesion', European Commission, Luxembourg.

Enyedi, Gy. (1996), 'Regionális folyamatok Magyarországon az átmenet időszakában' (Hungarian Regional Trends in the Transition Period), *Hilscher Rezső Szociálpolitikai Egyesület*, Budapest.

Enyedi, Gy. (2004), 'Process of regional development in Hungary', in Gy. Enyedi and I. Tózsa (eds), *The Region. Regional Development, Policy, Administration, and E-Government*, Akadémiai Kiadó, Budapest, pp. 21-35.

Feser, E.J. (2003), 'What regions do rather than make: a proposed set of knowledge-based occupation clusters', *Urban Studies*, 10, 1937-1958.

Gardiner, B., R. Martin and P. Tyler (2004), 'Competitiveness, productivity and economic growth across the European regions', *Regional Studies*, 9, 1045-1068.

HCSO (1999), *Territorial Statistical Yearbook*, Hungarian Central Statistical Office, Budapest.

HCSO (2004), *Territorial Statistical Yearbook*, Hungarian Central Statistical Office, Budapest.

Horváth, Gy. (2005), 'Decentralization, regionalism and the modernization of the regional economy in Hungary', in Gy. Barta, É.G. Fekete, I.Sz. Kukorelli and J. Timár (eds), *Hungarian Spaces and Places: Patterns of Transition*, Centre for Regional Studies of HAS, Pécs, 50-63.

IMD (2003), *The World Competitiveness Yearbook*, Lausanne, Switzerland.

Inzelt, A.M. (2004), 'The evolution of university-industry-government relationships during transition', *Research Policy*, 33. 975-995.

Kitson, M., R. Martin and P. Tyler (2004), 'Regional competitiveness: an elusive yet key concept?', *Regional Studies*, 9, 991-1000.

Krugman, P. (1994), 'Competitiveness: a dangerous obsession', *Foreign Affairs*, 2, 28-44.

Lengyel, B. (2005), 'Knowledge creation inside and among organisations: networks and spaces of regional innovation', in F. Farkas (ed.), *Current Issues in Change Management: Challenges and Organisational Responses*, University of Pécs, Pécs, pp. 215-224.

Lengyel, I. (1993), 'Development of local government finance in Hungary', in J. Bennett (ed.), *Local Government in the New Europe*, Belhaven Press, London, pp. 225-245.

Lengyel, I. (2000), 'A regionális versenyképességről' (On regional competitiveness), *Közgazdasági Szemle*, 12, 962-987.

Lengyel, I. (2003), *Verseny és területi fejlődés: térségek versenyképessége Magyarországon* (Competition and Territorial Development: The competitiveness of regions in Hungary), JATEPress, Szeged.

Lengyel, I. (2004), 'The pyramid model: enhancing regional competitiveness in Hungary', *Acta Oeconomica*, 54 (3), 323-342.

Lengyel, I. (2006a), 'A Szegedi Tudományegyetem lehetőségei a tudásalapú helyi gazdaságfejlesztésben' (Opportunities of the University of Szeged in the knowledge-based local economic development), in B. Máder and B. Rácz (eds), *85 éves a szegedi felsőoktatás*, University of Szeged, Szeged, pp. 45-52.

Lengyel, I. (2006b), 'A regionális versenyképesség értelmezése és piramismodellje' (Interpretation of Regional Competitiveness and the Pyramid Model), *Területi Statisztika*, 2, 131-147.

Lengyel, I. and M. Lukovics (2006), 'An attempt for the measurement of regional competitiveness in Hungary', Paper prepared for the 46th ERSA conference, Volos, 29.

Lukovics, M. (2006a), 'A possible method of measuring the competitiveness of Hungarian counties', *Gazdálkodás*, special edition, 17, 54-62.

Lukovics, M. (2006b), 'A magyar megyék és a főváros versenyképességének empirikus vizsgálata' (Empirical study of competitiveness of the counties and capital), *Területi Statisztika*, 2, 148-166.

Malecki, E.J. (2002), 'Hard and soft networks for urban competitiveness', *Urban Studies*, 5-6, 929-945.

Malecki, E.J. (2004), 'Jockeying for position: what it means and why it matters to regional development policy when places compete', *Regional Studies*, 9, 1101-1120.

Martin, R.L. (ed.) (2003), *A Study on the Factors of Regional Competitiveness*, A final report for the European Commission DG Regional Policy, University of Cambridge.

Nemes Nagy, J. (2001), 'New regional patterns in Hungary', in P. Meusburger and H. Jöns (eds), *Transformations in Hungary. Essays in Economy and Society*, Physica-Verlag, pp. 39-64.

Parkinson, M. et al. (2006), 'State of the English cities. A research study', 1, Office of the Deputy Prime Minister, London.

Patik, R. (2005), 'A regionális klaszterek feltérképezéséről' (On mapping of regional clusters), *Területi Statisztika*, 6, 519-541.

Patik R. and Sz. Deák (2005), 'Regionális klaszterek feltérképezése a gyakorlatban' (Mapping the regional clusters in practice), *Tér és Társadalom*, 3-4, 139-158.

Pike, A., T. Champion, M. Coombes, L. Humphrey and J. Tomaney (2006), 'New horizons programme. The economic viability and self-containment of geographical economies: a framework for analysis', Office of the Deputy Prime Minister, London.

Porter, M.E. (1990), *The Competitive Advantage of Nations*, The Free Press, New York.

Porter, M.E. (2003a), 'Building the microeconomic foundations of prosperity: findings from the microeconomic competitiveness index', in *The Global Competitiveness Report 2002-2003*, World Economic Forum, pp. 23-45.

Porter, M.E. (2003b), 'The economic performance of regions', *Regional Studies*, 6-7, 549-578.

Rechnitzer, J. (2000), 'The Features of the Transition of Hungary's Regional System', *Discussion Papers*, 32, Centre for Regional Studies of Hungarian Academy of Sciences, Pécs.

Rechnitzer, J., Z. Csizmadia and A. Grosz (2005), 'Knowledge-based innovation potential of the Hungarian urban network at the turn of the millennium', in Gy. Barta, É.G. Fekete, I.Sz. Kukorelli and J. Timár (eds), *Hungarian Spaces and Places: Patterns of Transition. Centre for Regional Studies of HAS*, Pécs, pp. 397-425.

Tötdling, F. and M. Trippl (2005), 'One size fits all? Towards a differentiated regional innovation policy approach', *Research Policy*, 1203-1219.

Varga, A. (2000), 'Local academic knowledge spillovers and the concentration of economic activity', *Journal of Regional Science*, 40, 289-309.

Varga, A. (2002), 'Knowledge transfers from universities and the regional economy: a review of the literature', in A. Varga and L. Szerb (eds), *Innovation, Entrepreneurship, Regions and Economic Development: International Experiences and Hungarian Challenges*, Press of University of Pécs, Pécs, pp. 147-171.

Varga, A. (2003), 'Agglomeration and the role of universities in regional economic development', in I. Lengyel (ed.), *Knowledge Transfer, Small and Medium-Sized Enterprises, and Regional Development in Hungary*, JATEPress, Szeged, pp. 15-31.

# 15. The care and feeding of high-growth businesses in rural areas: the role of universities

## Hugh D. Sherman, William B. Lamb and Kevin Aspegren

### 15.1 INTRODUCTION

Universities and related institutions play an essential role in the development of a knowledge-based economy. Kirchhoff et al. (2007), for example, demonstrate empirically that university R&D expenditures are positively associated with new business formation and employment growth in the surrounding region. While the economic development efforts of universities still have ample room for improvement (Kitagawa, 2004), it is clear that universities are a vital component of a region's innovation infrastructure (Varga, 2000). Given the special challenges facing entrepreneurial activity in rural regions (Barkema and Drabenstott, 2000; Shields, 2005), we argue that the economic development role of universities is even more important in rural settings than it is elsewhere. Universities have the potential to ameliorate many of the disadvantages associated with a rural business location. In this chapter we discuss the role universities can play in the care and feeding of an important sub-group among entrepreneurial firms: high-growth businesses (HGBs). HGBs, when properly supported, have the potential to make a substantial economic impact on a rural region (Westhead and Wright, 1998). Unfortunately, they are also among the most difficult businesses for economic developers to identify, understand, and assist. We conducted a four-region study of rural economic development efforts that included over 100 interviews of people who assist HGBs. Based on this information, we discuss the support needed by HGBs, much of which can be facilitated by university programs. In particular, we profile the successful HGB-support efforts of Ohio University's George Voinovich

School for Leadership and Public Affairs (GVS) in order to provide insight for future economic development initiatives.

## 15.2 UNIVERSITIES AND ECONOMIC DEVELOPMENT

Governments often use public universities as vehicles for direct investment, to encourage private sector collaboration, and to facilitate faculty participation in economic development activities. However, there is evidence that universities should become even more actively engaged in economic development. For example, some entrepreneurs have surprisingly mixed feelings about the effectiveness of universities' contributions. First, some entrepreneurs have criticized universities for not providing enough entrepreneurial training and assistance to students who want to start their own businesses. Second, entrepreneurs have expressed frustration over the obstacles they face when trying to commercialize the research produced at universities. Many of these criticisms relate to the fact that university-based economic development activities are sometimes handled in rather ad hoc ways, in part because academics and regional authorities do not always 'speak the same language' (Beer and Cooper, 2007). Historically the impetus to commercialize research, or to provide a firm with sophisticated technical assistance, has rested with individual faculty members or entrepreneurs most directly involved. If neither side sought the information, or if they failed to find each other, the economic development opportunity would never materialize.

Today, universities have become more actively involved in a variety of economic development efforts aimed at business attraction, creation, retention, and expansion. Governments use universities as bases for economic development activity due to their essential role in education, outreach, and community service (Tornatzky et al., 2002). Although we make suggestions later in this chapter about ways to expand and amplify these efforts, it is important to review some of the most important contributions universities already make.[1]

Universities help attract businesses to a community by making specific contributions to the local economic infrastructure. The quality of basic economic factors such as the labor force, customer base, and business advisory services tends to be higher near universities. Mueller (2006) finds evidence that, when firms draw knowledge from universities in a given region, the region experiences increased economic growth. She argues that this is due to the favorable impact of universities both on the quality of knowledge and on knowledge flows in the region. In the German sample she studies, this university effect appears to be strongest for engineering, technological, and scientific knowledge. Benneworth and Charles (2005) provide anecdotal evidence that university spin-off companies can spark greater innovative activity, even in remote rural settings.

Universities also enhance economic development by creating research centers, industry councils, and community-outreach programs (Wortman, 1996). For example, for decades US universities have made significant investments in agricultural extension. In some cases, universities make extra efforts to build local infrastructure in specific ways. For example they might provide workforce training programs specifically targeted to assist new plants that enter a region. Over the past 20 years more colleges and universities in the US have attracted businesses and government agencies to their region by developing research parks on or near their campuses (Appold, 2004). In addition to increasing economic activity in the region, these parks encourage links between companies and the faculty performing relevant research on campus (Löfsten and Lindelöf, 2002). Löfsten and Lindelöf (2002) find evidence that new-technology-based firms generate higher levels of job creation when brought into research parks. The high incidence of university/HGB relationships encouraged by research parks can help firms expand their business, promote regional development, and increase the likelihood that universities will receive more benefits from research commercialization.

One of the most important economic development activities of any university is research, and universities are growing increasingly sophisticated in their approaches to tapping the economic potential of the research their faculty, staff, and students conduct (Etzkowitz et al., 2000). In the US it is now commonplace for universities to have extensive programs to manage and help commercialize intellectual property generated on campus. For example Georgia Tech has established the Georgia Tech Research Corporation, which maintains ownership of intellectual property created at the university and helps inventors to start businesses or generate licensing income based on intellectual property (Tornatzky et al., 2002). In those regions of the world where entrepreneurial commercialization of university-generated knowledge is not yet commonplace, governments are trying to identify ways to spur such activity (e.g., Wong et al., 2007).

Universities in the US also have been actively involved in creating business incubators (Rothaermel and Thursby, 2005). Incubators are designed to alleviate some of the early survival pressures new firms face, particularly when the founders have little or no experience in running their own firms. Typically, member firms receive favorable terms on rent, basic office support, and access to business and technical advice through the university. Technology incubators are specifically designed to help firms cope with the stresses associated with technology commercialization. O'Neal (2005) describes a number of ways that a university connection is beneficial to successful new venture incubation, including improved access to faculty and student labor, access to laboratories and equipment, and enhanced credibility. Rothaermel and Thursby (2005) find that technology incubators with a university link enjoy a lower new venture failure rate, and that incubators with stronger university links have even lower failure rates.

It is important to remember that an HGB is by far the most difficult entrepreneurial venture for economic developers to support. As we will discuss, HGBs require more intensive and sophisticated technical assistance as well as access to a wide range of financing options. In rural areas, where HGBs face so many additional hurdles, universities can play a crucial role in improving access to many of the resources that HGBs need to launch, survive, and thrive.

## 15.3  HIGH-GROWTH BUSINESSES

According to a report from the US Bureau of Labor Statistics (Helfand et al., 2007) between 1990 and 2005 small firms created approximately 64 percent of the country's new jobs. Researchers across a variety of nations have shown empirically that small firms make significant contributions to economic growth as measured by net new job creation (Kirchhoff, 1994; Storey, 1994; Baldwin, 1995; Wennekers and Thurik, 1999). Findings indicate that, among small firms, newly formed firms create the largest share of net new jobs (Kirchhoff, 1994; Baldwin, 1995; Wennekers and Thurik, 1999). Also, there is evidence that a small subset of rapidly-growing firms generate the lion's share of new jobs (Storey et al., 1987; Birch, 1987; Pages and Poole, 2003). Moreover, highly innovative new firms create a disproportionately greater share of net new jobs compared with those of lesser innovation intensity (Kirchhoff, 1994). Therefore, highly innovative new firms are a major source of economic growth and should be a key area of focus for economic developers.

Tregear (2005) distinguishes between types of entrepreneurs based upon their goals for the enterprise: lifestyle or commercial growth. Likewise, the National Commission on Entrepreneurship (2002) uses the founding entrepreneur's personal goals to distinguish between two types of small businesses: 'lifestyle' businesses and 'entrepreneurial' or high-growth businesses. Lifestyle businesses provide family income or support a desired lifestyle. Once a lifestyle business reaches a certain size, founders typically forego additional growth. In contrast, founders of HGBs tend to have more ambitious, open-ended goals for the business. HGB founders are more likely to pursue high levels of visibility, wealth and job creation through their business. Taking the firm public is also an important goal for many HGB founders.

HGBs offer significant potential for economic growth, but they also require far more care and feeding than other kinds of startups. High-growth-potential businesses need two critical inputs: (a) equity capital financing, and (b) sophisticated business technical assistance. Encouraging the birth and growth of HGBs can be difficult in any circumstances, but HGBs are simply less likely to be founded in rural areas. For example, Kolko (2000) reports that from 1989 to 1995, even though rural job growth significantly outpaced urban job growth, information-technology-intensive industries were less likely than low- and medium-IT industries to locate and create jobs in rural regions. As McDaniel (2002) observes, growth in rural regions of the US has not been of the same scope and quality as the growth in metropolitan areas. McDaniel also finds that rural entrepreneurs are less likely to build HGBs. They tend to build smaller firms and generate lower incomes. For example, in the US in 2001, 5.5 percent of the rural self-employed worked in firms of more than 100 employees, while approximately 11 percent of the urban self-employed worked in such firms (McDaniel, 2002).

In addition to needing more support, rural HGBs pose an added risk for economic developers. Once a new HGB gets off the ground, the survival pressures it faces will often pull it towards urban locations where customers, financing, and other means of assistance are more readily available (Stam, 2007). Helping entrepreneurs start HGBs is not enough – rural economic developers must find ways to help HGBs founded in their region remain in the region.

## 15.4 CHALLENGES OF RURAL ENTREPRENEURIAL DEVELOPMENT

Rural entrepreneurial development is a topic that has been widely discussed in the literature in recent years, but which still requires further investigation. In the US, entrepreneurial firms have made a significant impact on the growth of urban economies, while evidence suggests that rural economies have not enjoyed comparable entrepreneurial gains (McDaniel, 2002). Rural regions pose a variety of special challenges to economic developers, and existing programs have had only mixed effects on rural entrepreneurial activity (Malecki, 1994; Wortman, 1996). In spite of the challenges, there is evidence that rural entrepreneurship is showing more promise as an economic development strategy (Lin et al., 1990). Smallbone et al. (1999) also demonstrate that some types of rural entrepreneurial firms are finding success.

Entrepreneurs who start, or want to start, a business in a rural region face a host of obstacles. Low population density and low levels of industrial activity in rural areas provide businesses with a relatively small base of nearby customers (Drabenstott and Meeker, 1999). Shields (2005) finds that rural entrepreneurs report particular difficulty related to the marketability of their products or services. Remote locations also increase transportation costs and reduce access to important customers and suppliers (Barkema and Drabenstott, 2000). Rural areas often have the advantage of lower labor costs, although the workers in these regions are less likely to have the most up-to-date training and job skills (Hoy and Vaught, 1980). One of the biggest problems faced in rural regions is how to attract and retain a younger, more educated workforce (Innovation and Information Consultants, Inc., 2006). Rural HGBs also have fewer opportunities than their urban counterparts to benefit from agglomeration economies. In a rural area there will tend to be less access to advice and support from the business network and fewer opportunities to benefit from knowledge spillovers.

Westhead and Wright (1998) point out that, while policymakers have created different types of economic development programs for rural and urban settings, our understanding of the differences between these two contexts is still a work in progress. Research suggests that limited understanding of the special needs of rural entrepreneurs has made attempts at support less successful than hoped (Meccheri and Pelloni, 2006). Skuras et al. (2000) argue that the emphasis on general grant-aid programs, rather than targeted, integrated assistance programs has been partly to blame for this lackluster track record.

## 15.5  EDA RESEARCH STUDY: FACILITATING HGB
##       GROWTH IN RURAL REGIONS

As part of a US Economic Development Administration research and technical assistance project, we interviewed approximately 100 entrepreneurs, state economic policymakers, financiers of entrepreneurial businesses and assistance-providers in four rural regions of the US. The goal for the interviews was to gain a better understanding of the challenges HGBs face from the perspective of those involved in rural economic development activities.[2] In the interviews there was wide agreement across the regions that HGBs were increasingly important to rural economic development efforts. Interviewees believed that HGBs can create significant new wealth and new employment opportunities. In the sections below we summarize key findings related to three main concerns expressed by respondents: (1) HGB technical assistance needs, (2) HGB financing needs, and (3) attitudes towards government support of HGBs.

### 15.5.1  Technical Assistance Needs

HGBs require much more sophisticated hands-on business and technical assistance than the typical small firm. In the interviews there was strong agreement among economic development officials, venture capitalists, angel investors, and bankers that rural entrepreneurs often lack the knowledge, management skills, experience, and network necessary to start or expand their businesses. Respondents reported that rural entrepreneurs are less likely to understand the importance of professional networking and access to outside counsel. Traditional venture capitalists and others who are providing business technical assistance reported that most, if not all, entrepreneurs starting HGBs in rural areas need intensive, hands-on, long-term assistance before their businesses becomes investment-ready. Technical assistance engagements can range from six to thirty-six months in length.

According to the respondents most rural entrepreneurs, even if they have a great product or service, do not have a clear growth strategy. They often do not understand how to reach a scalable level of revenues. While business incubators and small business development centers assist many rural entrepreneurs, there was consensus that such programs are more appropriate and helpful for lifestyle businesses than for HGBs. People who manage such programs can provide basic services such as accounting, marketing, and legal support. Generally, however, they lack experience at growing businesses and preparing them to attract venture capital. HGBs often require extensive organizational and management restructuring. HGB founders need assistance from people who understand what is necessary to grow the business and can provide the knowledge and experience needed to put together an equity deal.

### 15.5.2 Financing Needs

Contrary to the common perception, venture capitalists rarely invest in start-up firms. The majority of VC investments are follow-on funding, and these investments are nearly always placed in business sectors where rapid growth is expected. Venture capitalists look for high rates of return with an expected exit strategy of cashing out in less than eight years through an initial public offering or a merger with, or acquisition by, an established firm.

One difference between urban and rural contexts that is unequivocal is the relative difficulty rural entrepreneurs have obtaining angel investments[3] and venture capital. One recent report (Henderson, 2002) notes that two-thirds of all venture capital investments in the US go to just five states and that nearly all of these investments are made in metropolitan firms. Rural entrepreneurs, especially in remote areas, are at a distinct disadvantage when trying to raise capital for starting or growing their business (Henderson, 2002).

The venture capitalists interviewed identified two obstacles to equity investment in rural regions. First, relative to other deals they could work on, identifying attractive rural deals is very time-consuming and costly. Second, those who have found potential equity deals in rural areas report that it was more difficult than usual to help owners make the businesses ready for investment. Most venture capitalists operate across regional or state boundaries and therefore have a choice of many potential deals. They often reject entrepreneurs who have an interesting business model but need more assistance. This assistance costs extra money and time, introducing an added layer of uncertainty into the deal.

In our interviews, a broad group of respondents reported that people in rural areas tend not to be willing to provide financing to start up entrepreneurial firms. They have less knowledge about risk capital and tend to be relatively more conservative with their investments. Similarly, rural entrepreneurs often lack understanding of equity financing. They tend to have few friends or acquaintances who have experience with equity capital. Therefore, it is not surprising that, according to respondents, rural entrepreneurs tend to focus on the downsides of equity financing. Respondents indicated that many rural entrepreneurs resist giving up ownership of their business. They are less likely than their urban counterparts to understand the compensating advantages of equity financing, such as improved access to a network of contacts and sophisticated business assistance.

### 15.5.3 Concerns Related to Government Assistance

Interview responses indicated a need for major efforts to raise public understanding and support of entrepreneurship for rural economic growth and development. This includes efforts to raise the consciousness of politicians, investors, business owners, potential entrepreneurs (including university faculty) and community leaders about the financing and technical assistance needs of HGBs.

One issue respondents cited often was the tendency of politicians to focus on attracting jobs rather than encouraging entrepreneurship. Respondents noted that, although attracting jobs from elsewhere is very compelling from a political standpoint, it is not a sustainable economic development strategy. In fact, over the past decade most communities have lost more branch plants than they have been able to attract.

Respondents also believed politicians wanted to see quick, big-impact results. The entrepreneurial development process is slow and usually only creates a few jobs, initially. Some cited examples of government-sponsored programs that were terminated before they had a chance to work. Finally, some respondents believed that money provided by the state usually comes with too many strings attached, dampening the interest of potential investors. Respondents voiced strong belief that equity funds are best managed privately – that they should not be controlled by government. They worry that political factors will cloud sound business judgment, jeopardizing the opportunity for the fund to make successful investments.

## 15.5  CASE STUDY: THE GEORGE VOINOVICH SCHOOL FOR LEADERSHIP AND PUBLIC AFFAIRS AT OHIO UNIVERSITY

Ohio University is a major, comprehensive university located in the rural Appalachian Ohio region. The University has nearly 30,000 students located in Athens, Ohio, and on five regional campuses. The University's George Voinovich School for Leadership and Public Affairs (GVS) has created a multifaceted program to facilitate creation and expansion of HGBs.

GVS is Ohio University's largest public service organization, engaged in providing technical assistance and research services to local government, nonprofits and business organizations throughout the 29 counties of Appalachian Ohio. GVS has four major initiatives that include public service, energy and environment, an executive leadership institute, and a business development group. GVS employs more than 65 full-time professional staff members, works with more than 35 faculty members and engages more than 200 graduate and undergraduate students. GVS has completed hundreds of projects in conjunction with local development organizations, counties, municipalities, businesses, nonprofit organizations and state agencies.

The Business Development Group (BDG) – with its 11 full-time professional staff members, 60 part-time student interns and five faculty members – offers an integrated set of business technical-assistance services to new, existing, and fast growing businesses throughout Appalachian Ohio. The BDG has partnered with Adena Ventures, the nation's first New Markets Venture Capital Company.[4] Adena Ventures is a for-profit $32 million equity fund, created to provide equity financing to high-growth potential firms located in southern Ohio, eastern Kentucky, West Virginia, and parts of Maryland. The BDG provides high-end business consulting services to firms that might apply to Adena Ventures for equity investment or to firms in which Adena has already invested. GVS and Ohio University participated in the founding of Adena Ventures by helping to get local and state political support for the venture firm. In addition, the Ohio University Foundation provided $2 million to become the first investor in the fund.

The BDG also has developed a partnership with the Ohio University College of Business's full-time MBA program. Each MBA student works with the BDG for 12-15 hours per week as part of their 15-month MBA experience. Students obtain real world, project-based consulting experience to enhance their own professional development. Client firms in the economically distressed region receive valuable, in-depth consulting assistance free of charge. Under the guidance of the professional staff, MBA students conduct market research, develop marketing and business plans, compile financial projections, and develop information technology solutions. The majority of the professional staff, all of whom have significant business experience, has been hired for their ability to teach, coach, and mentor these students. It has become evident that having the professional staffers work with students and faculty is vital. Faculty have too many conflicting demands on their time, are not normally rewarded for this type of work, and in some cases do not have the in-depth business consulting experience necessary to work with HGBs. A summary of firms that received BDG assistance for two years (2005-06) is provided in Table 15.1.

*Table 15.1  Business development group high-growth business consulting engagements for the years 2005 and 2006*

| Type of engagement | Number of firms | Financing |
|---|---|---|
| Continuing intensive engagement | 23 | 1 Venture capital<br>5 Angel<br>6 Debt financing |
| Periodic contact and use of third party assistance | 46 | 1 Venture capital<br>2 Angel<br>2 Debt financing |
| No contact – business no longer in existence or client no longer wants assistance | 14 | |
| Total | 83 | |

The BDG worked with over 1,000 businesses in this two-year period. Of these firms, 83 were high-growth businesses. Approximately 200 were identified as middle-market firms – well established businesses with 5 to 150 employees. The remainder, over 700 businesses, were categorized as lifestyle businesses. This program assisted businesses in securing $38 million in new loans, $2.7 million in funds from individual investors and $6.5 million in venture capital.

### 15.6.1 Successful Project Examples

We present three examples of firms that have received in-depth technical assistance from the BDG to demonstrate the range and depth of the ongoing consulting services being provided to potential HGBs in Appalachian Ohio. These activities occurred from 2002 to mid-2005. In each case the level of support provided to the firms went well beyond the type of assistance that is generally available from economic development organizations or venture capitalists located in rural regions.

### e-Learning company
In 2003 the BDG began working with an e-learning company based in

Appalachian Ohio. The primary activity for this firm is creating on-line training materials for large corporations all over the US. The primary benefit for client firms is that employees can receive periodic training without having to leave their work site for extended periods. This results in reduced training costs and greater convenience for firms and employees. The company grew its revenue from approximately $1 million in 2002 to approximately $3 million in 2003; it needed organizational development help in order to cope with this rapid growth and launch future growth.

Through GVS, the company received detailed assistance in operations, marketing, financing, and employee recruitment. Improvements included the creation of a more appropriate organizational structure, as well as repositioning the firm's products and services in the marketplace. It helped the company develop a detailed business plan in 2003 and significantly revised it in 2004. This plan included extensive development of marketing, sales, and client-service strategies. The company presented the plan to venture capitalists at Adena Ventures, which in 2004 agreed to invest $750,000 in support of the company's expansion efforts.

**Distance learning company**
This firm, also located in Appalachian Ohio, develops and manages specialized software used by clients in the education, corporate, and government markets. Examples of clients include community colleges and universities offering distance learning programs. The firm offers online management of course materials, and can also process and fulfill mail or phone orders for materials.

The BDG, working with the firm since 2002, has provided business planning assistance and helped with the development of pro forma financials. In 2002 the firm had approximately 20 employees and sales revenue of $1 million. In 2003 this planning helped the firm secure an additional $600,000 from its original investors along with a $5 million line of credit. Other activities on behalf of the firm include development of human resource manuals, evaluation of the firm's network and telecommunications system, market research to identify potential corporate customers, and assessment of the firm's distribution and fulfillment systems.

Most significantly, the BDG has provided the firm with technical assistance on its back-office and customer-interface software program. BDG consultants first helped the client map out a long-term approach to maintaining and improving the software package. The team's contributions included designing the layout and developing faculty and administrative functions of the software. Software refinements helped the firm to secure a new client that will represent 40 percent of the firm's revenue in 2005. The firm is currently on target to generate $38 million in revenues and has more than 80 employees.

### Engine maker

Our final case example is a small manufacturer of external-combustion engines and ventilation equipment. The firm's engines, developed in the early 1980s, provide power to homes and businesses and have been sold primarily to customers in the developing world. In the late 1980s the company began making energy-efficient home ventilation systems that are based on the same technology used in its engines. The ventilators increase the percentage of fresh air brought into a building without significantly increasing the costs to heat or cool the space.

This firm began working with the BDG in 2003. Services provided to this firm included industry research, market analysis, product comparison, and competitor analysis. The BDG team determined that the firm needed to consolidate its debt and obtain additional working capital to meet its aggressive growth targets. The BDG developed a comprehensive business plan and detailed financial projections to ready it for bank financing. The bank approved the application, providing $1.5 million for debt consolidation and $500,000 in new working capital. This resulted in a lower effective interest rate and lower debt payments for the firm while also helping it implement a new marketing and sales plan. Further consulting through the BDG focused on implementing a just-in-time manufacturing process, redesigning the shop floor. The BDG redesigned the firm's Web site, produced a promotional video, developed sales-support tools, and created an employee handbook and financial planning tools. As a result of this assistance, the firm's unit sales have increased as has interest from venture capitalists. Recently, the firm secured $500,000 from a South Korean investor.

These examples demonstrate the level of ongoing, sophisticated business technical assistance that is required to provide the support needed by businesses that have high-growth potential. For each of these HGBs, the benefits of sophisticated technical assistance were only realized because the program also addressed financing concerns. Likewise, these firms could not have fully leveraged their new financing without the technical assistance they received. The dramatic growth these firms experienced was only possible through the integration of these two essential ingredients.

## 15.7 CONCLUSION

As we have discussed, HGBs can be an important component of any rural region's economic development strategy. However, most rural communities have not been able to create the comprehensive and sophisticated infrastructure necessary to meet the needs of HGBs.

One of the most important recommendations we have for economic developers who seek to support HGB formation and growth in their region is to build a program that addresses needs for both financial and technical assistance. As shown in our interviews, these two elements pose a 'chicken-and-egg' problem for economic developers. Without ample sources of multistage venture funding, HGBs rarely can reach their growth potential. Likewise, without sophisticated and in-depth business assistance, rural HGBs struggle to refine their business model to the level that is attractive to venture capitalists. Compared with traditional economic development efforts, HGB support is more labor- and resource-intensive, involves greater risk, and takes longer to show tangible results. HGBs, however, also provide the opportunity to make a dramatic impact on the economic development of a region, sometimes at lower total cost than traditional economic development techniques (e.g., tax abatements). Only through enlightened leadership at the university, regional and state levels can such a program begin to achieve successful outcomes.

In the case of Appalachia Ohio, we believe that Ohio University has played, and will continue to play, a vital role in successfully developing HGBs. Large universities do have sophisticated technical expertise that, when properly channeled, can address the problems of HGBs. In partnership with local and state government organizations, universities can become important engines for regional economic growth.

## NOTES

1. For a set of case studies detailing many of the economic development 'best practices' of US universities, see Tornatzky et al. (2002).
2. In order to explore the subject, the research team conducted site visits and interviews in four areas: Iowa, Nebraska, West Virginia, and southeastern Ohio. In each area we conducted interviews with 3 individuals involved in developing state or region-wide economic development policy, 3-5 individuals involved in financing entrepreneurs (venture capital or angel fund managers or institutions involved in providing capital to those funds), at least 10 high-growth-business owners or CEOs and 5 business technical assistance providers. These interviews were taped, transcribed and analyzed for common themes.

3. Angel investors are groups of loosely-organized individuals who pool financial resources to provide start-up or early-stage funds to firms. After exhausting their own financial resources or funds from friends and family, entrepreneurs often turn to angel investors. Angel investors are willing to meet smaller financing needs that traditional venture capitalists will not provide.
4. The New Markets Capital Program, federal legislation passed in 2001 and administered by the Small Business Administration, provides up to $150 million in debenture guarantees and $30 million in technical assistance grants. The technical assistance grants make available technical and managerial assistance to portfolio companies. However the money provided to selected New Markets Venture companies must be matched dollar for dollar in cash or in-kind contributions. At least 80 percent of the businesses receiving this funding must be in small and low-income geographic areas.

# REFERENCES

Appold, S.J. (2004), 'Research parks and the location of industrial research laboratories: an analysis of the effectiveness of a policy intervention', *Research Policy*, 33, 225-243.

Baldwin, J.R. (1995), *The Dynamics of Industrial Competition*, Cambridge, Cambridge University Press.

Barkema, A. and M. Drabenstott (2000), 'How rural America sees its future', *The Main Street Economist*, December, 1-4.

Beer, A. and J. Cooper (2007), 'University-regional partnership in a period of structural adjustment: lessons from southern Adelaide's response to an automobile plant closure', *European Planning Studies*, 15 (8), 1063-1084.

Benneworth, P. and D. Charles (2005), 'University spin-off policies and economic development in less successful regions: learning from two decades of policy practice', *European Planning Studies*, 13 (4), 537-557.

Birch, D. (1987), *Job Creation in America: How our Smallest Companies Put the Most People to Work*, New York, Free Press.

Drabenstott, M. and L. Meeker (1999), 'Equity for rural America: from Wall Street to Main Street; a conference summary', *Economic Review – Federal Reserve Bank of Kansas City*, 84 (2), 77-85.

Etzkowitz, H., A. Webster, C. Gebhardt and B.R.C. Terra (2000), 'The future of the university and the university of the future: evolution of ivory tower to entrepreneurial paradigm', *Research Policy*, 29, 313-330.

Helfand, J., A. Sadeghi and D. Talan (2007), 'Employment dynamics: small and large firms over the business cycle', *Monthly Labor Review*, March, 39-50.

Henderson, J.V. (2002), 'Are high growth entrepreneurs building rural economy?', Kansas City, Kan., Federal Reserve Bank of Kansas City.

Hoy, F. and B. Vaught (1980), 'The rural entrepreneur: a study in frustration', *Journal of Small Business Management*, 18 (1), 19-25.

Innovation and Information Consultants, Inc. (2006), 'An empirical approach to characterize rural small business growth and profitability', *Small Business Research Summary*, February, 271.

Kirchhoff, B.A. (1994), *Entrepreneurship and Dynamic Capitalism: The Economics of Business Firm Formation and Growth*, Westport, CT, Praeger.

Kirchhoff, B.A., S.L. Newbert, I. Hasan and C. Armington (2007), 'The influence of university R&D expenditures on new business formations and employment growth', *Entrepreneurship Theory and Practice*, 31, 543-559.

Kitagawa, F. (2004), 'Universities and regional advantage: higher education and innovation policies in English regions', *European Planning Studies*, 12 (6), 835-852.

Kolko, J. (2000), 'The high-tech rural renaissance? Information technology, firm size and rural employment growth', *Small Business Research Summary*, September, 201.

Lin, X., T.F. Buss and M. Popovich (1990), 'Entrepreneurship is alive and well in rural America: a four-state study', *Economic Development Quarterly*, 4, 254-259.

Löfsten, H. and P. Lindelöf (2002), 'Science parks and the growth of new technology-based firms – academic-industry links, innovation, and markets', *Research Policy*, 31, 859-876.

Malecki, E.J. (1994), 'Entrepreneurship in regional and local development', *International Regional Science Review*, 16 (1-2), 119-153.

McDaniel, K. (2002), 'Venturing into rural America. The Main Street Economist: Commentary on the Rural Economy', Kansas City, Center for the Study of Rural America, Federal Reserve Bank of Kansas City.

Meccheri, N. and G. Pelloni (2006), 'Rural entrepreneurs and institutional assistance: an empirical study from mountainous Italy', *Entrepreneurship & Regional Development*, 18 September, 371-392.

Mueller, P. (2006), 'Exploring the knowledge filter: how entrepreneurship and university-industry relationships drive economic growth', *Research Policy*, 35, 1499-1508.

National Commission on Entrepreneurship, (2002), 'Entrepreneurship, a candidate's guide: creating good jobs in your community', Kansas City, MO, Kauffman Center for Entrepreneurial Leadership.

O'Neal, T. (2005), 'Evolving a successful university-based incubator: lessons learned from the UCF technology incubator', *Engineering Management Journal*, 17 (3), 11-25.

Pages, E. and K. Poole (2003), 'Understanding entrepreneurship promotion as an economic development strategy: a three-state survey', Washington, DC, National Commission on Entrepreneurship and the Center for Regional Economic Competitiveness.

Rothaermel, F.T. and M. Thursby (2005), 'Incubator firm failure or graduation? The role of university linkages', *Research Policy*, 34, 1076-1090.

Shields, J.F. (2005), 'Does rural location matter? The significance of a rural setting for small businesses', *Journal of Developmental Entrepreneurship*, 10 (1), 49-63.

Skuras, D., E. Dimara and A. Vakrou (2000), 'The day after grant-aid: business development schemes for small rural firms in lagging areas of Greece', *Small Business Economics*, 14, 125-136.

Smallbone, D., D. North and C. Kalantaridis (1999), 'Adapting to peripherality: a study of small rural manufacturing firms in Northern England', *Entrepreneurship & Regional Development*, 11, 109-127.

Stam, E. (2007), 'Why butterflies don't leave: locational behavior of entrepreneurial firms', *Economic Geography*, 83 (1), 27-50.

Storey, D.J. (1994), *Understanding the Small Business Sector*, London, Routledge.

Storey, D.J., K. Keasey, R. Watson and P. Wynarczyk (1987), *The Performance of Small Firms: Profits, Jobs and Failures*, London, Croom Helm.

Tornatzky, L.G., P.G. Waugaman and D.O. Gray (2002), *Innovation U.: New University Roles in a Knowledge Economy*, Research Triangle Park, NC, Southern Growth Policies Board.

Tregear, A. (2005), 'Lifestyle, growth, or community involvement? The balance of goals of UK artisan food producers', *Entrepreneurship & Regional Development*, 17, 1-15.

Varga, A. (2000), 'Universities in local innovation systems', in Z. Ács (ed.), *Regional Innovation, Knowledge and Global Change*, London, Pinter.

Wennekers, S.R. and B. Thurik (1999), 'Linking entrepreneurship and economic growth', *Small Business Economics*, 13 (1), 27-55.

Westhead, P. and M. Wright (1998), 'Novice, portfolio, and serial founders in rural and urban areas', *Entrepreneurship Theory and Practice*, 22 (4), 63-100.

Wong, P.K., Y.P. Ho and A. Singh (2007), 'Towards an "entrepreneurial university" model to support knowledge-based economic development: the case of the National University of Singapore', *World Development*, 35 (6), 941-958.

Wortman, M.S. (1996), 'The Impact of Entrepreneurship Upon Rural Development', in T.D. Rowley, D.W. Sears, G.L. Nelson, J.N. Reid and M.J. Yetley (eds), *Rural Development Research: A Foundation for Policy, Contributions in Economics and Economic History no. 170*, Westport, CT, Greenwood Press.

# Index

Abramson, P. 22
absorptive capacity 26, 112, 175, 298–9,
302, 303, 312, 314, 315
academic careers 166–8, 177–9, 182
academic entrepreneurship 3, 5, 90,
163–85
and firm creation 172–82
academic knowledge as a non-
tradeable asset 174–6
business creation vs. patent
licensing 179–80
incentive problems rediscovered
176–9
intended and unintended
consequences of academic
entrepreneurship 180–82
as an individual feature 166–8
as an institutional feature 168–72, 183
policy implications 182–3
'straightforward' definition of 164–6
see also academic spin-offs;
entrepreneurship
academic knowledge as a non-tradeable
asset 174–6
academic knowledge transfer
definition of 1
geographical dimension of 2, 3
mechanisms for 1–2
non-geographical dimension of 2–3
recent international literature on 3
role of academic spin-offs in 192,
212–13
academic spin-offs 15, 90
academic entrepreneurship and
172–82
academic knowledge as a non-
tradeable asset 174–6
business creation vs. patent
licensing 179–80
incentive problems rediscovered
176–9

intended and unintended
consequences of academic
entrepreneurship 180–82
allocation of intellectual property
rights (IPR) 203, 204
drivers of success or failure 5, 191–6,
198–215
assumptions relating to knowledge-
intensive firm formation and
economic development 192–6,
212–15
effects of spin-off support on new
firm development 206–9
impact of locational factors related
to parent institute 209–10
implications for regional
development 210–12
research result allocation and
transfer process 203–6
study design and methodology
198–9
success indicators 206
summary of findings 212–15
support strategies of parent
organizations 199–203
and innovative activity 352
location of 193–5, 209–10, 214
public funding of 195, 210–11, 214
support by parent organization 27–8,
164, 180, 196, 199–209, 211–12
in Szeged subregion of Hungary 340
in Thailand 311
university technology development
measured by number of 21
see also academic entrepreneurship
Acemoglu, D. 49
Ács, Z.J. 2, 19, 25, 36, 37, 38, 40, 41, 43,
45, 47, 48, 50–51, 110, 139, 151,
271
Act on Local Governments (Hungary,
1990) 328

University of Szeged students from
330
*see also* Szeged; University of Szeged
Spain, rankings of universities in 234
spatial competition among universities
245–50
    empirical analysis for Austrian
        universities 251–64
    with heterogeneous goods 248–50
    with homogeneous goods 247–8
    summary and conclusions 264–5
spatial computable general equilibrium
    (SCGE) models 52
spatial econometrics 40
spatial equilibrium 44
spatial monopolies 248, 249
    characteristics of 251, 256
    Austrian universities as spatial
        monopolists or product
        differentiating suppliers? 251–65
'spatialized' explanation of economic
    growth 43–6
spatially segmented labour markets 274,
    281
Spencer, J.W. 139
spin-offs *see* academic spin-offs
Stahl, H. 29, 92
Stahlecker, T. 213
Stam, E. 354
standard competitiveness 323–5, 344
Standard Industrial Classification (SIC)
    60, 64
Stanford University 231, 237
start-ups
    factors determining 49
    innovations from 48–9, 50–51
Statistics Sweden (SCB) 96, 97
steel industry 302
Steffensen, M. 15
Stephan, P.E. 2, 170, 175, 178, 181
Sternberg, R. 194
Stevenson-Wydler Act 165, 184
Stokes, D. 18
Storey, D.J. 353
Storper, M. 25
Stracke, S. 300
Strambach, S. 192
Strathclyde University 269
Strobl, E. 102
strong market domination 252–5, 260

Stuart, T.E. 90, 174, 175, 178
student preferences, university ranking
    lists based on 224–6
sustainable development 341, 342
Sutz, J. 240
Sweden
    academic patents in 172
    control of private R&D in 86
    determinants of industry R&D location
        in 95–102
    conclusions 101–2
    empirical analysis 99–101
    empirical model 95–8
Swedish paradox 184
Swedish Road Administration (SRA) 96,
    97
Sylos-Labini, M. 184
Szatan, J. 22
Szeged
    BIOPOLIS programme in 340–43,
        345
    competitiveness of 333–5
    geography and demographics 329–30
    as a knowledge transfer region 334–5,
        336, 340, 345
    research institutes in 321, 340
    university-industry relations in the
        Csongrád county and Szeged
        subregion 335–9
    University of Szeged students from
        330, 344
    *see also* University of Szeged
Szerb, L. 3, 151

tacit knowledge
    as a component of industrial
        knowledge production 113
    as an input to knowledge production
        function 39, 88
    licensing of patents characterized by
        180
    in social sciences 193
Taylor, J. 268
teaching
    competition in business education
        between Austrian universities
        251–65
    cooperation resulting from 309, 310
    economic development impact of 16,
        17–18